ALGEBRA AND GEOMETRY

This text gives a basic introduction and a unified approach to algebra and geometry. It covers the ideas of complex numbers, scalar and vector products, determinants, linear algebra, group theory, permutation groups, symmetry groups and various aspects of geometry including groups of isometries, rotations and spherical geometry. The emphasis is always on the interaction between these topics, and each one is constantly illustrated by using it to describe and discuss the others. Many of the ideas are developed gradually throughout the book. For example, the definition of a group is given in Chapter 1 so that it can be used in a discussion of the arithmetic of real and complex numbers; however, many of the properties of groups are given later, and at a time when the importance of the concept has become clear. The text is divided into short sections, with exercises at the end of each one.

sla

ALGEBRA AND GEOMETRY

ALAN F. BEARDON

CAMBRIDGE
UNIVERSITY PRESS

CAMBRIDGE UNIVERSITY PRESS
Cambridge, New York, Melbourne, Madrid, Cape Town, Singapore, São Paulo

Cambridge University Press
The Edinburgh Building, Cambridge CB2 2RU, UK

Published in the United States of America by Cambridge University Press, New York

www.cambridge.org
Information on this title: www.cambridge.org/9780521813624

© Cambridge University Press 2005

This publication is in copyright. Subject to statutory exception
and to the provisions of relevant collective licensing agreements,
no reproduction of any part may take place without
the written permission of Cambridge University Press.

First published 2005
Reprinted 2006

Printed in the United Kingdom at the University Press, Cambridge

A catalogue record for this publication is available from the British Library

Library of Congress Cataloguing in Publication data

ISBN-13 978-0-521-81362-4 hardback
ISBN-10 0-521-81362-X hardback

ISBN-13 978-0-521-89049-6 paperback
ISBN-10 0-521-89049-7 paperback

Cambridge University Press has no responsibility for the persistence or accuracy of URLs for external or third-party internet websites referred to in this publication, and does not guarantee that any content on such websites is, or will remain, accurate or appropriate.

To Dylan, Harry, Fionn and Fenella

Study plan:
start 6/7/17. do one sub chapter a day.
so this book will be done in 97 days, roughly

Contents

	Preface	*page* xi
1	**Groups and permutations**	1
1.1	Introduction	1
1.2	Groups	2
1.3	Permutations of a finite set	6
1.4	The sign of a permutation	11
1.5	Permutations of an arbitrary set	15
2	**The real numbers**	22
2.1	The integers	22
2.2	The real numbers	26
2.3	Fields	27
2.4	Modular arithmetic	28
3	**The complex plane**	31
3.1	Complex numbers	31
3.2	Polar coordinates	36
3.3	Lines and circles	40
3.4	Isometries of the plane	41
3.5	Roots of unity	44
3.6	Cubic and quartic equations	46
3.7	The Fundamental Theorem of Algebra	48
4	**Vectors in three-dimensional space**	52
4.1	Vectors	52
4.2	The scalar product	55
4.3	The vector product	57
4.4	The scalar triple product	60

4.5	The vector triple product	62
4.6	Orientation and determinants	63
4.7	Applications to geometry	68
4.8	Vector equations	72
5	**Spherical geometry**	**74**
5.1	Spherical distance	74
5.2	Spherical trigonometry	75
5.3	Area on the sphere	77
5.4	Euler's formula	79
5.5	Regular polyhedra	83
5.6	General polyhedra	85
6	**Quaternions and isometries**	**89**
6.1	Isometries of Euclidean space	89
6.2	Quaternions	95
6.3	Reflections and rotations	99
7	**Vector spaces**	**102**
7.1	Vector spaces	102
7.2	Dimension	106
7.3	Subspaces	111
7.4	The direct sum of two subspaces	115
7.5	Linear difference equations	118
7.6	The vector space of polynomials	120
7.7	Linear transformations	124
7.8	The kernel of a linear transformation	127
7.9	Isomorphisms	130
7.10	The space of linear maps	132
8	**Linear equations**	**135**
8.1	Hyperplanes	135
8.2	Homogeneous linear equations	136
8.3	Row rank and column rank	139
8.4	Inhomogeneous linear equations	141
8.5	Determinants and linear equations	143
8.6	Determinants	144
9	**Matrices**	**149**
9.1	The vector space of matrices	149
9.2	A matrix as a linear transformation	154
9.3	The matrix of a linear transformation	158

9.4	Inverse maps and matrices	163
9.5	Change of bases	167
9.6	The resultant of two polynomials	170
9.7	The number of surjections	173

10 Eigenvectors — 175

10.1	Eigenvalues and eigenvectors	175
10.2	Eigenvalues and matrices	180
10.3	Diagonalizable matrices	184
10.4	The Cayley–Hamilton theorem	189
10.5	Invariant planes	193

11 Linear maps of Euclidean space — 197

11.1	Distance in Euclidean space	197
11.2	Orthogonal maps	198
11.3	Isometries of Euclidean n-space	204
11.4	Symmetric matrices	206
11.5	The field axioms	211
11.6	Vector products in higher dimensions	212

12 Groups — 215

12.1	Groups	215
12.2	Subgroups and cosets	218
12.3	Lagrange's theorem	223
12.4	Isomorphisms	225
12.5	Cyclic groups	230
12.6	Applications to arithmetic	232
12.7	Product groups	235
12.8	Dihedral groups	237
12.9	Groups of small order	240
12.10	Conjugation	242
12.11	Homomorphisms	246
12.12	Quotient groups	249

13 Möbius transformations — 254

13.1	Möbius transformations	254
13.2	Fixed points and uniqueness	259
13.3	Circles and lines	261
13.4	Cross-ratios	265
13.5	Möbius maps and permutations	268
13.6	Complex lines	271
13.7	Fixed points and eigenvectors	273

13.8	A geometric view of infinity	276
13.9	Rotations of the sphere	279

14 Group actions — 284

14.1	Groups of permutations	284
14.2	Symmetries of a regular polyhedron	290
14.3	Finite rotation groups in space	295
14.4	Groups of isometries of the plane	297
14.5	Group actions	303

15 Hyperbolic geometry — 307

15.1	The hyperbolic plane	307
15.2	The hyperbolic distance	310
15.3	Hyperbolic circles	313
15.4	Hyperbolic trigonometry	315
15.5	Hyperbolic three-dimensional space	317
15.6	Finite Möbius groups	319

Index — 320

Preface

> Nothing can permanently please, which does not contain in itself the reason why it is so, and not otherwise
>
> S.T. Coleridge, 1772–1834

The idea for this text came after I had given a lecture to undergraduates on the symmetry groups of regular solids. It is a beautiful subject, so why was I unhappy with the outcome? I had covered the subject in a more or less standard way, but as I came away I became aware that I had assumed Euler's theorem on polyhedra, I had assumed that every symmetry of a polyhedron extended to an isometry of space, and that such an isometry was necessarily a rotation or a reflection (again due to Euler), and finally, I had not given any convincing reason why such polyhedra did actually exist. Surely these ideas are at least as important (or perhaps more so) than the mere identification of the symmetry groups of the polyhedra?

The primary aim of this text is to present many of the ideas and results that are typically given in a university course in mathematics in a way that emphasizes the coherence and mutual interaction within the subject as a whole. We believe that by taking this approach, students will be able to support the parts of the subject that they find most difficult with ideas that they can grasp, and that the unity of the subject will lead to a better understanding of mathematics as a whole. Inevitably, this approach will not take the reader as far down any particular road as a single course in, say, group theory might, but we believe that this is the right approach for a student who is beginning a university course in mathematics. Increasingly, students will be taking more and more courses outside mathematics, and the pressure to include a wide spread of mathematics within a limited time scale will increase. We believe that the route advocated above will, in addition to being educationally desirable, help solve this problem.

To illustrate our approach, consider once again the symmetries of the five (regular) Platonic solids. These symmetries may be viewed as examples of permutations (acting on the vertices, or the faces, or even on the diagonals) of the solid, but they can also be viewed as finite groups of rotations of Euclidean 3-space. This latter point of view suggests that the discussion should lead into, or away from, a discussion of the nature of isometries of 3-space, for this is fundamental to the very definition of the symmetry groups. From a different point of view, probably the easiest way to identify the Platonic solids is by means of Euler's formula for the sphere. Now Euler's formula can be (and here is) proved by means of spherical geometry and trigonometry, and the requisite formulae here are simple (and important) applications of the standard scalar and vector product of the 'usual' vectors in 3-space (as studied in applied mathematics). Next, by studying rotation groups acting on the unit sphere in 3-space one can prove that the symmetry groups of the regular solids are the only finite groups of rotations of 3-space, a fact that it not immediately apparent from the geometry. Finally, by using stereographic projection (as appears in any complex analysis course that acknowledges the point at infinity) the symmetry groups of the regular solids appear as the only finite groups of Möbius transformations acting in hyperbolic space. Moreover in this guise one can also introduce rotations of 3-space in terms of quaternions which then appear as 2-by-2 complex matrices.

The author firmly believes that this is the way mathematics should be introduced, and moreover that it can be so introduced at a reasonably elementary level. In many cases, students find mathematics difficult because they fail to grasp the initial concepts properly, and in this approach preference is given to understanding and reinforcing these basic concepts from a variety of different points of view rather than moving on in the traditional way to provide yet more theorems that the student has to try to cope with from a sometimes uncertain base.

This text includes the basic definitions, and some early results, on, for example, groups, vector spaces, quaternions, eigenvectors, the diagonalization of matrices, orthogonal groups, isometries of the complex plane and of Euclidean space, scalar and vector products in 3-space, Euclidean, spherical and (briefly) hyperbolic geometries, complex numbers and Möbius transformations. Above all, it is these basic concepts and their mutual interaction which is the main theme of this text.

Finally an earlier version of this book can be freely downloaded as an html file from http://www.cambridge.org/0521890497. This file is under development and the aim is to create a fully linked electronic textbook.

1
Groups and permutations

1.1 Introduction

This text is about the interaction between algebra and geometry, and central to this interaction is the idea of a group. Groups are studied as abstract systems in algebra; they help us to describe the arithmetic structure of the real and complex numbers, and modular arithmetic, and they provide a framework for a discussion of permutations of an arbitrary set. Groups also arise naturally in geometry; for example, as the set of translations of the plane, the rotations of the plane about the origin, the symmetries of a cube, and the set of all functions of the plane into itself that preserve distance. We shall see that geometry provides many other interesting examples of groups and, in return, group theory provides a language and a number of fundamental ideas which can be used to give a precise description of geometry. In 1872 Felix Klein proposed his *Erlangen Programme* in which, roughly speaking, he suggested that we should study different geometries by studying the groups of transformations acting on the geometry. It is this spirit that this text has tried to capture.

We shall assume familiarity with the most basic facts about elementary set theory. We recall that if X is any set, then $x \in X$ means that x is an *element*, or *member*, of X, and $x \notin X$ means that x is not an element of X. The *union* $X \cup Y$ of two sets X and Y is the set of objects that are in at least one of them; the *intersection* $X \cap Y$ is the set of objects that are in both. The *difference set* $X \setminus Y$ is the set of objects that are in X but not in Y. The *empty set* $\emptyset$ is the set with no elements in it; for example, $X \setminus X = \emptyset$ for every set X. We say that X is *non-empty* when $X \neq \emptyset$.

In this chapter we shall define what we mean by a group, and then show that every non-empty set X has associated with it a group, which is known as the group of permutations of X. This basic fact underpins almost everything in this book. We shall also carry out a detailed study of the group of permutations of the finite set $\{1, 2, \ldots, n\}$ of integers. In Chapter 2 we review the algebraic properties of the real numbers in terms of groups, but in order to give concrete examples of groups now, we shall assume (in the examples) familiarity with the real numbers. Throughout the book we use $\mathbb{Z}$ for the set of integers, $\mathbb{Q}$ for the set of rational numbers, and $\mathbb{R}$ for the set of real numbers.

1.2 Groups

There are four properties that are shared by many mathematical systems and that have proved their usefulness over time, and any system that possesses these is known as a *group*. It is difficult to say when groups first appeared in mathematics for the ideas were used long before they were synthesized into an abstract definition of a group. Euler (1761) and Gauss (1801) studied modular arithmetic (see Section 2.4), and Lagrange (1770) and Cauchy (1815) studied groups of permutations (see Section 1.3). Important moves towards a more formal, abstract theory were taken by Cauchy (1845), von Dyck (1882) and Burnside (1897), thus group theory, as we know it today, is a relative newcomer to the history of mathematics.

First, we introduce the notion of a binary operation on a set X. A *binary operation* $*$ on X is a rule which is used to combine any two elements, say x and y, of X to obtain a third object, which we denote by $x*y$. In many cases $x*y$ will also be in X, and when this is so we say that X is *closed* with respect to $*$. We can now say what we mean by a group.

Definition 1.2.1 A *group* is a set G, together with a binary operation $*$ on G which has the following properties:

(1) for all g and h in G, $g*h \in G$;
(2) for all f, g and h in G, $f*(g*h) = (f*g)*h$;
(3) there a unique e in G such that for all g in G, $g*e = g = e*g$;
(4) if $g \in G$ there is some h in G such that $g*h = e = h*g$. □

A set X may support many different binary operations which make it a group so, for clarification, we often use the phrase 'X is a group with respect to $*$'. Property (1) is called the *closure axiom* for it says that G is closed with respect to $*$. Property (2) is the *associative law*, and this says that $f * g * h$ is uniquely defined regardless of which of the two operations $*$ we choose to do first. The point here is that as $*$ only combines *two* objects at a time, we have to apply $*$ twice (in some order) to obtain $f*g*h$. There are exactly two ways to do this, and (2) says that these two ways must yield the same result. Obviously, this idea extends to more elements, and reader should now use (2) to verify that the element $f*g*h*i$ is defined independently of the order in which the three applications of $*$ are carried out. It is important to understand that the associative law is not self-evident; indeed, if $a*b = a/b$ for for positive numbers a and b then, in general, $(a*b)*c \neq a*(b*c)$.

The element e in (3) is the *identity element* of G, and the reader should note that (3) requires that *both* $e*g$ and $g*e$ are g. In the example just considered (where $a*b = a/b$) we have $a*1 = a$ but $1*a \neq a$ (unless $a = 1$). We also note that in conjunction with (1) and (2), we could replace (3) by the weaker statement that there exists some e in G such that $g*e = g = e*g$ for every g in G. Indeed, suppose that G contains elements e and e' such that, for all g in G, $g*e = g = e*g$ and $g*e' = g = e'*g$. Then $e' = e*e' = e$ so that $e' = e$ (so that such an element is necessarily unique). It follows that when we need to prove that, say, G is a group we need only prove the existence of some element e in G such that $e*g = g = g*e$ for every g (and it is not necessary to prove the uniqueness of e). However, we cannot replace (3) by this weaker version of (3) in the definition of a group without making (4) ambiguous.

The element h in (4) is the *inverse of g*, and henceforth will be written as g^{-1}. However, before we can legitimately speak of *the* inverse of g, and use the notation g^{-1}, we need to show that each g has only one inverse.

Lemma 1.2.2 *Let G be any group. Then, given g in G, there is only one element h that satisfies (4). In particular, $(g^{-1})^{-1} = g$.*

Proof Take any g and suppose that h and h' satisfy $h*g = e = g*h$ and $h'*g = e = g*h'$. Then

$$h = h*e = h*(g*h') = (h*g)*h' = e*h' = h'$$

as required. As $g*g^{-1} = e = g^{-1}*g$, it is clear that $(g^{-1})^{-1} = g$. □

The next three results show that one can manipulate expressions, and solve simple equations, in groups much as one does for real numbers.

Lemma 1.2.3 *Suppose that a, b and x are in a group G. If $a*x = b*x$ then $a = b$. Similarly, if $x*a = x*b$ then $a = b$.*

Proof If $a*x = b*x$ then $(a*x)*x^{-1} = (b*x)*x^{-1}$. Now $(a*x)*x^{-1} = a*(x*x^{-1}) = a*e = a$, and similarly for b instead of a; thus $a = b$. The second statement follows in a similar way. For obvious reasons, this result is known as the *cancellation law*. □

Lemma 1.2.4 *Suppose that a and b are in a group G. Then the equation $a*x = b$ has a unique solution in G, namely $x = a^{-1}*b$. Similarly, $x*a = b$ has a unique solution, namely $b*a^{-1}$.*

Proof As $a*(a^{-1}*b) = (a*a^{-1})*b = e*b = b$, we see that $a^{-1}*b$ is a solution of $a*x = b$. Now let y_1 and y_2 be any solutions. Then $a*y_1 = b = a*y_2$ so that, by Lemma 1.2.3, $y_1 = y_2$. The second statement follows in a similar way. □

Lemma 1.2.5 *In any group G, e is the unique solution of $x*x = x$.*

Proof As $y*e = y$ for every y, we see that $e*e = e$. Thus e is one solution of $x*x = x$. However, if $x*x = x$ then $x*x = x*e$ so that from Lemma 1.2.3, $x = e$. □

The reader should note that the definition of a group does *not* include the assumption that $f*g = g*f$; indeed, there are many interesting groups in which equality does not hold. However, this condition is so important that it carries its own terminology.

Definition 1.2.6 Let G be a group with respect to $*$. We say that f and g in G *commute* if $f*g = g*f$. If $f*g = g*f$ for every f and g in G, we say that G is an *abelian*, or a *commutative*, group. We often abbreviate this to 'G is abelian'. □

Several straightforward examples of groups are given in the Exercises. We end this section with an example of a non-commutative group.

Example 1.2.7 Let G be the set of functions of the form $f(x) = ax + b$, where a and b are real numbers and $a \neq 0$. It is easy to see that G is a group with respect to the operation $*$ defined by making $f*g$ the function $f(g(x))$. First, if $g(x) = ax + b$ and $h(x) = cx + d$, then $g*h$ is in G because $(g*h)(x) = g(h(x)) = acx + (ad + b)$. It is also easy (though tedious) to check that for any f, g and h in G, $f*(g*h) = (f*g)*h$. Next, the function $e(x) = 1x + 0$ is in G and satisfies $e*f = f = f*e$ for every f in G. Finally, if $g(x) = ax + b$ then $g*g^{-1} = e = g^{-1}*g$, where $g^{-1}(x) = x/a - b/a$. We have shown that G is a group, but it is *not* abelian as $f*g \neq g*f$ when $f(x) = x + 1$ and $g(x) = -x + 1$. In the same way we see that the set of functions of the form $f(x) = ax + n$, where $a = \pm 1$ and n is an integer is also a non-abelian group. □

1.2 Groups

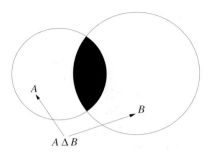

Figure 1.2.1

Exercise 1.2

1. Show that the set $\mathbb{Z}$ is a group with respect to addition. Show also that the set of positive real numbers is a group with respect to multiplication.
2. Let $\mathbb{Q}$ be the set of rational numbers (that is, numbers of the form m/n where m and n are integers and $n \neq 0$), and let $\mathbb{Q}^+$ and $\mathbb{Q}^*$ be the set of positive, and non-zero, rational numbers, respectively. Show that $\mathbb{Q}$, but not $\mathbb{Q}^+$, is a group with respect to addition. Show that $\mathbb{Q}^+$ and $\mathbb{Q}^*$ are groups with respect to with respect to multiplication, but that $\mathbb{Q}$ is not. Is the set of rational numbers of the form p/q, where p and q are positive odd integers, a group with respect to multiplication?
3. Show that $\mathbb{Z}$, with the operation $*$ defined by $m*n = m+n+1$, is a group. What is the identity element in this group? Show that the inverse of n is $-(n+2)$.
4. Show that $\mathbb{Z}$, with the operation $m*n = m + (-1)^m n$, is a group. Show that in this group the inverse n^{-1} of n is $(-1)^{n+1}n$. For which n is $n^{-1} = n$?
5. Let $G = \{x \in \mathbb{R} : x \neq -1\}$, where $\mathbb{R}$ is the set of real numbers, and let $x*y = x + y + xy$, where xy denotes the usual product of two real numbers. Show that G with the operation $*$ is a group. What is the inverse 2^{-1} of 2 in this group? Find $(2^{-1})*6*(5^{-1})$, and hence solve the equation $2*x*5 = 6$.
6. For any two sets A and B the *symmetric difference* $A \triangle B$ of A and B is the set of elements in *exactly one* of A and B; thus

$$A \triangle B = \{x \in A \cup B : x \notin A \cap B\} = (A \cup B)\backslash(A \cap B)$$

(see Figure 1.2.1). Let Ω be a non-empty set and let G be the set of subsets of Ω (note that G includes both the empty set $\emptyset$ and Ω). Show that G with the operation $\triangle$ is a group with $\emptyset$ as the identity element of G. What is A^{-1}? Now let $\Omega = \{1, 2, \ldots, 7\}$, $A = \{1, 2, 3\}$, $B = \{3, 4, 5\}$ and

$C = \{5, 6, 7\}$. By considering A^{-1} and B^{-1}, solve the two equations $A \triangle X = B$, and $A \triangle X \triangle B = C$.

1.3 Permutations of a finite set

We shall now discuss permutations of a non-empty set X. We shall show (in Section 1.5) that the permutations of X form a group, and we shall use this to examine the nature of the permutations. This is most effective when X is a finite set, and we shall assume that this is so during this and the next section. Before we can consider permutations we need to understand what we mean by a function and, when it exists, its inverse function. As (for the moment) we are only considering functions between finite sets, we can afford to take a fairly relaxed view about functions; a more detailed discussion of functions (between arbitrary sets) is given in Section 1.5.

A *function* $f : X \to X$ from a finite set X to itself is a rule which assigns to each x in X a unique element, which we write as $f(x)$, of X. We can define such a function by giving the rule explicitly; for example, when $X = \{a, b, c\}$ we can define $f : X \to X$ by the rule $f(a) = b$, $f(b) = c$ and $f(c) = a$. Note that f cyclically permutes the elements a, b and c, and this is our first example of a permutation. Two functions, say $f : X \to X$ and $g : X \to X$ are *equal* if $f(x) = g(x)$ for every x in X, and in this case we write $f = g$. The *identity function* $I : X \to X$ on X is the function given by the rule $I(x) = x$ for all x in X.

Suppose now that we have two functions f and g from X to itself. Then for every x in X there is a unique element $g(x)$ in X, and for every y in X there is a unique element $f(y)$ in X. If we choose x first, and then take $y = g(x)$, we have created a rule which takes us from x to the element $f(g(x))$. This rule defines a function which we denote by $fg : X \to X$. We call this function the *composition* (or sometimes the *product*) of f and g, and it is obtained by *applying g first, and then f*. This function is sometimes denoted by $f \circ g$, but it is usual to use the less cumbersome notation fg.

Given a function $f : X \to X$, the function $g : X \to X$ is the *inverse* of f if, for every x in X, we have $f(g(x)) = x$ and $g(f(x)) = x$, or, more succinctly, if $fg = I = gf$, where I is the identity function on X. It is important to note that not every function $f : X \to X$ has an inverse function. Indeed, f *has an inverse function precisely when, for every y in X, there is exactly one x in X such that $f(x) = y$*, for then the inverse function is the rule which takes y back to x. We say that a function $f : X \to X$ is *invertible* when the inverse of f exists, and then we denote the inverse by f^{-1}. Note that if f is invertible, then so is f^{-1}, and $(f^{-1})^{-1} = f$. We are now ready to define what we mean by a permutation of a set X.

Definition 1.3.1 A *permutation* of X is an invertible map $f : X \to X$. The set of permutations of X is denoted by $\mathcal{P}(X)$. □

Theorem 1.3.2 *The set $\mathcal{P}(X)$ of permutations of a finite non-empty set X is a group with respect to the composition of functions.*

We remark that it is usual to speak of the *product of permutations* rather than the composition of permutations.

Proof We must show that the operation $*$ defined on $\mathcal{P}(X)$ by $f*g = fg$ (the composition) satisfies the requirements of Definition 1.2.1. First, we show that $*$ is associative. Let f, g and h be any functions, and let $u = gf$ and $v = hg$. Then, for every x in X,

$$\begin{aligned}(h(gf))(x) &= (hu)(x) \\ &= h(u(x)) \\ &= h(g(f(x))) \\ &= v(f(x)) \\ &= (vf)(x) \\ &= ((hg)f)(x).\end{aligned} \quad (1.3.1)$$

This shows that $h(gf) = (hg)f$ and, as a consequence of this, we can now use the notation hgf (without brackets) for the composition of three (or more) functions in an unambiguous way.

Next, the identity map $I : X \to X$ is the identity element of $\mathcal{P}(X)$ because if f is any permutation of X, then $fI = f = If$; explicitly, for every x, $fI(x) = f(x) = I(f(x))$. Next, if f is any permutation of X, then f is invertible, and the inverse function f^{-1} is also a permutation of X (because it too is invertible). Moreover, f^{-1} is the inverse of f in the sense of groups because $ff^{-1} = I = f^{-1}f$. Finally, suppose that f and g are permutations of X. Then fg is invertible (and so is a permutation of X) with inverse $g^{-1}f^{-1}$; indeed

$$(fg)(g^{-1}f^{-1}) = f(gg^{-1})f^{-1} = fIf^{-1} = ff^{-1} = I,$$

and similarly, $(g^{-1}f^{-1})(fg) = I$. This completes the proof. □

Examples of permutation groups will occur throughout this text. However, for the rest of this and the next section we shall focus on the group of permutations of the finite set $\{1, 2, \ldots, n\}$ of integers.

Definition 1.3.3 The *symmetric group* S_n is the group of permutations of $\{1, \ldots, n\}$. □

As a permutation ρ is a function we can use the usual notation $\rho(k)$ for the image of an integer k under ρ. However, it is customary, and convenient, to write ρ in the form

$$\rho = \begin{pmatrix} 1 & 2 & \cdots & n \\ \rho(1) & \rho(2) & \cdots & \rho(n) \end{pmatrix},$$

where the image $\rho(k)$ of k is placed in the second row underneath k in the first row; for example, the permutation β of $\{1, 2, 3, 4\}$ such that $\beta(1) = 4$, $\beta(2) = 2$, $\beta(3) = 1$ and $\beta(4) = 3$ is denoted by

$$\begin{pmatrix} 1 & 2 & 3 & 4 \\ 4 & 2 & 1 & 3 \end{pmatrix}.$$

It is not necessary to order the columns according to the natural order of the top row, and we may use any order that we wish; for example,

$$\rho = \begin{pmatrix} 1 & \cdots & n \\ a_1 & \cdots & a_n \end{pmatrix}, \quad \rho^{-1} = \begin{pmatrix} a_1 & \cdots & a_n \\ 1 & \cdots & n \end{pmatrix}.$$

A permutation ρ is said to *fix* k, and k is a *fixed point* of ρ, if $\rho(k) = k$. By convention, we may omit any integers in the expression for ρ that are fixed by ρ (and any integers that are omitted in this expression may be assumed to be fixed by ρ). For example, if ρ is a permutation of $\{1, \ldots, 9\}$, and if

$$\rho = \begin{pmatrix} 1 & 8 & 3 & 7 \\ 8 & 1 & 7 & 3 \end{pmatrix},$$

then ρ interchanges 1 and 8, and 3 and 7, and it fixes 2, 4, 5, 6 and 9.

If α and β are permutations of $\{1, \ldots, n\}$ then $\alpha\beta$ is the permutation obtained by applying β first and then α. The following simple example illustrates a purely mechanical way of computing this composition: if

$$\alpha = \begin{pmatrix} 1 & 2 & 3 & 4 \\ 2 & 4 & 3 & 1 \end{pmatrix}, \quad \beta = \begin{pmatrix} 1 & 2 & 3 & 4 \\ 2 & 3 & 4 & 1 \end{pmatrix}$$

then (re-arranging α so that its top row coincides with the bottom row of β, and remembering that we apply β first) we have

$$\alpha\beta = \begin{pmatrix} 2 & 3 & 4 & 1 \\ 4 & 3 & 1 & 2 \end{pmatrix} \begin{pmatrix} 1 & 2 & 3 & 4 \\ 2 & 3 & 4 & 1 \end{pmatrix} = \begin{pmatrix} 1 & 2 & 3 & 4 \\ 4 & 3 & 1 & 2 \end{pmatrix}.$$

Note that $\alpha\beta \neq \beta\alpha$ (that is, α and β do not commute). We shall now define what we mean by disjoint permutations, and then show that *disjoint permutations commute*.

Definition 1.3.4 We say that two permutations α and β are *disjoint* if, for every k in $\{1, \ldots, n\}$, either $\alpha(k) = k$ or $\beta(k) = k$. $\square$

Theorem 1.3.5 *If α and β are disjoint permutations then $\alpha\beta = \beta\alpha$.*

Proof Take any k in $\{1, \ldots, n\}$. As either α or β fixes k we may suppose that $\alpha(k) = k$. Let $k' = \beta(k)$; then $\alpha(\beta(k)) = \alpha(k')$ and $\beta(\alpha(k)) = \beta(k) = k'$ so we need to show that α fixes k'. This is true (by assumption) if β does not fix k', so we may suppose that β fixes k'. But then $\beta(k) = k' = \beta(k')$, and applying β^{-1}, we see that $k = k'$, so again α fixes k'. $\square$

A permutation that cyclically permutes some set of integers is called a cycle. More precisely, we have the following definition.

Definition 1.3.6 The *cycle* $(n_1 \ldots n_q)$ is the permutation

$$\begin{pmatrix} n_1 & n_2 & \cdots & n_{q-1} & n_q \\ n_2 & n_3 & \cdots & n_q & n_1 \end{pmatrix}.$$

Explicitly, this maps n_j to n_{j+1} when $1 \leq j < q$, and n_q to n_1, and it fixes all other integers in $\{1, \ldots, n\}$. We say that this cycle has *length q*, or that it is a *q-cycle*. $\square$

Notice that we can write a cycle in three different ways; for example,

$$(1\,3\,5) = \begin{pmatrix} 1 & 3 & 5 \\ 3 & 5 & 1 \end{pmatrix} = \begin{pmatrix} 1 & 2 & 3 & 4 & 5 \\ 3 & 2 & 5 & 4 & 1 \end{pmatrix}.$$

To motivate the discussion that follows, observe that if

$$\sigma = \begin{pmatrix} 1 & 2 & 3 & 4 & 5 & 6 & 7 \\ 5 & 7 & 2 & 1 & 4 & 3 & 6 \end{pmatrix},$$

then (by inspection) $\sigma = (1\,5\,4)(2\,7\,6\,3)$ and so, by Theorem 1.3.5,

$$\sigma = (1\,5\,4)(2\,7\,6\,3) = (2\,7\,6\,3)(1\,5\,4).$$

We shall now show that this is typical of *all* permutations. Take any permutation ρ of $\{1, \ldots, n\}$, and any integer k in this set. By applying ρ repeatedly we obtain the points $k, \rho(k), \rho^2(k), \ldots$, and as two of these points must coincide, we see that there are integers p and q with $\rho^p(k) = \rho^q(k)$ where, say, $q < p$. As ρ^{-1} exists, $\rho^{p-q}(k) = k$. Now let u be the smallest positive integer with the property that $\rho^u(k) = k$; then the distinct numbers $k, \rho(k), \rho^2(k), \ldots, \rho^{u-1}(k)$ are cyclically permuted by ρ. We call

$$O(k) = \{k, \rho(k), \rho^2(k), \ldots, \rho^{u-1}(k)\}. \quad (1.3.2)$$

the *orbit* of k under ρ. Now every point m in $\{1, \ldots, n\}$ lies in some orbit (which will have exactly one element if and only if ρ fixes m), and it is evident that two orbits are either identical or disjoint. Thus we can write

$$\{1, \ldots, n\} = O(k_1) \cup \cdots \cup O(k_m), \qquad (1.3.3)$$

where the orbits $O(k_i)$ are pairwise disjoint sets, and where each of these sets is cyclically permuted by ρ. We call (1.3.3) the *orbit-decomposition* of $\{1, \ldots, n\}$.

Each orbit $O(k)$ in (1.3.2) provides us with an associated cycle

$$\rho_0 = (k \ \rho(k) \ \rho^2(k) \ \cdots \ \rho^{u-1}(k)).$$

Note that ρ and ρ_0 have exactly the same effect on the integers in $O(k)$, but that ρ_0 fixes every integer that is not in $O(k)$. Now consider the decomposition (1.3.3) of $\{1, \ldots, n\}$ into mutually disjoint orbits, and let ρ_j be the cycle associated to the orbit $O(k_j)$. Then it is clear that the cycles ρ_j are pairwise disjoint (because their corresponding orbits are); thus they commute with each other. Finally, if $x \in O_j$, then $\rho_j(x) = \rho(x)$, and $\rho_i(x) = x$ if $i \neq j$, so that $\rho = \rho_1 \cdots \rho_m$. We summarize this result in our next theorem.

Theorem 1.3.7 *Let ρ be a permutation of $\{1, \ldots, n\}$. Then ρ can be expressed as a product of disjoint (commuting) cycles.*

It is evident that the expression $\rho = \rho_1 \cdots \rho_m$ that was derived from the orbit decomposition (1.3.3) is unique up to the order of the 'factors' ρ_j. Indeed if $\rho = \mu_1 \cdots \mu_v$, where the μ_i are pairwise disjoint cycles, then the set of points not fixed by μ_i, constitutes an orbit for ρ, so that μ_i must be some ρ_j. In particular, the number m of factors in this product is uniquely determined by ρ, and we shall return to this later. We pause to name this representation of ρ.

Definition 1.3.8 *The representation $\rho = \rho_1 \cdots \rho_m$ which is derived from the orbit decomposition (1.3.3), and which is unique up to the order of the factors ρ_j, is called the standard representation of ρ as a product of cycles.* □

Let us illustrate these ideas with an example. Consider

$$\rho = \begin{pmatrix} 1 & 2 & 3 & 4 & 5 & 6 & 7 & 8 & 9 \\ 7 & 1 & 8 & 4 & 6 & 9 & 2 & 3 & 5 \end{pmatrix}$$

as a permutation of $\{1, \ldots, 9\}$. The orbits of ρ are $\{1, 7, 2\}$, $\{3, 8\}$, $\{4\}$ and $\{5, 6, 9\}$, and the standard representation of ρ as a product of disjoint cycles is $(1\ 7\ 2)(3\ 8)(4)(5\ 6\ 9)$.

There is an interesting corollary of Theorem 1.3.7. First, if μ is a cycle of length k, then μ^k (that is, μ applied k times) is the identity map. Suppose now that $\rho = \rho_1 \cdots \rho_m$ is the standard representation of ρ, and let d be any positive integer. As the ρ_j commute, we have

$$\rho^d = (\rho_1 \cdots \rho_m)^d = \rho_1^d \cdots \rho_m^d.$$

It follows that if d is the least common multiple of $q_1, \ldots, q_m$, where q_j is the length of the cycle ρ_j, then $\rho^d = I$. For example if $\rho = (1\,3\,4)(2\,9\,5\,6)(7\,8)$, then $\rho^{12} = I$. In fact, it is not difficult to see that the least common multiple d of the q_j is the smallest positive integer t for which $\rho^t = I$. As d divides $n!$, this shows that $\rho^{n!} = I$ for every permutation ρ of $\{1, \ldots, n\}$.

Exercise 1.3

1. Show that
$$\begin{pmatrix} 1 & 2 & 3 & 4 & 5 & 6 & 7 & 8 & 9 \\ 4 & 7 & 9 & 2 & 6 & 8 & 1 & 5 & 3 \end{pmatrix} = (1\,4\,2\,7)(3\,9)(5\,6\,8).$$

2. Show that $(1\,2\,3\,4) = (1\,4)(1\,3)(1\,2)$. Express $(1\,2\,3\,4\,5)$ as a product of 2-cycles. Express $(1\,2\,\ldots\,n)$ as a product of 2-cycles.

3. Express the permutation
$$\rho = \begin{pmatrix} 1 & 2 & 3 & 4 & 5 & 6 & 7 & 8 & 9 & 10 \\ 8 & 7 & 10 & 9 & 4 & 3 & 6 & 5 & 1 & 2 \end{pmatrix}$$
as a product of cycles, and hence (using Exercise 1.3.2) as a product of 2-cycles. Use this to express ρ^{-1} as a product of 2-cycles.

4. Show that the set $\{I, (1\,2)(3\,4), (1\,3)(2\,4), (1\,4)(2\,3)\}$ of permutations is a group.

5. Suppose that the permutation ρ of $\{1, \ldots, n\}$ satisfies $\rho^3 = I$. Show that ρ is a product of 3-cycles, and deduce that if n is not divisible by 3 then ρ fixes some k in $\{1, \ldots, n\}$.

1.4 The sign of a permutation

A 2-cycle $(r\,s)$ (which interchanges the distinct integers r and s and leaves all other integers fixed) is called a *transposition*. Notice that $(r\,s) = (s\,r)$, and that $(r\,s)$ is its own inverse. Common experience tells us that any permutation can be achieved by a succession of transpositions, and this suggests the following result.

Theorem 1.4.1 *Every permutation is a product of transpositions.*

Proof As every permutation is a product of cycles, and as for distinct integers a_i we have (by inspection)

$$(a_1\,a_2\,\cdots\,a_p) = (a_1\,a_p)\cdots(a_1\,a_3)(a_1\,a_2), \qquad (1.4.1)$$

the result follow immediately. □

In fact (1.4.1) leads to the following quantitative version of Theorem 1.4.1.

Theorem 1.4.2 *Let ρ be a permutation acting on $\{1, \ldots, n\}$, and suppose that ρ partitions $\{1, \ldots, n\}$ into m orbits. Then ρ can be expressed as a composition of $n - m$ transpositions.*

Proof Let $\rho = \rho_1 \cdots \rho_m$ be the standard representation of ρ as a product of disjoint cycles, and let n_j be the length of the cycle ρ_j. Thus $\sum_j n_j = n$. If $n_j \geq 2$ then, from (1.4.1), ρ_j can be written as a product of $n_j - 1$ transpositions. If $n_j = 1$ then ρ_j is the identity, so that no transpositions are needed for this factor. However, in this case $n_j - 1 = 0$. It follows that we can express ρ as a product of $\sum_j (n_j - 1)$ transpositions, and this number is $n - m$. □

We come now to the major result of this section, namely the *number of transpositions used to express a permutation ρ as a product of transpositions.* Although this number is not uniquely determined by ρ, we will show that its *parity* (that is, whether it is even or odd) is determined by ρ. First, however, we prove a preliminary result.

Lemma 1.4.3 *Suppose that the identity permutation I on $\{1, 2, \ldots, n\}$ can be expressed as a product of m transpositions. Then m is even.*

Proof The proof is by induction on n, and we begin with the case $n = 2$. In this case we write $I = \tau_1 \cdots \tau_m$, where each τ_j is the transposition $(1\ 2)$. As $(1\ 2)^m = (1\ 2)$ if m is odd, we see that m must be even, so the conclusion is true when $n = 2$.

We now suppose that the conclusion holds when the permutations act on $\{1, 2, \ldots, n-1\}$, and consider the situation in which $I = \tau_1 \cdots \tau_m$, where each τ_j is a transposition acting on $\{1, \ldots, n\}$. Clearly, $m \neq 1$, thus $m \geq 2$. Suppose, for the moment, that τ_m does not fix n. Then, for a suitable choice of a, b and c, we have one of the following situations:

$$\tau_{m-1}\tau_m = \begin{cases} (n\ b)(n\ a) = (a\ b\ n) = (n\ a)(a\ b); \\ (a\ b)(n\ a) = (a\ n\ b) = (n\ b)(a\ b); \\ (b\ c)(n\ a) = (n\ a)(b\ c); \\ (n\ a)(n\ a) = I = (a\ b)(a\ b). \end{cases}$$

It follows that we can now write I as a product of m transpositions in which the first transposition to be applied fixes n (this was proved under the assumption that $\tau_m(n) \neq n$, and I is already in this form if $\tau_m(n) = n$). In other words, we may assume that $\tau_m(n) = n$. We can now apply the same argument to $\tau_1 \cdots \tau_{m-1}$ (providing that $m - 1 \geq 2$), and the process can be continued to the point where we can write $I = \tau_1 \cdots \tau_m$, where each of $\tau_2, \ldots, \tau_m$ fixes n. But then τ_1 also fixes n, because

$$\tau_1(n) = \tau_1 \cdots \tau_m(n) = I(n) = n.$$

Thus, we can now write $I = \tau_1 \cdots \tau_m$, where each τ_j is a transposition *acting on* $\{1, \ldots, n-1\}$. The induction hypothesis now implies that m is even and the proof is complete. □

The main result now follows.

Theorem 1.4.4 *Suppose that a permutation ρ can be expressed both as a product of p transpositions, and also as a product of q transpositions. Then p and q are both even, or both odd.*

Proof Suppose that $\tau_1 \cdots \tau_p = \sigma_1 \cdots \sigma_q$, where each τ_i and each σ_j is a transposition. Then $\sigma_q \sigma_{q-1} \cdots \sigma_1 \tau_1 \cdots \tau_p = I$, so that, by Lemma 1.4.3, $p + q$ is even. It follows from this that p and q are both even, or both odd. □

As an example, consider the permutation $\rho = (1\,3\,5)(2\,4\,6\,8)(7)$ acting on $\{1, \ldots, 8\}$. Here, $n = 8$ and $N(\rho) = 3$ so that, by Lemma 1.4.3, ρ can be expressed as a product of five transpositions. Theorem 1.4.4 now implies that if we write ρ as a product of tranpositions *in any way whatsoever*, then there will necessarily be an odd number of transpositions in the product. This discussion suggests the following definition.

Definition 1.4.5 The *sign* $\varepsilon(\rho)$ of a permutation ρ is $(-1)^q$, where ρ can be expressed as a product of q transpositions. We say that ρ is an *even permutation* if $\varepsilon(\rho) = 1$, and an *odd permutation* if $\varepsilon(\rho) = -1$. □

Observe from (1.4.1) that if ρ is a p-cycle then $\varepsilon(\rho) = (-1)^{p+1}$; thus *a cycle of even length is odd*, and *a cycle of odd length is even*. If the permutations α and β can be expressed as products of p and q transpositions, respectively, then the composition $\alpha\beta$ can be expressed as a product of $p + q$ transpositions; thus the next two results are clear.

Theorem 1.4.6 *If α and β are permutations, then $\varepsilon(\alpha\beta) = \varepsilon(\alpha)\varepsilon(\beta)$. In particular, $\varepsilon(\alpha) = \varepsilon(\alpha^{-1})$.*

Theorem 1.4.7 *The product of two even permutations is an even permutation. The inverse of an even permutation is an even permutation. More generally, the set of even permutations in S_n is a group.*

Definition 1.4.8 The *alternating group* A_n is the group of all even permutations in S_n.

It is easy to find the number of elements in the symmetric group S_n and in the alternating group A_n.

Theorem 1.4.9 *The symmetric group S_n has $n!$ elements, and the alternating group A_n has $n!/2$ elements.*

Proof Elementary combinatorial arguments show that S_n has exactly $n!$ elements for, in order to construct a permutation of $\{1, \ldots, n\}$, there are n ways to choose the image of 1, then $n - 1$ ways to choose the image of 2 (distinct from the image of 1), and so on. Thus S_n has $n!$ elements.

Now let σ be the transposition $(1\ 2)$ and let $f : S_n \to S_n$ be the function defined by $f(\rho) = \sigma\rho$. We note that f is invertible, with $f^{-1} = f$, because, for every ρ, $f(f(\rho)) = f(\sigma\rho) = \sigma\sigma\rho = \rho$. It is clear that f maps even permutations to odd permutations, and odd permutations to even permutations and, as f is invertible, there are the same number of even permutations in S_n as there are odd permutations. Thus there are exactly $n!/2$ even permutations in S_n. □

Theorem 1.4.1 says that every permutation is a product of 2-cycles. Are there any other values of m with the property that every permutation a product of m-cycles? The answer is given in the next theorem.

Theorem 1.4.10 *Let ρ be a permutation of $\{1, \ldots, n\}$, and let m be an integer satisfying $2 \leq m \leq n$. Then ρ is a product of m-cycles if and only if either ρ is an even permutation, or m is an even integer.*

Proof Take any integer m with $2 \leq m \leq n$. Suppose first that ρ is an even permutation. The identity

$$(a_1\ a_2)(a_1\ a_3) = (a_1\ a_2\ a_3\ a_4 \cdots a_m)(a_m \cdots a_4\ a_3\ a_1\ a_2),$$

where the a_i are distinct (and which can be verified by inspection) shows that it suffices to express ρ as a product of terms $\tau_i \tau_j$, where τ_i and τ_j are transpositions with exactly one entry in common. Now as ρ is even it can certainly be be written as a product of terms of the form $\tau_i \tau_j$, where each τ_k is a transposition, and clearly we may assume that $\tau_i \neq \tau_j$. If τ_i and τ_j have no elements in common then we can use the identity

$$(a\ b)(c\ d) = (a\ b)(a\ c)(a\ c)(c\ d),$$

to obtain ρ as a product of the desired terms.

Let us now suppose that ρ is an odd permutation. If ρ is a product of m-cycles, say, $\rho = \rho_1 \cdots \rho_t$, then

$$-1 = \varepsilon(\rho) = \varepsilon(\rho_1) \cdots \varepsilon(\rho_t) = [(-1)^{m-1}]^t,$$

so that m is even. Finally, take any even m, and let $\sigma_0 = (1\,2\,3\cdots m)$. As σ_0 is odd, we see that $\sigma_0\rho$ is even. It follows that $\sigma_0\rho$, and hence ρ itself, can be written as a product of m-cycles. □

Exercise 1.4

1. Show that the permutation $\begin{pmatrix} 1 & 2 & 3 & 4 & 5 & 6 & 7 \\ 5 & 7 & 2 & 1 & 4 & 3 & 6 \end{pmatrix}$ is odd.
2. Find all six elements of S_3 and determine which are even and which are odd. Find all twelve even permutations of S_4.
3. The *order* of a permutation ρ is the smallest positive integer m such that ρ^m (that is, ρ applied m times) is the identity map.
   (a) What is the order of the permutation $(1\,2\,3\,4)(5\,6\,7\,8\,9)$?
   (b) Which element of S_9 has the highest order, and what is this order?
   (c) Show that every element of order 14 in S_{10} is odd.
4. (i) By considering $(1\,a)(1\,b)(1\,a)$, show that any permutation in S_n can be written as a product of the transpositions $(1, 2), (1\,3), \ldots, (1\,n)$, each of which may be used more than once.
   (ii) Use (i) to show that any permutation in S_n can be written as a product of the transpositions $(1, 2), (2\,3), \ldots, (n-1\,n)$, each of which may be used more than once.
   [This is the basis of bell-ringing, for a bell-ringer can only 'change places' with a neighbouring bell-ringer.]
5. Show that any subgroup of S_n (that is, a subset of S_n that is a group in its own right) which is not contained in A_n contains an equal number of even and odd permutations.

1.5 Permutations of an arbitrary set

This section is devoted to a careful look at functions between arbitrary sets. The reader will have already met functions defined by algebraic rules (for example, $x^2 + 3x + 5$), but we need to understand what one means by a function between sets in the absence of any arithmetic. We can say that a function $f : X \to Y$ is a rule that assigns to each x in X a unique y in Y and this seems clear enough, but what do we actually mean by a rule, and why should it be easier to define a 'rule' than a function? In fact, it is easier to think about a function in terms of its graph, and this is what we shall do next. As an example, the graph $G(f)$ of the function $f(x) = x^2$, where $x \in \mathbb{R}$, is the set

$$G(f) = \{(x, x^2) : x \in \mathbb{R}\} = \{(x, f(x)) \in \mathbb{R}^2 : x \in \mathbb{R}\}$$

in the plane. Now $G(f)$ contains all the information that there is about the function f, and therefore, conceptually, it is *equivalent* to f. However, $G(f)$ is a *set* (not a 'rule'), and if we can characterize sets of this form that arise from functions we can then base our definition of functions on sets and thereby avoid the problem of defining what we mean by a 'rule'. With this in mind, we note that $G(f)$ has the important property that *for every x in X there is exactly one y in Y such that $(x, y) \in G(f)$*. We now define what we mean by a function, but note that the words 'function', 'map' and 'mapping' are used interchangeably.

Definition 1.5.1 Let X and Y be non-empty sets. A *function, map*, or *mapping*, f from X to Y is a set $G(f)$ of ordered pairs (x, y) with the properties (a) if $(x, y) \in G(f)$ then $x \in X$ and $y \in Y$, and (b) for every x in X there is exactly one y in Y such that (x, y) is in $G(f)$. □

Suppose that we are given a function in this sense. Then to each x in X we can assign the unique y in Y such that $(x, y) \in G(f)$; in other words, the set $G(f)$ can be used in this natural way to define a function from X to Y. We emphasize, however, the subtle and important point that a function given by Definition 1.5.1 is based on set theory, and not on a vague idea of a 'rule'. As usual, we write $f : X \to Y$ to mean that f is a function from X to Y, and we write the unique element y that the rule f assigns to x as $f(x)$. We can now revert to the more informal and common use of functions safe in the knowledge that the foundation of this idea is secure. In the case when a function is given by a simple algebraic rule, for example $f(x) = x^2$, it is often convenient to use the notation $f : x \mapsto x^2$, or simply $x \mapsto x^2$.

Suppose now that we have two functions $g : X \to Y$ and $f : Y \to Z$. Then for every x in X there is a unique element $g(x)$ in Y, and for every y in Y there is a unique element $f(y)$ in Z. If we choose x, and then take $y = g(x)$ here we have created a rule which takes us from x in X to the unique element $f(g(x))$ in Z. We have therefore created a function which we denote by $fg : X \to Z$; this function is the *composition* of f and g and it is obtained by *applying g first, and then f*. It is a simple but vitally important fact that *the composition of functions is associative*; thus given functions $f : X \to Y, g : Y \to W$ and $h : W \to Z$, we have $h(gf) = (hg)f$. To prove this we need only check that the argument in (1.3.1) remains valid in this more general situation. As a consequence, we can now use the notation hgf (without brackets) for the composition of three (or more) functions in an unambiguous way.

There is one special case of composition that is worth mentioning, and which we have already used for permutations. If $f : X \to X$ is any function, then we can form composition ff to give the function $x \mapsto f(f(x))$. Naturally we denote this by f^2. We can then form the composition $f^2 f$ (or ff^2); we denote this by f^3 so that $f^3(x) = f(f(f(x)))$. More generally, if $f : X \to X$ is any function, we can apply f repeatedly and f^n is the function obtained by applying f exactly n times; this is the *n*-th *iterate* of f. Notice that $f^n(x)$ is the effect of starting with x and applying f exactly n times; it is not the *n*-th power of a number $f(x)$ which we would write as $(f(x))^n$.

For any non-empty set X, we can form the function $f : X \to X$ consisting of all pairs of the form (x, x), where $x \in X$. This function is the rule that takes every x to itself, and in the usual notation we would write $f(x) = x$. This function will play a special role in what follows (it will be the identity element in a group of transformations of X) so we give it a special symbol, namely I_X. Notice that for any function $g : X \to Y$, we have $gI_X = g$ and $I_Y g = g$.

We now discuss conditions under which a function $f : X \to Y$ has an inverse function $f^{-1} : Y \to X$ which 'reverses' the action of f. Explicitly, we want to find a function $g : Y \to X$ such that for all x in X, $gf(x) = x$ and for all y in Y, $fg(y) = y$. These two conditions are equivalent to $gf = I_X$ and $fg = I_Y$.

Definition 1.5.2 Let $f : X \to Y$ be any function. Then a function $g : Y \to X$ is an *inverse* of f if $gf = I_X$ and $fg = I_Y$. If an inverse of f exists we say that f is *invertible*. □

It is clear that there can be at most one inverse of f, for if $g : Y \to X$ and $h : Y \to X$ are both inverses of f, then $fg = I_Y = fh$ so that $g = I_X g = gfg = gfh = I_X h = h$. Henceforth, when the inverse of f exists, we shall denote it by f^{-1}. We can now give conditions that guarantee that f^{-1} exists.

Definition 1.5.3 A function $f : X \to Y$ is *injective* if, for each y in Y, $f(x) = y$ for at most one x in X. □

Definition 1.5.4 A function $f : X \to Y$ is *surjective* if, for each y in Y, $f(x) = y$ for at least one x in X. □

Definition 1.5.5 A function $f : X \to Y$ is *bijective*, or is a *bijection*, if it is both injective and surjective; that is, if, for each y in Y, there is exactly one x in X such that $f(x) = y$. □

Sometimes the term *one-to-one* is used to mean injective, and f is said to map X onto Y when $f : X \to Y$ is surjective. Note that to show that f is injective it is sufficient to show that $f(x_1) = f(x_2)$ implies that $x_1 = x_2$.

Theorem 1.5.6 *A function $f : X \to Y$ is invertible if and only if it is a bijection. If this is so then $f^{-1} : Y \to X$ is also a bijection.*

Proof Suppose first that $f : X \to Y$ is bijective; thus, for each y in Y, there is exactly one x in X with $f(x) = y$. Let us write this x, which depends on y, by $g(y)$. Then, by definition, $f(g(y)) = y$. Next, take any x_0 in X and let $y_0 = f(x_0)$. Then, by the definition of g, we have $g(y_0) = x_0$ so that $g(f(x_0)) = x_0$. Thus f in invertible with inverse g.

Now suppose that $f^{-1} : Y \to X$ exists. To check that f is injective we assume that $f(x_1) = f(x_2)$ and then show that $x_1 = x_2$. This, however, is immediate for $x_1 = f^{-1}(f(x_1)) = f^{-1}(f(x_2)) = x_2$. To show that $f : X \to Y$ is surjective, consider any y in Y. We want to show that there is some x such that $f(x) = y$. As $f^{-1}(y)$ is in X, we can take this as our x and this gives us what we want as $f(x) = f(f^{-1}(y)) = y$. We have now shown that f is invertible if and only if it is a bijection.

It remains to show that if f is bijective then so is f^{-1}, but this is clear for, applying what we have just proved about f to the function f^{-1}, we see that f^{-1} is bijective if and only if it is invertible. However, it is obvious that f^{-1} is invertible if and only f is, and hence if and only if f is bijective. □

The following definition is consistent with Definition 1.3.1, and the basic result about permutations is that they form a group.

Definition 1.5.7 Let X be a non-empty set. A *permutation of X* is a bijection of X onto itself. □

Theorem 1.5.8 *The set $\mathcal{P}(X)$ of permutations of a non-empty set X is a group with respect to the composition of functions.*

Proof Let f and g be permutations of X; then, by Theorem 1.5.6, their inverses f^{-1} and g^{-1} exist. It is easy to check that the composition fg has inverse $g^{-1}f^{-1}$, so that fg is also a permutation of X. Thus $\mathcal{P}(X)$ is closed under the composition of functions. Next, we have seen that the composition of functions is associative. The identity map $I_X : X \to X$ defined by $I_X(x) = x$ is the identity element of $\mathcal{P}(X)$ for $I_X f = f = f I_X$ for every f in $\mathcal{P}(X)$. Finally, the inverse function f^{-1} is a permutation of X (Theorem 1.5.6), and it is indeed the inverse of f in the sense of group theory because $ff^{-1} = I_X = f^{-1}f$. □

We end this chapter by giving the number of different types of functions from a set X with m elements to a set Y with n elements.

1.5 Permutations of an arbitrary set

Theorem 1.5.9 *Suppose that X and Y have m and n elements, respectively. Then there are*

(a) *n^m functions from X to Y;*
(b) *$n!$ bijections from X to Y when $m = n$, and none otherwise;*
(c) *$n!/(n-m)!$ injections from X to Y when $n \geq m$, and none otherwise;*
(d) *$\sum_{k=1}^{n}(-1)^{n-k}\binom{n}{k}k^m$ surjections from X to Y when $m \geq n$, and none otherwise.*

Proof Each element of X can be mapped to any element of Y; thus there are exactly n^m functions from X to Y. Clearly, if there is a bijection from X to Y then $m = n$. If $m = n$ we write $X = \{x_1, \ldots, x_n\}$ and we can construct the general bijection f by taking any one of n choices for $f(x_1)$, then any one of $n-1$ remaining choices for $f(x_2)$, and so on, and this gives (b). There are no injections from X to Y unless $n \geq m$, so suppose that $n \geq m$. Then there are $\binom{n}{m}$ subsets of Y that have exactly m elements, and hence there are $m!\binom{n}{m}$ injections from X to Y; thus (c) holds.

Finally, we prove (d). It is clear that no such surjective maps exist if $m < n$, so we suppose that $m \geq n \geq 1$. Let $S(m, k)$ be the number of surjective maps from X onto a set with exactly k elements, and note that $S(m, 0) = 0$. As there are $\binom{n}{k}$ distinct subsets of Y with exactly k elements, and as any map from X to Y is a surjective map onto some subset of Y with k elements (for some k), we see that

$$n^m = \sum_{k=1}^{n}\binom{n}{k}S(m,k) = \sum_{k=0}^{n}\binom{n}{k}S(m,k).$$

We now want a way of solving this system of equations in the 'unknowns' $S(m, k)$, and this is given by the following 'inversion formula'. □

Lemma 1.5.10 *Given any sequence of numbers $A_0, A_1, \ldots$, let*

$$B_n = \sum_{k=0}^{n}\binom{n}{k}A_k, \quad n = 1, 2, \ldots.$$

Then

$$A_n = \sum_{k=0}^{n}\binom{n}{k}(-1)^{n+k}B_k.$$

If we apply this with $B_k = k^m$ and $A_k = S(m, k)$, (d) follows. □

The proof of Lemma 1.5.10 We consider the second sum, write each B_k in terms of the A_j, and then interchange the order of summation. Thus

$$\sum_{k=0}^{n} \binom{n}{k}(-1)^{n+k} B_k = \sum_{k=0}^{n} \binom{n}{k}(-1)^{n+k} \left(\sum_{r=0}^{k} \binom{k}{r} A_r \right)$$

$$= \sum_{k=0}^{n} \sum_{r=0}^{k} \binom{n}{k}\binom{k}{r}(-1)^{n+k} A_r$$

$$= \sum_{r=0}^{n} \sum_{k=r}^{n} \binom{n}{k}\binom{k}{r}(-1)^{n+k} A_r.$$

It is clear that the coefficient of A_n on the right is one; thus we only have to show that the coefficients of $A_0, \ldots, A_{n-1}$ are zero. Now for $r = 0, \ldots, n-1$, the coefficient of A_r is λ_r, say, where

$$\lambda_r = \sum_{k=r}^{n} \binom{n}{k}\binom{k}{r}(-1)^{n+k}$$

$$= \sum_{k=r}^{n} \binom{n}{r}\binom{n-r}{k-r}(-1)^{n+k}$$

$$= (-1)^n \binom{n}{r} \sum_{j=0}^{n-r} \binom{n-r}{j}(-1)^{r+j}$$

$$= (-1)^{n+r} \binom{n}{r} [1 + (-1)]^{n-r}$$

$$= 0.$$

This completes the proof of Lemma 1.5.10 and Theorem 1.5.9. □

Exercise 1.5

1. Show that the map $f : \mathbb{R} \to \mathbb{R}$ defined by $f(x) = x^n$, where n is a positive integer, is a bijection if and only if n is odd.
2. Show that the map $f : \mathbb{R} \to \mathbb{R}$ defined by $f(x) = 1/x$ when $x \neq 0$, and $f(0) = 0$, is a bijection.
3. Let $f : X \to Y$ be any function. Show that
   (a) f is injective if and only if there is a function $g : Y \to X$ such that $gf = I_X$, and
   (b) f is surjective if and only if there is a function $h : Y \to X$ such that $fh = I_Y$.

4. Let $f : X \to Y$ and $g : Y \to Z$ be given functions. Show that if f and g are injective, then so is $gf : X \to Z$. Show also that if f and g are surjective, then so is $gf : X \to Z$.
5. Consider the (infinite) set $\mathbb{Z}$ of integers. Show that there is a function $f : \mathbb{Z} \to \mathbb{Z}$ that is injective but not surjective, and a function $g : \mathbb{Z} \to \mathbb{Z}$ that is surjective but not injective. Now let X be any finite set, and let $f : X \to X$ be any function. Show that the following statements are equivalent:
   (a) $f : X \to X$ is injective;
   (b) $f : X \to X$ is surjective;
   (c) $f : X \to X$ is bijective.

2
The real numbers

2.1 The integers

This chapter contains a brief review of the algebraic properties of the real numbers. As we usually think of the real numbers as the coordinates of points on a straight line we often refer to the set $\mathbb{R}$ of real numbers as the *real line*. The set $\mathbb{R}$ carries (and is characterized by) three important structures, namely an *algebraic structure* (addition, subtraction, multiplication and division), an *order* (positive numbers, negative numbers and zero), and the *least upper bound property* (or its equivalent). We shall take the existence of real numbers, and many of their properties (in particular, their order, and the existence of the n-th root of a positive number) for granted, and we concentrate on their algebraic properties. The rest of this section is devoted to a discussion of the set $\mathbb{Z}$ of integers. We review the algebraic structure of $\mathbb{R}$ (using groups) in Section 2.2, and we introduce the idea of a *field* (an algebraic structure that has much in common with $\mathbb{R}$) in Section 2.3. In Section 2.4 we discuss modular arithmetic.

We now discuss the set $\mathbb{N} = \{1, 2, 3, \ldots\}$ of *natural numbers*, and the the set $\mathbb{Z} = \{\ldots, -2, -1, 0, 1, 2, \ldots\}$ of *integers*. One of the most basic facts about the integers is the

The Well-Ordering Principle *Any non-empty subset of $\mathbb{N}$ has a smallest member.*

This apparently obvious result justifies the use of induction.

The Principle of Induction I *Suppose that $A \subset \mathbb{N}$, $1 \in A$, and for every m, $m \in A$ implies that $m + 1 \in A$. Then $A = \mathbb{N}$.*

This is the set-theoretic version of the (presumably) familar result that *if a statement $\mathcal{P}(n)$ about n is true when $n = 1$, and if the truth of $\mathcal{P}(m)$ implies the truth of $\mathcal{P}(m + 1)$, the $\mathcal{P}(n)$ is true for all n*. Indeed, if we let A be the set of n

for which $\mathcal{P}(n)$ is true, the two versions are easily seen to be equivalent to each other.

Proof Let B be the set of positive integers that are not in A; thus $A \cap B = \emptyset$ and $A \cup B = \mathbb{N}$. We want to prove that $B = \emptyset$ for then, $A = \mathbb{N}$. Suppose not, then, by the Well-Ordering Principle, B has a smallest element, say b. As $1 \in A$ we see that $1 \notin B$; thus $b \geq 2$ so that $b - 1 \geq 1$ (that is, $b - 1 \in \mathbb{N}$). As b is the smallest element in B, it follows that $b - 1 \notin B$, and hence that $b - 1 \in A$. But then, by our hypothesis, $(b - 1) + 1 \in A$, so that $b \in A$. This is a contradiction (to $A \cap B = \emptyset$), so that $B = \emptyset$, and $A = \mathbb{N}$. □

There is an alternative version of induction that is equally important (and the difference between the two versions will be illustrated below).

The Principle of Induction II *Suppose that $A \subset \mathbb{N}$, $1 \in A$, and for every m, $\{1, \ldots, m\} \subset A$ implies that $m + 1 \in A$. Then $A = \mathbb{N}$.*

Proof Let B be the set of positive integers that are not in A. Suppose that $B \neq \emptyset$; then, by the Well-Ordering Principle, B has a smallest element, say b. As before, $b \geq 2$, so that now $\{1, \ldots, b - 1\} \subset A$. With the new hypothesis, this implies that $b \in A$ which is again a contradiction. Thus (as before) $B = \emptyset$, and $A = \mathbb{N}$. □

We remark that there is a slight variation of each version of induction in which the process 'starts' at an integer k instead of 1. For example, the first version becomes: *if $A \subset \mathbb{N}$, $k \in A$, and for every $m \geq k$, $m \in A$ implies that $m + 1 \in A$, then $A = \{k, k + 1, \ldots\}$*.

To illustrate the difference between these two versions of induction, let $\mathcal{P}(n)$ be the number of prime factors of n, and let us try to prove, by induction, that $\mathcal{P}(n) < n$ for $n \geq 2$. For the moment, we shall assume familiarity with the primes, and the factorization of the integers. It is clear that $\mathcal{P}(2) = 1 < 2$. Now consider $\mathcal{P}(m + 1)$. If $m + 1$ is prime then $\mathcal{P}(m + 1) = 1 < m + 1$ as required. If not, then we can write $m + 1 = ab$ and so (obviously) $\mathcal{P}(m + 1) = \mathcal{P}(a)\mathcal{P}(b)$. Now there will be many cases in which neither a nor b is m, so knowing only that $\mathcal{P}(m)$ is true is of no help. However, if we know that $\mathcal{P}(\ell) < \ell$ for all ℓ in $\{2, \ldots m\}$, then

$$\mathcal{P}(m + 1) = \mathcal{P}(a)\mathcal{P}(b) < ab = m + 1$$

as required. Thus, in some circumstances, we do need the second form of induction.

One of the earliest recorded uses of induction is in the book *Arithmeticorum Libri Duo*, written by Francesco Maurolico in 1575, in which he uses induction to show that

$$1 + 3 + 5 + \cdots (2n - 1) = n^2.$$

The term 'mathematical induction' was suggested much later (in 1838) by Augustus DeMorgan. The two forms of induction, and the Well-Ordering Principle are all, in fact, equivalent to each other. The logical foundation for the natural numbers was given, in 1889, by Giuseppe Peano who reduced the theory to five simple axioms, one of which was the Principle of Induction.

Let us now consider the divisibility properties of integers. An integer m *divides* an integer n, and m is a *divisor*, or *factor*, of n, if $n = mk$ for some integer k. The integer n, where $n \geq 2$, has 1 and n as factors, and n is said to be a *prime*, or is a *prime number*, if these are its only positive factors. Suppose, for example, that a is a positive factor of 2. Then $2 = ab$ for some b, so that $2 = ab \geq a$. Thus a is 1 or 2, so that 2 is a prime. As 2 is not a factor of 3 (else for some $a \neq 1$, $3 = 2a \geq 4$), we see that 3 is also prime. The following important result is proved using the *second* form of induction.

The Fundamental Theorem of Arithmetic *Every integer n, where $n \geq 2$, can be factorized into a product of primes in only one way apart from the order of the factors.*

Proof It is worthwhile to consider this proof in detail. First, we show that every integer can be expressed as a product of primes. Let A be the set of integers n, with $n \geq 2$, that can be expressed in this way. Obviously, A contains every prime (for each prime can be considered as a product with just one term in the product). In particular, $2 \in A$. Now suppose that $2, \ldots n$ are all in A, and consider $n + 1$. If $n + 1$ is prime, then it is in A. If not, then we can write $n + 1 = ab$, where $1 < a < n + 1$. It follows that $1 < b < n + 1$, and hence that both a and b are in A. Thus both a and b can be written as a product of primes, and hence so too can their product $n + 1$. We deduce that $n + 1 \in A$ so, by the second form of induction, $A = \{2, 3, \ldots\}$. In other words, every integer n with $n \geq 2$ can be written as a product of primes.

It remains to prove that each n, where $n \geq 2$, can be expressed as such a product in only one way, up to the order of the factors, and we shall also prove this by induction. Let A be the set of n ($n \geq 2$) that can be factorized into a product of primes in only one way apart from the order of the factors. Clearly $2 \in A$. Now suppose that $2, \ldots, n$ are all in A, and consider $n + 1$. If $n + 1$ is prime then $n + 1 \in A$. If not (as we shall now assume), consider two possible prime factorizations of $n + 1$, say $n + 1 = p_1 \ldots p_r = q_1 \ldots q_s$, where r and s are at least two. Clearly, we may assume that $2 \leq p_1 \leq \cdots \leq p_r$, and similarly for the q_i, and that $p_1 \leq q_1$. There are two cases to consider, namely (i) $p_1 = q_1$, and (ii) $p_1 < q_1$.

In case (i), $p_2 \ldots p_r = q_2 \ldots q_s = m$, say, and as $2 \leq m \leq n$ we see that $m \in A$. This means that the two factorizations of m are the same (up to order), and hence the same is true of $n + 1$. This in this case, $n + 1 \in A$.

In case (ii), $p_1 < q_1$ and we shall now show that this cannot happen. Assume that $p_1 < q_1$; then

$$p_1(p_2 \ldots p_r - q_2 \ldots q_s) = q_2 \ldots q_s(q_1 - p_1).$$

As the term on the right is strictly less than $n + 1$, the induction hypothesis implies that both sides of this equation have the same prime factors. As $p_1 < q_j$ for every j, and the q_i are primes, this means that p_1 must be a prime factor of $q_1 - p_1$. Thus for some m, $q_1 - p_1 = mp_1$; hence p_1 divides q_1. As q_1 is prime, and as $2 \leq p_1 < q_1$ this is a contradiction. Finally, as case (ii) cannot occur, we must have $n + 1 \in A$ and so $A = \{2, 3, \ldots\}$. □

The following two results on integers will be of use later.

Theorem 2.1.1 *Let G be a set of integers that is a group with respect to addition. Then, for some integer k, $G = \{kn : n \in \mathbb{Z}\}$.*

Proof Let e be the identity in G. Then $x + e = x$ for every x in G, and taking $x = e$ we see that $e = 0$. If $G = \{0\}$ then the conclusion holds with $k = 0$. Suppose now that $G \neq \{0\}$. Then G contains some non-zero x and its inverse $-x$; thus G contains some positive integer. Let k be the smallest positive integer in G, and let $K = \{kn : n \in \mathbb{Z}\}$. Obviously $k + \cdots + k \in G$; thus $K \subset G$. Now take any g in G and write $g = ak + b$, where $0 \leq b < k$. It follows that b (which is $g - ak$) is in G, and hence, from the definition of k, $b = 0$. Thus $g \in K$ so that $G \subset K$. We conclude that $G = K$. □

We say that two integers a and b are *coprime* if they have no common factor except 1.

Theorem 2.1.2 *Suppose that a and b are coprime integers. Then there are integers u and v such that $au + bv = 1$.*

Proof Let $G = \{ma + nb : m, n \in \mathbb{Z}\}$. It is easy to see that G is a group with respect to addition (we leave the reader to check this) so, by Theorem 2.1.1, there is a positive integer k such that

$$\{ma + nb : m, n \in \mathbb{Z}\} = G = \{kn : n \in \mathbb{Z}\}.$$

As a is in the set on the left (take $m = 1$ and $n = 0$) we see that $a = kn$ for some n. Thus k divides a and, similarly, k divides b. By assumption, a and b are coprime; thus $k = 1$ and $G = \mathbb{Z}$. It follows that $1 \in G$ and this is the desired conclusion. □

Finally, we comment on rational and irrational numbers. The *rational numbers* are numbers of the form m/n, where m and n are integers with $n \neq 0$, and $\mathbb{Q}$ denotes the set of rational numbers. A real number that is not rational is said to be *irrational*. Not every real number is rational; for example, $\sqrt{2}$ is not. More generally, suppose that n is a positive integer and that $\sqrt{n}$ is rational. Then we can write $\sqrt{n} = p/q$, where p and q are non-zero integers. By cancelling common factors we may assume that p and q have no common prime factor. Now $p^2 = nq^2$ so that every prime factor of q divides p^2, and hence (from the Fundamental Theorem of Arithmetic) it also divides p. We deduce that q has no prime factor; thus $q = 1$ and $n = p^2$. This shows that *if n is a positive integer then $\sqrt{n}$ is either an integer or is irrational*. Once the concept of *length* has been defined it can be shown that $\mathbb{Q}$ has length zero; as $\mathbb{R}$ has infinite length, we see that (in some sense) 'most' real numbers are irrational.

Exercise 2.1

1. Show that $\sqrt{2/3}$ is irrational. Use the prime factorization of integers to show that if $\sqrt{p/q}$ is rational, where p and q are positive integers with no common factors, then $p = r^2$ and $q = s^2$ for some integers r and s.
2. Show that if x is rational and y is irrational, then $x + y$ is irrational. Show that if, in addition, $x \neq 0$, then xy is irrational.
3. Find two irrational numbers whose sum is rational. Find two irrational numbers whose sum is irrational.
4. Let a and b be real numbers with $a < b$. Show that there are infinitely many rational numbers x with $a < x < b$, and infinitely many irrational numbers y with $a < y < b$. Deduce that there is no smallest positive irrational number, and no smallest positive rational number.

2.2 The real numbers

Consider the set $\mathbb{R}$ of real numbers with the usual operations of addition $x + y$ and multiplication xy. We recognize immediately that $\mathbb{R}$ is an abelian group with respect to $+$; explicitly,

(1a) if x and y are in $\mathbb{R}$, then so is $x + y$;
(2a) if x, y and z are in $\mathbb{R}$, then $x + (y + z) = (x + y) + z$;
(3a) there is a number 0 such that for all x in $\mathbb{R}$, $x + 0 = x = 0 + x$;

(4a) for each x there is some $-x$ such that $x + (-x) = 0 = (-x) + x$;
(5a) for all x and y, $x + y = y + x$.

Notice that in (4a) we have replaced the symbol g^{-1} for the inverse of g in an abstract group by the usual symbol $-x$.

We leave the reader to verify that the set $\mathbb{R}^{\#}$ of *non-zero* real numbers is an abelian group with respect to multiplication. Here, the identity element is 1, and the inverse of x is $1/x$ (which is also written x^{-1}). The operations of addition and multiplication in $\mathbb{R}$ are not independent of each other for they satisfy the *Distributive Laws*:

$$x(y + z) = xy + xz, \quad (y + z)x = yx + zx.$$

One of the fundamental properties of $\mathbb{R}$ is that $0x = 0 = x0$ for every x. To see this, note first that as 0 is the additive identity, $0 + 0 = 0$. Thus, from the Distributive Law,

$$x0 + x0 = x(0 + 0) = x0 = x0 + 0. \tag{2.2.1}$$

If we subtract $x0$ from both sides of this equation we see that $x0 = 0$ and, by commutativity, $0x = x0 = 0$. As $0 \neq 1$ this means that there is no real number y such that $y0 = 1 = 0y$; thus 0 has no inverse with respect to multiplication, and *this is why we cannot divide by zero*. In particular, $\mathbb{R}$ is not a group with respect to multiplication.

Exercise 2.2

1. Show that the set $\mathbb{R}^{\#}$ of non-zero real numbers is a group with respect to multiplication.
2. Show that the set of non-zero numbers of the form $a + b\sqrt{2}$, where a and b are rational, is a group with respect to multiplication.

2.3 Fields

The algebraic properties of $\mathbb{R}$ described in Section 2.2 are so important that we give a name to any system that possesess them.

Definition 2.3.1 A *field* $\mathbb{F}$ is a set with two binary operations $+$ and $\times$, which we call addition and multiplication, respectively, such that

(1) $\mathbb{F}$ is an abelian group with respect to $+$ (with identity $0_{\mathbb{F}}$);
(2) $\{x \in \mathbb{F} : x \neq 0_{\mathbb{F}}\}$ is an abelian group with respect to $\times$ (with identity $1_{\mathbb{F}}$);
(3) the distributive laws hold. $\square$

We have seen that $\mathbb{R}$ with the usual definitions of $+$ and $\times$ is a field. The reader should check that the set $\mathbb{Q}$ of rational numbers is also a field. The set $\mathbb{Z}$ of integers is a group with respect to $+$, but it not a field because ± 1 are the only integers whose multiplicative inverse is an integer. The only sets of integers that are groups with respect to multiplication are $\{1\}$ and $\{-1, 1\}$. A slightly less trivial example of a field is the set $\{a + b\sqrt{2} : a, b \in \mathbb{Q}\}$, again with the usual definitions of addition and multiplication. We leave the reader to verify this, with the hint that

$$(a + b\sqrt{2})^{-1} = \left(\frac{a}{a^2 - 2b^2}\right) + \left(\frac{-b}{a^2 - 2b^2}\right)\sqrt{2}.$$

Note that $a^2 - 2b^2 \neq 0$ here because otherwise, $\sqrt{2}$ would be rational.

One important property that is shared by all fields (and which we have seen that $\mathbb{R}$ has) is that for all x in $\mathbb{F}$, $x \times 0_\mathbb{F} = 0_\mathbb{F} = 0_\mathbb{F} \times x$. The proof for a general field is exactly the same as for $\mathbb{R}$; see (2.2.1).

Exercise 2.3

1. Let $\mathbb{F}$ be a field. Show that the only solutions of the equation $x^2 = x$, where x is in $\mathbb{F}$, are $0_\mathbb{F}$ and $1_\mathbb{F}$.
2. Prove that in any field, $x^2 - y^2 = (x + y) \times (x - y)$. You should state at each step of your proof which property of the field is being used. Deduce that in any field, $x^2 = y^2$ if and only if $x = \pm y$.
3. Show that in any field, $x^3 - y^3 = (x - y)(x^2 + xy + y^2)$.

2.4 Modular arithmetic

We end this chapter with a discussion of *modular arithmetic* which provides us with a rich source of examples. Let n be a positive integer, and let

$$\mathbb{Z}_n = \{0, 1, \ldots, n - 1\};$$

this is the set of all possible 'remainders' after any integer is divided by n. Addition modulo n, which we denote by $\oplus_n$, is defined for any integers x and y as follows. Given x and y, we can write $x + y = an + b$, where the integers a and b are uniquely determined by the condition that $0 \leq b < n$, in other words, by the condition that $b \in \mathbb{Z}_n$. We now define $x \oplus_n y = b$ so that in all cases $x \oplus_n y \in \mathbb{Z}_n$. As examples, we have $3 \oplus_6 7 = 4$ and $0 \oplus_n (-1) = n - 1$. *Multiplication modulo* n, which we denote by $\otimes_n$ is defined similarly, namely we write $xy = an + b$, where $b \in \mathbb{Z}_n$, and then $x \otimes_n y = b$. As examples, $6 \otimes_7 5 = 2$ and $2 \otimes_4 8 = 0$.

2.4 Modular arithmetic

The definitions of $\oplus_n$ and $\otimes_n$ provide an addition and multiplication on $\mathbb{Z}_n$ in which $x \oplus_n y$ and $x \otimes_n y$ are in $\mathbb{Z}_n$ whenever x and y are. While it is a straightforward exercise to show that $\mathbb{Z}_n$ is a group with respect to $\oplus$ (with identity 0, and $n - x$ as the inverse of x), the multiplication $\otimes_n$ in $\mathbb{Z}_n$ has different properties depending on whether or not n is a prime number. We investigate this in more detail.

If n is not a prime then $n = \ell m$, say, where ℓ and m are non-zero elements of $\mathbb{Z}_n$, and then $\ell \otimes_n m = 0$. For example, $2 \otimes_6 3 = 0$. We have just shown that *if n is not a prime, then $\mathbb{Z}_n$ contains two non-zero elements whose product is zero.* As this cannot be so in any field (see Section 2.3), we conclude that *for non-prime n, $\mathbb{Z}_n$ is not a field.* By contrast, we have the following result.

Theorem 2.4.1 *Let p be a prime; then $\mathbb{Z}_p$ is a field.*

Notice that Theorem 2.4.1 implies that *a field need not be an infinite set*; for example, {0, 1, 2} with addition $\oplus_3$ and multiplication $\otimes_3$ is a field (the reader should verify this directly).

Proof of Theorem 2.4.1 Suppose that p is a prime and that x is a non-zero number in $\mathbb{Z}_p$. Then x and p are coprime so, by Theorem 2.1.2, there are integers u and v with $xu + pv = 1$. Now write $u = ap + b$, where $b \in \mathbb{Z}_p$. Then $xb + (xa + v)p = 1$, so that $x \otimes_p b = 1$ and hence $x^{-1} = b$. This shows that every nonzero element of $\mathbb{Z}_p$ has an inverse with respect to the multiplication $\otimes_p$. The rest of the proof is straightforward, and is left to the reader to complete. □

Exercise 2.4

1. Show that $134 \oplus_{71} 928 = 68$, and that $46 \otimes_{17} 56 = 9$.
2. Show that $x \oplus_n y = (x + y) - n[(x + y)/n]$, where $[x]$ denotes the integral part of x.
3. Find integers u and v such that $31u + 17v = 1$, and hence find the multiplicative inverse of 17 in $\mathbb{Z}_{31}$.
4. Does 179 have a multiplicative inverse in $\mathbb{Z}_{971}$? If so, find it.
5. Show that the integer m in $\mathbb{Z}_n$ has a multiplicative inverse in $\mathbb{Z}_n$ if and only if m and n are coprime.
6. Show that if $x \in \mathbb{Z}_4$ then x^2 is 0 or 1 in $\mathbb{Z}_4$. Deduce that no integer of the form $4k + 3$ can be written as the sum of two squares.
7. Show that if 3 divides $a^2 + b^2$ then it divides both a and b.
8. Show that for some integer m, $3^m = 1$ in $\mathbb{Z}_7$. Deduce that 7 divides $1 + 3^{2001}$.

9. Find all solutions of the equation $x^2 = x$ in $\mathbb{Z}_{12}$. (Note that there are more than two solutions.)
10. Show that the equation $x^{-1} + y^{-1} = (x + y)^{-1}$ has no solutions x and y in $\mathbb{R}$. Show, however, that this equation does have a solution in $\mathbb{Z}_7$.
11. Show that the set $\{2, 4, 8\}$ with multiplication modulo 14 is a group. What is the identity element? Is $\{n, n^2, n^3\}$ a group with respect to multiplication in $\mathbb{Z}_m$, where $m = n + n^2 + n^3$?
12. Let f and g be the functions from $\mathbb{Z}_8$ to itself defined by $f(x) = x \oplus_8 2$ and $g(x) = 5 \otimes_8 x$ (that is, $f(x) = x + 2$ and $g(x) = 5x$, both modulo 8). Show that f and g are permutations of $\mathbb{Z}_8$. Show that f and g commute. What is the smallest positive integer n such that g^n is the identity map?
13. Let $X = \{0, 1, 2, \ldots, 16\}$. Express each of the following permutations of X as a product of disjoint cycles:
    (a) the function f_1 defined by $f_1(x) \equiv x + 5 \bmod 17$;
    (b) the function f_2 defined by $f_2(x) \equiv 2x \bmod 17$;
    (c) the function f_3 defined by $f_3(x) \equiv 3x + 1 \bmod 17$.

3
The complex plane

3.1 Complex numbers

Ordered pairs (x, y) of real numbers x and y arise naturally as the coordinates of a point in the Euclidean plane, and we shall adopt the view that the plane *is* the set of ordered pairs of real numbers. Complex numbers arise by denoting the point (x, y) by a new symbol $x + iy$, and then introducing simple algebraic rules for the numbers $x + iy$ with the assumption that $i^2 = -1$. A *complex number*, then, is a number of the form $x + iy$, and we stress that *this is no more than an alternative notation* for (x, y). Thus we see that $x + iy = u + iv$ if and only if $(x, y) = (u, v)$; that is, if and only if $x = u$ and $y = v$. Complex notation has enormous benefits, not least that while a real polynomial need not have any real roots, it always has complex roots; for example, $x^2 + 1$ has no real roots but it has complex roots i and $-i$. As we identify the point $(x, 0)$ in the plane with the real number x, so we also identify the complex number $x + 0i$ with x. We denote the set of complex numbers by $\mathbb{C}$.

Historically, the very existence of a number i with $i^2 = -1$ was in doubt, and for this reason complex numbers were called 'imaginary numbers'. In 1545 Cardan published his book *Ars magna* in which he discussed solutions of the simultaneous equations $x + y = 10$ and $xy = 40$. He used complex numbers in a purely symbolic way to obtain the 'complex' solutions $5 \pm \sqrt{-15}$ of these equations, and later he went on to study cubic equations (see Section 3.6). Wallis (1616–1703) realized that real numbers could be represented on a line, and he made the first real attempt to represent complex numbers as points in the plane. Later, Wessel (1797), Gauss (around 1800) and Argand (1806) all successfully represented complex numbers as points in the plane. We call x the *real part*, and y the *imaginary part*, of the complex number $x + iy$; these terms were introduced by Descartes (1596–1650) whose name later gave rise to the term *cartesian coordinates*. Gauss introduced the term 'complex number' in 1832.

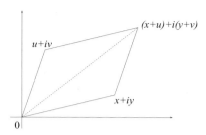

Figure 3.1.1

For us, the complex number i is simply the ordered pair $(0, 1)$, which certainly does exist, and after we have defined multiplication of complex numbers we shall be able to check without difficulty that $i^2 = -1$.

We shall now introduce the arithmetic of complex numbers. Many readers will be familiar with vector addition in the plane, namely

$$(x, y) + (u, v) = (x + u, y + v),$$

and the addition of complex numbers is defined simply by rewriting this in our new notation; thus

$$(x + iy) + (u + iv) = (x + u) + i(y + v);$$

see Figure 3.1.1. With this, it is easy to see that the complex numbers form a group with respect to addition. The identity is $0 + i0$, the additive inverse of $x + iy$ is $-x + i(-y)$, and subtraction is the addition of the additive inverse.

To motivate the definition of multiplication of complex numbers we compute a product of complex numbers in a purely formal way, treating i as any other number, and assuming that $ia = ai$ for every real a; thus we obtain

$$(x + iy)(u + iv) = xu + i^2 yv + i(xv + yu).$$

To convert this product into a complex number we have to specify the value of i^2 and the whole theory rests on the specification that $i^2 = -1$. Putting these tentative steps aside, we start afresh and now *define* the product of two complex numbers by the rule

$$(x + iy)(u + iv) = (xu - yv) + i(xv + yu). \qquad (3.1.1)$$

It is natural to adopt the convention of ignoring 0's and 1's in the obvious way so that, for example, we write $0 + i0, 0 + iy, x + i0$ and $x + i1$ as $0, iy, x$ and $x + i$, respectively. Strictly speaking, we should check that this convention will not lead to any inconsistencies in our arithmetic; it does not and we shall take this for granted. Now, *as a consequence* of definition (3.1.1), we have

$$i^2 = (0 + i1)(0 + i1) = -1 + i0 = -1.$$

In fact, $(x + iy)^2 = -1$ if and only if $x^2 - y^2 = -1$ and $xy = 0$; that is, if and only if $x + iy = \pm i$. In short, the quadratic equation $z^2 + 1 = 0$ has exactly two roots, namely i and $-i$.

To find the multiplicative inverse z^{-1} of a complex number z we adopt a common strategy in mathematics, namely we first *assume* that the inverse exists and then find an explicit expression for it. It is then quite legitimate to start with this expression (which does exist) and then verify that it has the desired properties. Given the complex number $z = x + iy$, then, we assume that z^{-1} exists, write $z^{-1} = u + iv$, and then impose the condition $zz^{-1} = 1$. This yields

$$(xu - yv) + i(xv + yu) = 1,$$

so that $xu - yv = 1$ and $xv + yu = 0$. Thus $u = x/(x^2 + y^2)$ and $v = -y/(x^2 + y^2)$. So, for any non-zero z, say $z = x + iy$ (where $x^2 + y^2 \neq 0$), we define

$$z^{-1} = \frac{x}{x^2 + y^2} + i\left(\frac{-y}{x^2 + y^2}\right) = \frac{x - iy}{x^2 + y^2}. \quad (3.1.2)$$

It is now easy to verify that $zz^{-1} = 1 = z^{-1}z$, so that z^{-1} is indeed the multiplicative inverse of z; thus *every non-zero complex number has a multiplicative inverse*. Finally, we define division by

$$\frac{z}{w} = zw^{-1}. \quad (3.1.3)$$

We have now defined addition and multiplication in $\mathbb{C}$, and these obey the same rules as in $\mathbb{R}$ with the added convention that $i^2 = -1$. In fact, with these operations the complex numbers form a field. The verification of this is tedious but elementary, and we omit the details.

Theorem 3.1.1 *The set $\mathbb{C}$ of complex numbers is a field with respect to the addition and multiplication defined above.*

The *complex conjugate* $\bar{z}$ of z is given by $\bar{z} = x - iy$, where $z = x + iy$, and geometrically, z and $\bar{z}$ are mirror images of each other in the real axis. We leave the reader to verify that

$$\overline{z + w} = \bar{z} + \bar{w}, \quad \overline{(zw)} = \bar{z}\bar{w}, \quad z\bar{z} = x^2 + y^2,$$

so, in particular, $z\bar{z}$ is real and non-negative. If $z \neq 0$, and $w = 1/z$, then $\bar{z}\bar{w} = \overline{zw} = \bar{1} = 1$, so that $\bar{w} = 1/\bar{z}$. More generally, if $w \neq 0$ then

$$\overline{z/w} = \bar{z}/\bar{w}.$$

Next, for any complex numbers a and b with $b \neq 0$, we can always write a/b as $a\bar{b}/b\bar{b}$, where the denominator $b\bar{b}$ of the second quotient is real and positive. As an illustration,

$$\frac{2-i}{1+i} = \frac{(2-i)(1-i)}{(1+i)(1-i)} = \tfrac{1}{2}(1-3i).$$

Let $z = x + iy$, and recall that x and y are the *real part* and *imaginary part* of z, respectively. We denote these by Re[z], and Im[z], and we leave the reader to check that

$$\text{Re}[z] = x = \tfrac{1}{2}(z+\bar{z}), \quad \text{Im}[z] = y = \tfrac{1}{2i}(z-\bar{z}).$$

The *modulus* $|z|$ of z is defined by

$$|z| = \sqrt{x^2 + y^2}, \tag{3.1.4}$$

where we take the non-negative square root of the real number $x^2 + y^2$. The basic properties of the modulus are

$$|\text{Re}[z]| \leq |z|, \quad |\text{Im}[z]| \leq |z|; \tag{3.1.5}$$

$$|\bar{z}| = |z|, \quad z\bar{z} = |z|^2; \tag{3.1.6}$$

$$|zw| = |z||w|; \tag{3.1.7}$$

$$|z+w| \leq |z| + |w| \tag{3.1.8}$$

and these are easily proved. The first inequality in (3.1.5) is true because

$$|\text{Re}[z]| = |x| = \sqrt{x^2} \leq \sqrt{x^2 + y^2} = |z|,$$

and the second inequality is proved similarly. Next, (3.1.6) follows immediately from the definition $\bar{z} = x - iy$, and (3.1.7) holds because, by (3.1.6),

$$|zw|^2 = (zw)(\overline{zw}) = (zw)(\bar{z}\bar{w}) = (z\bar{z})(w\bar{w}) = |z|^2|w|^2.$$

Finally, (3.1.8) is obvious if $z + w = 0$. If not, then, from (3.1.5), we have

$$1 = \text{Re}\left[\frac{z+w}{z+w}\right]$$

$$= \text{Re}\left[\frac{z}{z+w}\right] + \text{Re}\left[\frac{w}{z+w}\right]$$

$$\leq \frac{|z|}{|z+w|} + \frac{|w|}{|z+w|},$$

3.1 Complex numbers

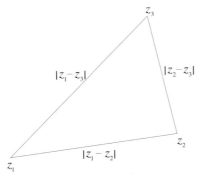

Figure 3.1.2

which is (3.1.8). Of course, repeated applications of (3.1.8) show that

$$|z_1 + \cdots + z_n| \leq |z_1| + \cdots + |z_n|.$$

If $z = x + iy$ and $w = u + iv$, then

$$|z - w|^2 = (x - u)^2 + (y - v)^2,$$

so that, by Pythagoras' theorem, $|z - w|$ is the distance between the points z and w in the plane. Our geometric intuition tells us that the length of any side of a triangle is at most the sum of the lengths of the other two sides. A formal statement of this result, which is known as the *triangle inequality*, and a proof of it (that does not rely on our intuition) now follows; see Figure 3.1.2.

Theorem 3.1.2 *For all complex numbers z_1, z_2 and z_3,*

$$|z_1 - z_3| \leq |z_1 - z_2| + |z_2 - z_3|. \tag{3.1.9}$$

Proof We simply put $z = z_1 - z_2$ and $w = z_2 - z_3$ in (3.1.8). □

Finally, we mention a useful inequality that is related to (3.1.8), and that gives a *lower bound* of $|z + w|$.

Theorem 3.1.3 *For all z and w, $|z \pm w| \geq \big||z| - |w|\big|.$*

Proof For all z and w we have

$$|z| = |(z + w) + (-w)| \leq |z + w| + |-w| = |z + w| + |w|.$$

This, and a second inequality obtained by interchanging z and w in this, gives the stated inequality for $z + w$. If we now replace w by $-w$ we obtain the inequality for $z - w$. This second inequality is illustrated in Figure 3.1.3 in the case when $|z| > |w|$. □

The complex plane

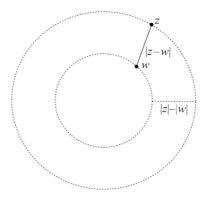

Figure 3.1.3

Exercise 3.1

1. Show that $i^{-1} = -i$, $(1+i)^{-1} = \frac{1}{2}(1-i)$, and $(1+i)^2 = 2i$.
2. Show that $z^2 = 2i$ if and only if $z = \pm(1+i)$.
3. Show that $\overline{zw} = \bar{z}\bar{w}$.
4. Verify directly that $zw = 0$ if and only if $z = 0$ or $w = 0$.
5. Suppose that $zw \neq 0$. Show that the segment joining 0 to z is perpendicular to the segment joining 0 to w if and only if $\text{Re}[z\bar{w}] = 0$.
6. Let T be a triangle in $\mathbb{C}$ with vertices at 0, w_1 and w_2. By applying the mapping $z \mapsto \bar{w}_2 z$, show that the area of T is $\frac{1}{2}|\text{Im}[w_1\bar{w}_2]|$.
7. Show that for any positive integer n,
$$z^n - w^n = (z-w)(z^{n-1} + z^{n-2}w + \cdots + zw^{n-2} + w^{n-1}).$$
Deduce that $z^3 - w^3 = (z-w)^3 + 3zw(z-w)$.
8. Prove (by induction) the *binomial theorem*: for any positive integer n, and any complex numbers z and w,
$$(z+w)^n = \sum_{k=0}^{n} \binom{n}{k} z^k w^{n-k}.$$

3.2 Polar coordinates

Given a non-zero complex number z, the *modulus* $|z|$ of z is the length of the line segment from 0 to z. The *argument* $\arg z$ of z is the angle θ between the positive real axis and the segment from 0 to z measured in the anti-clockwise direction: see Figure 3.2.1. It is clear that if $z = x + iy$, then

$$x = |z|\cos\theta, \quad y = |z|\sin\theta, \tag{3.2.1}$$

3.2 Polar coordinates

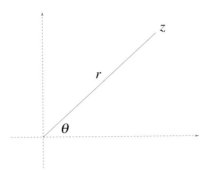

Figure 3.2.1

where $\theta = \arg z$. It is customary to write $r = |z|$ and $\theta = \arg z$; then (r, θ) are the *polar coordinates* of z. If $z = 0$ then $r = 0$ but $\arg z$ is not defined. The transition between the real and imaginary parts x and y of z and the polar coordinates (r, θ) of z is given by (3.1.4) and (3.2.1).

It is important to realize that $\arg z$ is determined by z from (3.2.1) only *to within an integer multiple of* 2π; that is, if θ is one value of $\arg z$, then $\theta + 2n\pi$ is another value for any integer n. We can, if we wish, insist that θ is chosen so as to satisfy $0 \leq \theta < 2\pi$ or, if we prefer, $-\pi \leq \theta < \pi$, and in both cases, the resulting θ would be unique. However, *there is no choice of θ that is universally advantageous*, so it is better not to prejudice our thinking by giving prominence to one choice over any other. We shall agree, then, to leave $\arg z$ determined only up to the addition of an integer multiple of 2π. Despite this ambiguity (which causes no problems), the values $\cos(\arg z)$ and $\sin(\arg z)$ are uniquely determined by z because the trigonometric functions $\cos$ and $\sin$ (whose properties we assume here) are periodic with period 2π.

Neither one of the equations in (3.2.1) is by itself sufficient to determine $\arg z$ (even to within an integral multiple of 2π). Moreover, although these two equations give the single equation

$$\arg(z) = \tan^{-1}\left(\frac{y}{x}\right),$$

this equation is also insufficient to determine $\arg z$ to within a multiple of 2π for, as $\tan(\theta + \pi) = \tan \theta$, it will only determine θ to within an integral multiple of π. As there is often confusion about this matter, we give pause to give a single formula for $\arg z$.

Suppose that we restrict z to lie in the complex plane from which the negative real axis (including 0) has been is removed (the condition for this is that $x + |z| \neq 0$); see Figure 3.2.2. Then we can always choose a *unique* value θ of $\arg z$

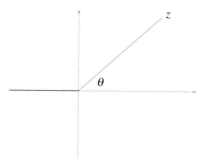

Figure 3.2.2

so that $-\pi < \theta < \pi$, and this value of θ is given by a single formula which, incidentally, confirms that $\arg z$ varies continuously with z.

Theorem 3.2.1 *Suppose that z is not real and negative, or zero, and let θ be the unique value of $\arg z$ that satisfies $-\pi < \theta < \pi$. Then*

$$\theta = 2\tan^{-1}\left(\frac{y}{x+|z|}\right).$$

In particular, θ varies continuously with z.

Proof The given formula is an immediate consequence of the identity $\tan(\theta/2)(1+\cos\theta) = \sin\theta$, and the equations $x = |z|\cos\theta$ and $y = |z|\sin\theta$. The main point here is that we must consider $\theta/2$ rather than θ because $\tan^{-1}$ is single valued on the interval $(-\pi/2, \pi/2)$ but not on $(-\pi, \pi)$. We shall not give a formal proof that θ varies continuously with z, but it is clear from the formula for θ that this must be so with any reasonable definition of continuity. □

We have already assumed that the reader is familiar with the trigonometric functions, and it is now convenient to write

$$\cos\theta + i\sin\theta = e^{i\theta} = \exp(i\theta), \qquad (3.2.2)$$

where the last two expressions are *alternative notations for the left-hand side*. With this, (3.1.4) and (3.2.1) yield $z = re^{i\theta}$, and we call this the *polar form* of the complex number z. At some stage the reader will learn that

$$\exp z = \sum_{n=0}^{\infty} \frac{z^n}{n!},$$

and this gives an alternative interpretation of (3.2.2), but we do not need this here. In this context, (3.2.2) was proved in 1748 by Euler, although it was apparently known to Cotes in 1714, and it produces what is probably the most striking formula in mathematics, namely

$$e^{i\pi} = -1.$$

The fundamental property of $e^{i\theta}$ is as follows.

Theorem 3.2.2 $e^{i\theta} = 1$ *if and only if for some integer n*, $\theta = 2n\pi$.

Proof By definition, $e^{i\theta} = 1$ if and only if $\cos\theta = 1$ and $\sin\theta = 0$. Now $\sin\theta = 0$ if and only if $\theta = m\pi$ for some integer m, and $\cos m\pi = 1$ if and only if m is even. □

Next, we derive the polar form of the product zw in terms of the polar forms of z and w.

Theorem 3.2.3 *If* $z = r_1 e^{i\theta_1}$, $w = r_2 e^{i\theta_2}$ *then* $zw = (r_1 r_2) e^{i(\theta_1 + \theta_2)}$.

Proof From trigonometry we have

$$\begin{aligned} zw &= (r_1 r_2) e^{i\theta_1} e^{i\theta_2} \\ &= (r_1 r_2)(\cos\theta_1 + i\sin\theta_1)(\cos\theta_2 + i\sin\theta_2) \\ &= (r_1 r_2)(\cos(\theta_1 + \theta_2) + i\sin(\theta_1 + \theta_2)) \\ &= (r_1 r_2) e^{i(\theta_1 + \theta_2)}. \end{aligned}$$

□

Observe that Theorem 3.2.3 shows that

$$\arg(zw) = \arg z + \arg w,$$

and, taking w to be $1/z$ and $\bar{z}$ in turn, we have

$$\arg(1/z) = -\arg z = \arg \bar{z},$$

where, of course, all of these terms are only defined to within an integral multiple of 2π. Finally, taking $r_1 = r_2 = 1$ in Theorem 3.2.3, we obtain

$$e^{i\theta_1} e^{i\theta_2} = e^{i(\theta_1 + \theta_2)}, \qquad (3.2.3)$$

a special case of which is $1/e^{i\theta} = e^{-i\theta}$. Combining (3.2.3) with an argument by induction, we see that for all integers n, $(e^{i\theta})^n = e^{in\theta}$. This proves the following result due to de Moivre and Euler.

Theorem 3.2.4 *For all n*, $(\cos\theta + i\sin\theta)^n = \cos(n\theta) + i\sin(n\theta)$.

Exercise 3.2

1. Show that $1 + i = \sqrt{2}e^{i\pi/4}$ and $\sqrt{3} - i = 2e^{-i\pi/6}$.
2. Show that if $w = re^{i\theta}$ and $w \neq 0$, then $z^2 = w$ if and only if $|z| = \sqrt{r}$ and $\arg z$ is $\theta/2$ or $\theta/2 + \pi$.
3. Show that $z^4 = 1$ if and only if $z \in \{1, i, -1, -i\}$.
4. (i) Show that $|z_1 + \cdots + z_n| \leq |z_1| + \cdots + |z_n|$.
   (ii) Show that if $|\arg z| \leq \pi/4$, then $x \geq 0$ and $|z| \leq \sqrt{2}x$, where $z = x + iy$. Deduce that if $|\arg z_j| \leq \pi/4$ for $j = 1, \ldots, n$, then
$$\frac{|z_1| + \cdots + |z_n|}{\sqrt{2}} \leq |z_1 + \cdots + z_n| \leq |z_1| + \cdots + |z_n|.$$
5. Show that $\cos(\pi/5) = \lambda/2$, where $\lambda = (1 + \sqrt{5})/2$ (the Golden Ratio). [Hint: As $\cos 5\theta = 1$, where $\theta = 2\pi/5$, we see from De Moivre's theorem that $P(\cos \theta) = 0$ for some polynomial P of degree five. Now observe that $P(z) = (1 - z)Q(z)^2$ for some quadratic polynomial Q.]
6. Use De Moivre's theorem and the binomial theorem to show that $\cos n\theta$ is a polynomial in $\cos \theta$. This means that there are polynomials $T_0, T_1, \ldots$ such that $\cos n\theta = T_n(\cos \theta)$. The polynomial T_n is called the n-th *Chebychev polynomial*. By considering appropriate trigonometric identities, show that $T_{n+1}(z) + T_{n-1}(z) = 2zT_n(z)$, and hence show that $T_3(z) = 4z^3 - 3z$.
7. Show that if θ is real then $|e^{i\theta} - 1| = 2\sin(\theta/2)$. Use this to derive Ptolemy's theorem: *if the four vertices of a quadrilateral Q lie on a circle, then $d_1 d_2 = \ell_1 \ell_3 + \ell_2 \ell_4$, where d_1 and d_2 are the lengths of the diagonals of Q, and ℓ_1, ℓ_2, ℓ_3 and ℓ_4 are the lengths of its sides taken in this order around Q.*

3.3 Lines and circles

Complex numbers provide an easy way to describe straight lines and circles in the plane. As $|z - a|$ is the distance between z and a, the circle C with centre a and radius r has equation $|z - a| = r$. This equation is equivalent to $r^2 = (z - a)\overline{(z - a)}$, so the equation of C is

$$z\bar{z} - (\bar{a}z + a\bar{z}) + |a|^2 - r^2 = 0.$$

More generally, the equation $z\bar{z} - (\bar{a}z + a\bar{z}) + k = 0$ has no solution if $k > |a|^2$, a single solution if $k = |a|^2$, and a circle of solutions if $k < |a|^2$.

Any straight line L is the set of points that are equidistant from two distinct points u and v. Thus L has equation $|z - u|^2 = |z - v|^2$ and, after simplification, this is seen to be of the form $\bar{a}z + a\bar{z} + b = 0$, where b is real. Not every equation of the form $az + b\bar{z} + c = 0$ has a solution. Indeed, by taking the real and imaginary parts of such an equation we obtain two linear equations in x and y. The solutions of each of these equations give rise to a line, say L_1 and L_2, respectively, and the set of solutions of the single complex equation is $L_1 \cap L_2$. Thus the set of solutions of the complex equation is either empty, a point, or a line. The three equations $z + \bar{z} = i$, $z + 2\bar{z} = 0$, and $z + \bar{z} = 0$ illustrate each of these cases. The following theorem describes the general situation (as we shall not need this we leave the proof, which is not entirely trivial, as an exercise for the reader).

Theorem 3.3.1 *Suppose that a and b are not both zero. Then the equation $az + b\bar{z} + c = 0$ has*

(1) *a unique solution if and only if $|a| \neq |b|$;*
(2) *no solution if and only if $|a| = |b|$ and $b\bar{c} \neq \bar{a}c$;*
(3) *a line of solutions if and only if $|a| = |b|$ and $b\bar{c} = \bar{a}c$.*

Exercise 3.3

1. Find the radius of the circle whose equation is $z\bar{z} + 5z + 5\bar{z} + 9 = 0$.
2. Find the equation of the line $y = x$ in the form $\bar{a}z + a\bar{z} = b$.
3. Suppose that $b \neq 0$. Show that the equation of the line that passes through the origin in the direction b is $\bar{b}z = b\bar{z}$. What is the line $\bar{b}z = -b\bar{z}$?
4. Suppose that $a \neq 0$. Show that the equation of the line that passes through a, and is in a direction perpendicular to the direction of a is $\bar{a}z + a\bar{z} = 2|a|^2$.
5. Suppose that $a \neq 0$. Show that the equation of the line that passes through z_0, and is in the direction a, is $z\bar{a} - \bar{z}a = z_0\bar{a} - \overline{z_0}a$.
6. Show that, *in general*, there are *exactly two solutions* of the equation $z\bar{z} + az + \bar{b}z + c = 0$, where a, b and c are complex numbers. When are there more than two solutions?

3.4 Isometries of the plane

An *isometry* of the complex plane is a function $f: \mathbb{C} \to \mathbb{C}$ that preserves the distance between points; that is, it satisfies $|f(z) - f(w)| = |z - w|$ for all z and w. Each translation, and each rotation, is an isometry. The next result describes all isometries.

Theorem 3.4.1 *Each of the maps*

$$z \mapsto az + b, \quad z \mapsto a\bar{z} + b, \qquad (3.4.1)$$

where $|a| = 1$, is an isometry, and any isometry is of one of these forms.

Proof It is clear that both of the maps in (3.4.1) are isometries; for example, $|(az+b) - (aw+b)| = |a| |z-w| = |z-w|$. Suppose now that f is an isometry such that $f(0) = 0$, $f(1) = 1$ and $f(i) = i$. If we write $z = x + iy$ and $f(z) = u + iv$ and consider the distances of z, and $f(z)$, from 0, 1 and i we see that

$$u^2 + v^2 = x^2 + y^2,$$
$$(u-1)^2 + v^2 = (x-1)^2 + y^2,$$
$$u^2 + (v-1)^2 = x^2 + (y-1)^2.$$

These equations imply that $u = x$ and $y = v$ so that $f(z) = z$ for all z. A similar argument shows that if f is an isometry and $f(0) = 0$, $f(1) = 1$ and $f(i) = -i$, then $f(z) = \bar{z}$ for all z.

Now suppose that F is any isometry, and let

$$F_1(z) = \frac{F(z) - F(0)}{F(1) - F(0)}.$$

Then $|F(1) - F(0)| = 1$ so that F_1 is an isometry with $F_1(0) = 0$ and $F_1(1) = 1$. This implies that $F_1(i)$ is i or $-i$, and we deduce (from above) that either $F_1(z) = z$ for all z, or $F_1(z) = \bar{z}$ for all z. Both of these cases imply that F is of one of the forms given in (3.4.1). □

Theorem 3.4.1 has the following corollary.

Theorem 3.4.2 *Each isometry f is an invertible map of $\mathbb{C}$ onto itself, and f^{-1} is also an isometry. Moreover, the isometries form a group with respect to the composition of functions.*

Proof First, if $f(z) = az + b$ then $f^{-1}(z) = \bar{a}z - \bar{a}b$, while if $f(z) = a\bar{z} + b$ then $f^{-1}(z) = a\bar{z} - a\bar{b}$. In each case f^{-1} is of one of the forms in (3.4.1), and so is an isometry. It is obvious that if f and g are isometries then fg is also an isometry, and we already know that the composition of functions is associative; see (1.3.1). As the identity map I is an isometry, the proof is complete. □

There are four types of isometries of $\mathbb{C}$, namely *translations*, *rotations*, *reflections* (across a line), and *glide reflections* (a reflection across a line L followed by a non-zero translation along L). The next result shows how to recognize each of these algebraically and, at the same time, shows that every isometry is of one of these types.

Theorem 3.4.3

(i) *Suppose that $f(z) = az + b$, where $|a| = 1$. If $a = 1$ then f is a translation; if $a \neq 1$ then f is a rotation.*

(ii) *Suppose that $f(z) = a\bar{z} + b$, where $|a| = 1$. If $a\bar{b} + b = 0$ then f is a reflection in some line; if $a\bar{b} + b \neq 0$ then f is a glide reflection. In particular, any isometry is of one of the four types listed above.*

Proof Assume that $f(z) = az + b$. If $a = 1$ then f is a translation. If $a \neq 1$, then $f(w) = w$, where $w = b/(1-a)$, and $f(z) - w = a(z - w)$. It is now clear that f is a rotation about w of angle θ, where $a = e^{i\theta}$.

Now assume that $f(z) = a\bar{z} + b$, where $a = e^{i\theta}$. If (ii) is true, then we must be able to write $f = tr$, where r is a reflection in a line L, t is a translation along L (possibly of zero translation length), and where r and t commute. Assuming this is so, then $f^2 = trtr = t^2 r^2 = t^2$, and this tells us how to find t, and also r as $r = t^{-1} f$. As $f^2(z) = z + a\bar{b} + b$, we now define maps t and r by

$$t(z) = z + \tfrac{1}{2}(a\bar{b} + b), \quad r(z) = t^{-1} f(z) = a\bar{z} + \tfrac{1}{2}(b - a\bar{b}).$$

It is clear that t is a translation, and as

$$\tfrac{1}{2}(a\bar{b} + b) = \tfrac{1}{2} e^{i\theta/2}(e^{i\theta/2}\bar{b} + e^{-i\theta/2} b),$$

we see that the translation is in the direction $e^{i\theta/2}$. Next, a simple computation shows that $r^2(z) = z$, and that $r(z) = z$ whenever $z = \tfrac{1}{2} b + \rho e^{i\theta/2}$, where ρ is any real number. As r is not the identity, we see that r is the reflection in the line

$$L = \{\tfrac{1}{2} b + \rho e^{i\theta/2} : \rho \in \mathbb{R}\},$$

and t is a translation of $\tfrac{1}{2}(a\bar{b} + b)$ along the direction of L. It follows that f is a reflection if $a\bar{b} + b = 0$, and a glide reflection if $a\bar{b} + b \neq 0$.

Finally, as any isometry is of one of the forms given in (3.4.1), it follows that any isometry is one of the four types listed above. □

Exercise 3.4

1. Show that if a is real and non-zero then
   (a) $z \mapsto \bar{z} + a$ is a glide reflection along the real axis, and
   (b) $z \mapsto -\bar{z} + ia$ is a glide reflection along the imaginary axis.
2. Find the formulae as in (3.4.1) for each of the following:
   (a) the rotation of angle $\pi/2$ about the point i;
   (b) the reflection in the line $y = x$;
   (c) a reflection in $x = y$ followed by a translation by $1 + i$.
3. If g is a reflection then there are infinitely many lines L satisfying $g(L) = L$. Show that if f is a glide reflection then there is only one line L such that $f(L) = L$; we call this line the *axis* of f. Show that if f is a glide

reflection with axis L, then $\frac{1}{2}(z + f(z))$ lies on L for every z. This shows how to find L (choose two different values of z).
4. Let $f(z) = az + b$ and $g(z) = \alpha z + \beta$, where neither is the identity. Show that $fgf^{-1}g^{-1}$ is a translation. Show also that f commutes with g if and only if either f and g are translations, or f and g have a common fixed point.
5. Suppose that f is a reflection in the line L, and that $f(z) = a\bar{z} + b$. Show that $f(z) = a(\bar{z} - \bar{b})$. As $|a| = 1$ we can write $a = e^{i\theta}$; let $c = e^{i\theta/2}$. By considering the fixed points of f, show that L is given by the equation $c\bar{z} - \bar{c}z = c\bar{b}$.

3.5 Roots of unity

For the remainder of this chapter we shall be concerned with the problem of finding the zeros of a complex polynomial. This section is devoted to the equation $z^n = 1$ and its solutions, namely the *n*-th *roots of unity*. We leave the reader to prove our first result.

Theorem 3.5.1 *The set* $\{z \in \mathbb{C} : |z| = 1\}$ *is a group with respect to multiplication.*

The next result gives the *n*-th roots of unity (see Figure 3.5.1 for the case $n = 8$).

Theorem 3.5.2 *Let n be a positive integer. The n-th roots of unity are the distinct complex numbers* $1, \omega, \omega^2, \ldots, \omega^{n-1}$, *where* $\omega = e^{2\pi i/n}$. *These points are equally spaced around the circle* $|z| = 1$ *starting at 1, and they form a group with respect to multiplication.*

Proof If $z = \omega^m$ then $z^n = (\omega^m)^n = \omega^{mn} = (\omega^n)^m = 1$ as $\omega^n = 1$. Conversely, if $z^n = 1$, write z in polar form as $re^{i\theta}$. Then $r^n e^{in\theta} = 1$, so that $r = 1$ and $n\theta = 2\pi m$ for some integer m. If $m = pn + q$, where $0 \leq q < n$, then $z = \omega^m = \omega^q$, so that $\{z : z^n = 1\} = \{1, \omega, \ldots, \omega^{n-1}\}$. Moreover, the points listed in this set are obviously distinct because $\arg \omega^k = 2\pi k/n$. We leave the proof that these points form a group to the reader. □

The next result is a slight generalization of Theorem 3.5.2.

Theorem 3.5.3 *Let w be any non-zero complex number. Then there are exactly n distinct solutions of the equation* $z^n = w$.

Proof Let $w = Re^{i\varphi}$, and $z_0 = R^{1/n}e^{i\varphi/n}$. Then $z^n = w$ if and only if $(z/z_0)^n = 1$ so the solutions are $z_0, \omega z_0, \ldots, \omega^{n-1}z_0$. □

3.5 Roots of unity

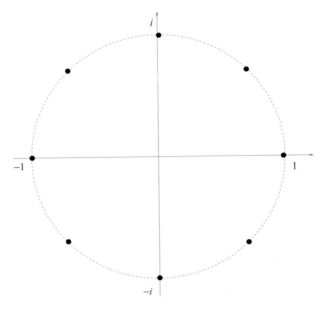

Figure 3.5.1

We end this section with a slight digression in which we apply these ideas to the geometry of a regular polygon. A regular n-gon is a polygon whose n vertices are evenly spaced around a circle, and the angle at a vertex of a regular n-gon is easily seen to be $(n-2)\pi/n$. This means that we can fit k regular n-gons together at a common vertex, filling the plane near the common vertex but without overlapping, precisely when $k(n-2)\pi/n = 2\pi$, and this simplifies to $(n-2)(k-2) = 4$. The solutions of this are easily seen to be $(n, k) = (3, 6), (4, 4)$ or $(6,3)$ and these solutions correspond to six equilateral triangles, four squares and three regular hexagons, respectively, meeting at a point.

Following this idea a little further, we can fit k regular n-gons together at a vertex to form a *non-planar* pyramid-like structure (in three dimensions) if and only if $k(n-2)\pi/n < 2\pi$ or, equivalently, if and only if $(n-2)(k-2) < 4$. The solutions of this are

$$(n, k) = (3, 3), \ (3, 4), \ (3, 5), \ (4, 3), \ (5, 3) \qquad (3.5.1)$$

which correspond to three, four or five triangles meeting at a point, three squares meeting at a point, and three pentagons meeting at a point, respectively. The significance of these solutions is that they are precisely the configurations that can occur at the vertex of a regular polyhedron; for example, exactly three (square) faces of a cube meet at any vertex of the cube. We shall return to this topic later (in Section 5.5, and also when we discuss the symmetry groups of regular polyhedra in Chapter 14).

Exercise 3.5

1. Show that the three cube roots of unity are 1, $(-1 + i\sqrt{3})/2$ and $(-1 - i\sqrt{3})/2$.
2. Suppose that the vertices of a regular pentagon lie on the circle $|z| = 1$. Show that the distance between any two distinct vertices is $2\sin(\pi/5)$ or $2\sin(2\pi/5)$.
3. If we place a unit mass at each vertex of a regular n-gon whose vertices are on the circle $|z| = 1$, the centre of gravity of the masses should be at the origin. Prove (algebraically) that this is so. Let $\omega = e^{2\pi i/n}$, and let k be a positive integer. Show that
$$1 + \omega^k + \cdots + \omega^{k(n-1)} = \begin{cases} n & \text{if } n \text{ divides } k, \\ 0 & \text{otherwise.} \end{cases}$$
4. Show that every arc of positive length on the circle $|z| = 1$ contains points which are roots of unity (for some n), and points which are not roots of unity (for any n).
5. Show that the set of roots of unity for all n (that is, the set z for which $z^n = 1$ for some n) is a group with respect to multiplication.
6. An n-th root of unity z is said to be *primitive* if $z^m \neq 1$ for $m = 1, 2, \ldots, n - 1$. Show that the primitive fourth roots of unity are i and $-i$. Show that there are only two primitive 6-th roots of unity and find them. Show that, for a general n, $e^{2\pi i k/n}$ is a primitive n-th root of unity if and only if k and n have no common divisor other than 1.

3.6 Cubic and quartic equations

Every quadratic equation has two solutions. Suppose now that we want to solve the cubic equation $p_1(z) = 0$, where $p_1(z) = z^3 + az^2 + bz + c$. Now $p_1(z - a/3)$ is a cubic polynomial with no term in z^2 so, by considering this polynomial and relabelling its coefficients, it is sufficient to find the zeros of p, where now

$$p(z) = z^3 + 3bz - c. \tag{3.6.1}$$

The advantage of using this form of the polynomial is that

$$p(z - b/z) = z^3 - \frac{b^3}{z^3} - c,$$

so that $p(z - b/z) = 0$ providing that $z^3 - b^3/z^3 = c$. This quadratic equation in z^3 can be solved to obtain z^3, and hence a value, say ζ, of z. As $\zeta^3 - b^3/\zeta^3 = c$ it follows that $p(\zeta - b/\zeta) = 0$, and we have found a solution of $p(z) = 0$. There is one matter that needs further discussion here. This algorithm apparently gives two values of z^3, and hence six values of ζ. However, the values of ζ are the roots of the equation $z^6 - cz^3 - b^3 = 0$, and if v is a root of this equation, then so (trivially) is $v_1 = -b/v$. As $v_1 - b/v_1 = v - b/v$, we see that these six values of ζ can only provide at most three distinct roots of p.

The Italian del Ferro (1465–1526) is usually credited with the first solution of the general cubic. It is said that he passed the secret onto Tartaglia who subsequently divulged it to Cardan, who then included it in his text *Ars magna* published in 1545. Cardan's student Ferrari then showed how to solve the quartic in the following way. Suppose that $p(z)$ is a quartic polynomial. By considering $p(z - z_0)$ for a suitable z_0, we may assume that $p(z)$ has no term in z^3. Thus it is sufficient to solve the equation $z^4 + az^2 + bz + c = 0$. Now if z is a solution of this equation then, for any w, we have

$$(z^2 + a + w)^2 = z^2(a + 2w) - bz + (a + w)^2 - c. \qquad (3.6.2)$$

Thus if we can choose w such that the right-hand side of this equation is of the form $(uz + v)^2$, then z satisfies

$$z^2 + a + w = \pm(uz + v),$$

and these two quadratic equations can be solved in the usual way. The condition that the right-hand side of (3.6.2) is of the form $(uz + v)^2$ is the familiar condition that this quadratic has repeated roots, and a closer look shows that this is equivalent to saying that w satisfies a certain cubic equation. As we can solve cubics, we can find an appropriate value of w, and hence solve the original quartic equation. Of course, these methods are rather involved, and in general the calculations will be cumbersome. Nevertheless, we have seen that it is possible to solve cubic and quartic equations without developing any further theory. Unfortunately, there is no simple way to solve the general quintic equation.

Exercise 3.6

1. Solve the equation $z^3 - z^2 + z - 1 = 0$ first by inspection, and then by the method described above.
2. Solve the equation $z^3 + 6z = 20$ (this was considered by Cardan in *Ars magna*).

3. Verify that (in the solution of the quartic described above) we need w to satisfy a certain cubic equation.

3.7 The Fundamental Theorem of Algebra

A *polynomial* of *degree* n is a function of z of the form

$$p(z) = a_0 + a_1 z + \cdots + a_n z^n \tag{3.7.1}$$

where $a_0, a_1, \ldots, a_n$ are given complex numbers and $a_n \neq 0$. Frequently we want to find the *zeros*, or *roots*, of p; that is, we wish to solve the equation $p(z) = 0$. We know that if we work only with real numbers, there need not be any roots. The existence of complex solutions of *any* equation $p(z) = 0$, where p is a non-constant polynomial, is an extremely important and non-trivial fact which is known as

The Fundamental Theorem of Algebra. *Let p be given by* (3.7.1), *where $n \geq 1$ and $a_n \neq 0$. Then there are complex numbers $z_1, \ldots, z_n$ such that, for all z,*

$$p(z) = a_n (z - z_1) \cdots (z - z_n). \tag{3.7.2}$$

Much work on the Fundamental Theorem of Algebra was done during the eighteenth century, principally by Leibniz, d'Alembert, Euler, Laplace and Gauss. The first proof for complex polynomials was given by Gauss in 1849. There are now many proofs of this result available, and the proof we sketch below is based on topological arguments. An analytic proof is usually given in a first course on complex analytic functions.

The first (and major) step in any proof of (3.7.2) is to show that *p has at least one root*. We may assume that $a_0 \neq 0$, since if $a_0 = 0$, then $p(0) = 0$. As $n \geq 1$ and $a_n \neq 0$, we may divide by a_n and so assume that $a_n = 1$; henceforth, then, we may assume that

$$p(z) = a_0 + a_1 z + \cdots + z^n, \quad n \geq 1, \quad a_0 \neq 0.$$

To understand why p has a root in this case, we examine the effect of applying the function p to circles centred at the origin. To be specific, let $z = re^{i\theta}$, where $0 \leq \theta \leq 2\pi$. As θ increases from 0 to 2π, so z moves once around the circle with centre the origin and radius r, and the point $p(z)$ traces out some curve C_r. If r is very large, then the term z^n in $p(z)$ has modulus much larger than the sum of all of the other terms and so C_r is not very different from the curve traced out by z^n, namely n revolutions around the circle centred the origin with

3.7 The Fundamental Theorem of Algebra

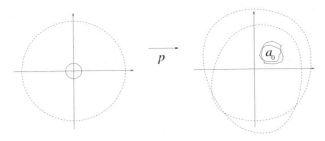

Figure 3.7.1

radius r^n. Thus if r is large, C_r is a curve which begins and ends at the point $p(r)$ (which has large modulus), and which winds n times around the origin (see Figure 3.7.1).

Now suppose that r is very small. In this case, as z traces out the circle of radius r, $p(z)$ stays close to a_0 which, by assumption, is non-zero. It follows that if r is small enough, then C_r lies in the vicinity of the non-zero number a_0 and so does *not* enclose the origin. Now imagine the curve C_r changing continuously as r varies from small to large values; the curve C_r varies continuously from a curve that does not enclose the origin to one that does, and this means that at some stage the curve C_r must pass through the origin. At this stage we have found a zero of p. This is not a formal proof (although it can be converted into one) for there is much that we have not verified; nevertheless, it does indicate why the result is true.

Taking the existence of a root of p for granted, it is easy to see (by induction) that every non-constant polynomial p factorises into n linear factors. Clearly, when $n = 1$ we can write p in the form (3.7.2). Next, suppose that any polynomial with degree $n - 1$ or less factorizes in the manner of (3.7.2) and consider p given by (3.7.1). According to the argument sketched above, there is some z_1 with $p(z_1) = 0$; thus

$$p(z) - p(z_1) = \sum_{k=1}^{n} a_k(z^k - z_1^k).$$

Now for any complex numbers u and v, we have

$$u^k - v^k = (u - v)(u^{k-1} + u^{k-2}v + \cdots + uv^{k-2} + v^{k-1}),$$

so that $z^k - z_1^k = (z - z_1)q_k(z)$, where q_k is a polynomial of degree $k - 1$. We deduce that $p(z) - p(z_1) = (z - z_1)q(z)$, where q is a polynomial of degree $n - 1$ and, by our induction hypothesis $q(z) = a(z - z_2) \cdots (z - z_n)$, where $a \neq 0$, and this gives (3.7.2). Finally, because $\mathbb{C}$ is a field, a product of complex numbers is zero if and only if one of the factors is zero; thus $p(z) = 0$ if and only if some z is some z_j. □

We remark that the problem of finding the zeros of a given polynomial is extremely difficult, and usually the best one can do is to use one of the many computer programs now available for *estimating* (not finding!) the zeros.

Suppose now that p is a *real polynomial*; that is, $p(z) = \sum a_j z^j$, where each a_j is real. Then, by taking conjugates and using the fact that $\bar{a}_j = a_j$, we see that $p(z_0) = 0$ if and only if $p(\bar{z}_0) = 0$. Thus a real polynomial has a number of real zeros, say $x_1, \ldots, x_m$, and a number of pairs of complex (non-real) zeros, say $z_1, \bar{z}_1, \ldots, z_k, \bar{z}_k$. As

$$p(z) = c(z - x_1) \cdots (z - x_m)(z - z_1)(z - \bar{z}_1) \cdots (z - z_k)(z - \bar{z}_k),$$

for some constant c, and as $(z - z_j)(z - \bar{z}_j)$ is a real quadratic polynomial, we see that *any real polynomial can be written as the product of real linear and real quadratic polynomials*. This is often used to obtain a real partial fraction expansion of a real rational function; however, it is no easier to prove this than it is to prove the Fundamental Theorem of Algebra (see Exercise 3.7.1).

Finally, notice that the Fundamental Theorem of Algebra has the following corollary.

Theorem 3.7.1 *Let p and q be polynomials of degree at most n. If $p(z) = q(z)$ at $n + 1$ distinct points then $p(z) = q(z)$ for all z.*

Proof As p and q are of degree at most n, so is $p - q$; thus we can write $p(z) - q(z) = a(z - z_1) \cdots (z - z_k)$, for some a, k and z_i, where $k \leq n$. Now Suppose that $p = q$ at the $n + 1$ distinct points $w_1, \ldots, w_{n+1}$. There is some w_j that is not any of the z_i, and as $p(w_j) - q(w_j) = 0$ we see that $a = 0$. Thus $p = q$. □

We end this chapter with an amusing application of these ideas to prove the following result (due to Cotes in 1716): let $A_1, \ldots, A_n$ *be equally spaced points on a circle of radius one and centre O, and let P be the point on the radius OA_1 at a distance x from O. Then*

$$PA_1.PA_2.\cdots.PA_n = 1 - x^n. \tag{3.7.3}$$

We may assume that the points A_j are the n-th roots of unity, so that in complex notation we have

$$PA_1.PA_2.\cdots.PA_n = \left|(x-1)(x-\omega)\cdots(x-\omega^{n-1})\right|.$$

3.7 The Fundamental Theorem of Algebra

However, we know that

$$(z-1)(z-\omega)\cdots(z-\omega^{n-1}) = z^n - 1,$$

as the two sides are polynomials of degree n with the same set of n distinct zeros, and the same coefficient of z^n. If we now put $z = x$ and equate the moduli of the two sides we obtain (3.7.3).

Exercise 3.7

1. Suppose that we know that any real polynomial is the product of real linear and real quadratic polynomials. Show that this implies that any real polynomial has a root (this is trivial). Now take any complex polynomial p, say $p(z) = \sum a_k z^k$, and let $q(z) = p(z)r(z)$, where $r(z) = \sum \bar{a}_k z^k$. Show that the polynomial q has real coefficients, and hence a root w say. Deduce that w or $\bar{w}$ is a root of p (and so we have 'proved' the Fundamental Theorem of Algebra).
2. Show that all roots of $a + bz + cz^2 + z^3 = 0$ lie inside the circle $|z| = \max\{1, |a|+|b|+|c|\}$.
3. Suppose that $n \geq 2$. Show that
   (i) all roots of $1 + z + z^n = 0$ lie inside the circle $|z| = 1 + 1/(n-1)$;
   (ii) all roots of $1 + nz + z^n = 0$ lie inside the circle $|z| = 1 + 2/(n-1)$.
4. Suppose that ζ is a solution of the $3 - 2z + z^4 + z^5 = 0$. Use the inequalities

$$3 = |2\zeta - \zeta^4 - \zeta^5| \leq 2|\zeta| + |\zeta|^4 + |\zeta|^5,$$
$$|\zeta|^5 = |-\zeta^4 + 2\zeta - 3| \leq 3 + 2|\zeta| + |\zeta|^4,$$

to show that $0{\cdot}89426 < |\zeta| < 1{\cdot}7265$.

4
Vectors in three-dimensional space

4.1 Vectors

A vector is sometimes described as an object having both magnitude and direction, but this is unnecessarily restrictive. The essential algebraic properties of vectors are that we can take real (and sometimes complex) multiples of a vector, and that we can add vectors, and that in each case we obtain another vector. In short, if $v_1, \ldots, v_n$ are vectors, and if $\lambda_1, \ldots, \lambda_n$ are real numbers, then we can form the *linear combination* $\lambda_1 v_1 + \ldots + \lambda_n v_n$ of vectors and *this is again a vector*. This ability to form the linear combinations of vectors is far more important, and more general, than ideas about length and direction. For example, the set of polynomials is a set of vectors (although polynomials have no magnitude or direction), so is the set of solutions of the differential equation $\ddot{x} + x = 0$, the set of sequences of real numbers, the set of functions $f : \mathbb{R} \to \mathbb{R}$, and so on.

In this chapter we shall consider vectors that lie in three-dimensional Euclidean space

$$\mathbb{R}^3 = \{(x_1, x_2, x_3) : x_1, x_2, x_3 \in \mathbb{R}\},$$

and we shall apply them to various problems about the geometry of $\mathbb{R}^3$. Unfortunately, there seems to be little agreement about what vectors in $\mathbb{R}^3$ actually are. Some say that they are points in $\mathbb{R}^3$, some that they are directed line segments, and for others they are classes of line segments, where two segments are in the same class if they determine the same 'displacement'. It seems reasonable to insist that we should not only define what we mean by a vector, but also that thereafter we should be consistent about its use. The cost of consistency, however, is that we cannot simultaneously embrace all of the suggestions made above, for they clearly are different types of objects. In any event, *whatever we choose as our definition of vectors*, we must be careful to distinguish between

4.1 Vectors

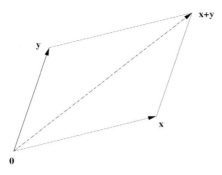

Figure 4.1.1

points and directed line segments. To see why, consider the statement that 'three vectors are coplanar'. Now any three points in $\mathbb{R}^3$ are coplanar, but the generic triple of directed line segments emanating from the origin in $\mathbb{R}^3$ are definitely not coplanar.

In this text a *vector* in $\mathbb{R}^3$ is simply a point of $\mathbb{R}^3$; that is, an ordered triple (x_1, x_2, x_3) of real numbers. As is customary, we shall call the real numbers *scalars* to distinguish them from vectors. To conform with common practice we shall write vectors in boldface type, for example **a**, **b**, ..., but later, when we discuss abstract vector spaces, we shall abandon this notation for it is both unnecessary and awkward. Throughout this text, **x** will be the triple (x_1, x_2, x_3), and the x_j are the *coordinates*, or *components*, of **x**. A similar notation will be used for vectors **a**, **b**, ... and so on. The *origin* **0** in $\mathbb{R}^3$ is the vector $(0, 0, 0)$.

A linear combination of two vectors is defined in the natural way, namely

$$\lambda \mathbf{x} + \mu \mathbf{y} = (\lambda x_1 + \mu y_1, \lambda x_2 + \mu y_2, \lambda x_3 + \mu y_3),$$

where λ and μ are real numbers, and this definition extends in the obvious way to give linear combinations of any finite number of vectors. The addition of vectors (which is defined by taking $\lambda = \mu = 1$ in the above expression) can be illustrated geometrically by noting that $\mathbf{x} + \mathbf{y}$ is the fourth vertex of the parallelogram whose other three vertices are **0**, **x** and **y**; this is often referred to as the *Parallelogram Law of Addition*: see Figure 4.1.1.

There are a variety of simple, and self-evident, rules that hold between vectors and scalars; for example

$$\lambda(\mathbf{x} + \mathbf{y}) = \lambda \mathbf{x} + \lambda \mathbf{y} \quad \lambda \mathbf{x} + \mu \mathbf{x} = (\lambda + \mu) \mathbf{x}, \quad \lambda(\mu \mathbf{x}) = (\lambda \mu) \mathbf{x}. \quad (4.1.1)$$

We shall take these and other equally simple facts for granted. Note that $1\mathbf{x} = \mathbf{x}$ and $0\mathbf{x} = \mathbf{0}$. We write $-\mathbf{x}$ instead of $(-1)\mathbf{x}$. Finally, it is easy to check that *vectors form an abelian group under addition*; the identity vector is **0** and the inverse of **x** is −**x**.

The distance $||\mathbf{x} - \mathbf{y}||$ between **x** and **y** in $\mathbb{R}^3$ is given by

$$||\mathbf{x} - \mathbf{y}||^2 = (x_1 - y_1)^2 + (x_2 - y_2)^2 + (x_3 - y_3)^2$$
$$= ||\mathbf{x}||^2 + ||\mathbf{y}||^2 - 2(x_1 y_1 + x_2 y_2 + x_3 y_3). \qquad (4.1.2)$$

We call $||\mathbf{x}||$, namely $(x_1^2 + x_2^2 + x_3^2)^{1/2}$, the *norm*, or *length*, of **x**. For every λ and every **x**, $||\lambda \mathbf{x}|| = |\lambda| \, ||\mathbf{x}||$, so that $||-\mathbf{x}|| = ||\mathbf{x}||$. A vector **x** is a *unit vector* if $||\mathbf{x}|| = 1$. If $\mathbf{x} \neq \mathbf{0}$, the vector $\mathbf{x}/||\mathbf{x}||$ is the unit vector *in the direction* **x**; the vector −**x** is in the *opposite direction* to **x**. In general, **x** and **y** are in the *same direction* if there are positive numbers λ and μ such that $\lambda \mathbf{x} = \mu \mathbf{y}$.

We shall need to consider directed line segments, and we denote the *directed line segment from the point* **a** *to the point* **b** by [**a**, **b**]. Specifically, [**a**, **b**] is the set of points $\{\mathbf{a} + t(\mathbf{b} - \mathbf{a}) : 0 \leq t \leq 1\}$, with *initial point* **a** and *final point* **b**. The two segments [**a**, **b**] and [**c**, **d**] are *parallel* if **b** − **a** and **d** − **c** are in the same direction.

Next, we recall the familar unit vectors

$$\mathbf{i} = (1, 0, 0), \quad \mathbf{j} = (0, 1, 0), \quad \mathbf{k} = (0, 0, 1);$$

these points lie at a unit distance along the three coordinate axes. Although the notation **i**, **j** and **k** is universal, we shall also use the notation $\mathbf{e}_1$, $\mathbf{e}_2$ and $\mathbf{e}_3$ in place of **i**, **j** and **k**, respectively, for this new notation adapts more easily to higher dimensions. Note that for all **x**, $\mathbf{x} = x_1 \mathbf{i} + x_2 \mathbf{j} + x_3 \mathbf{k}$. Finally, to emphasize a point made earlier, we note that the vectors **i**, **j** and **k** *are* coplanar (for they lie in the plane given by $x_1 + x_2 + x_3 = 1$), but the segments [**0**, **i**], [**0**, **j**] and [**0**, **k**] are not.

Exercise 4.1

1. Verify the rules given in (4.1.1), and show that the vectors form an abelian group with respect to addition.
2. Given vectors **a** and **b**, and positive numbers ℓ_1 and ℓ_2 such that $\ell_1 + \ell_2 = ||\mathbf{a} - \mathbf{b}||$, let **c** be the unique vector on [**a**, **b**] such that $||\mathbf{c} - \mathbf{a}|| = \ell_1$ and $||\mathbf{c} - \mathbf{b}|| = \ell_2$. By writing $\mathbf{c} - \mathbf{a} = t(\mathbf{b} - \mathbf{a})$, for some real t, show that

$$\mathbf{c} = \frac{\ell_2}{\ell_1 + \ell_2} \mathbf{a} + \frac{\ell_1}{\ell_1 + \ell_2} \mathbf{b}.$$

What is the mid-point of the segment [**a**, **b**]?

3. Suppose that $\mathbf{u} = s_1\mathbf{i} + s_2\mathbf{j}$ and $\mathbf{v} = t_1\mathbf{i} + t_2\mathbf{j}$, where s_1, s_2, t_1 and t_2 are real numbers. Find a necessary and sufficient condition on these real numbers such that every vector in the plane of $\mathbf{i}$ and $\mathbf{j}$ can be expressed as a linear combination of the vectors $\mathbf{u}$ and $\mathbf{v}$.

4.2 The scalar product

The formula (4.1.2) motivates the following definition.

Definition 4.2.1 The *scalar product* $\mathbf{x} \cdot \mathbf{y}$ of the two vectors $\mathbf{x}$ and $\mathbf{y}$ is given by $\mathbf{x} \cdot \mathbf{y} = x_1 y_1 + x_2 y_2 + x_3 y_3$. □

The following properties of the scalar product are immediate:

(1) $(\lambda \mathbf{x} + \mu \mathbf{y}) \cdot \mathbf{z} = \lambda(\mathbf{x} \cdot \mathbf{z}) + \mu(\mathbf{y} \cdot \mathbf{z})$;
(2) $\mathbf{x} \cdot \mathbf{y} = \mathbf{y} \cdot \mathbf{x}$;
(3) $||\mathbf{x} - \mathbf{y}||^2 = ||\mathbf{x}||^2 + ||\mathbf{y}||^2 - 2(\mathbf{x} \cdot \mathbf{y})$.
(4) $\mathbf{i} \cdot \mathbf{j} = \mathbf{j} \cdot \mathbf{k} = \mathbf{k} \cdot \mathbf{i} = 0$, and $\mathbf{i} \cdot \mathbf{i} = \mathbf{j} \cdot \mathbf{j} = \mathbf{k} \cdot \mathbf{k} = 1$.

To obtain a geometric interpretation of $\mathbf{x} \cdot \mathbf{y}$ in $\mathbb{R}^3$, consider the triangle with vertices $\mathbf{0}, \mathbf{x}$ and $\mathbf{y}$ and let β be the angle between the segments $[\mathbf{0}, \mathbf{x}]$ and $[\mathbf{0}, \mathbf{y}]$. Applying the cosine rule to this triangle we obtain

$$||\mathbf{x} - \mathbf{y}||^2 = ||\mathbf{x}||^2 + ||\mathbf{y}||^2 - 2||\mathbf{x}|| \, ||\mathbf{y}|| \cos \beta$$

which, by comparison with (3), yields

$$\mathbf{x} \cdot \mathbf{y} = ||\mathbf{x}|| \, ||\mathbf{y}|| \cos \beta. \qquad (4.2.1)$$

Note that (4.2.1) shows that if $\mathbf{y}$ is a unit vector then $|\mathbf{x} \cdot \mathbf{y}|$ *is the length of the projection of* $[\mathbf{0}, \mathbf{x}]$ *onto the line through* $\mathbf{0}$ *and* $\mathbf{y}$; see Figure 4.2.1. We say that $\mathbf{x}$ and $\mathbf{y}$ are *orthogonal*, or *perpendicular*, and write $\mathbf{x} \perp \mathbf{y}$, if the angle between $[\mathbf{0}, \mathbf{x}]$ and $[\mathbf{0}, \mathbf{y}]$ is $\pi/2$ and, from (4.2.1), this is so if and only if $\mathbf{x} \cdot \mathbf{y} = 0$; thus the scalar product gives a convenient test for orthogonality.

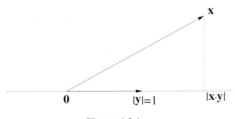

Figure 4.2.1

We give two examples in which the scalar product is used to calculate angles.

Example 4.2.2 We calculate the angle θ between the segments $[\mathbf{a}, \mathbf{b}]$ and $[\mathbf{a}, \mathbf{c}]$, where $\mathbf{a} = (1, 2, 3)$, $\mathbf{b} = (-1, 0, 1)$ and $\mathbf{c} = (1, -2, 5)$. Clearly, the translated segments $[\mathbf{0}, \mathbf{b} - \mathbf{a}]$ and $[\mathbf{0}, \mathbf{c} - \mathbf{a}]$ also meet at an angle θ, so that

$$(\mathbf{b} - \mathbf{a}) \cdot (\mathbf{c} - \mathbf{a}) = ||\mathbf{b} - \mathbf{a}||\,||\mathbf{c} - \mathbf{a}|| \cos \theta.$$

As $\mathbf{b} - \mathbf{a} = (-2, -2, -2)$ and $\mathbf{c} - \mathbf{a} = (0, -4, 2)$, $\cos \theta = 1/\sqrt{15}$. □

Example 4.2.3 Consider the cube in $\mathbb{R}^3$ with vertices $(\varepsilon_1, \varepsilon_2, \varepsilon_3)$, where each ε_j is 0 or 1, and let β be the angle between the diagonals $[\mathbf{0}, \mathbf{i} + \mathbf{k}]$ and $[\mathbf{0}, \mathbf{j} + \mathbf{k}]$ of two faces (the reader should draw a diagram). Each of these diagonals has length $\sqrt{2}$, so from (4.2.1) we see that

$$\sqrt{2}\sqrt{2} \cos \beta = (\mathbf{i} + \mathbf{k}) \cdot (\mathbf{j} + \mathbf{k}) = (\mathbf{i} \cdot \mathbf{j}) + (\mathbf{i} \cdot \mathbf{k}) + (\mathbf{k} \cdot \mathbf{j}) + (\mathbf{k} \cdot \mathbf{k}) = 1.$$

This shows that $\beta = \pi/3$ (and this can also be seen by noting that the triangle with vertices $\mathbf{0}, \mathbf{i} + \mathbf{k}$ and $\mathbf{j} + \mathbf{k}$ is an equilateral triangle). □

The inequality

$$(\mathbf{x} \cdot \mathbf{y})^2 \leq ||\mathbf{x}||^2\, ||\mathbf{y}||^2 \qquad (4.2.2)$$

is an immediate consequence of (4.2.1), and with (3) this shows that

$$||\mathbf{x} - \mathbf{y}||^2 \leq ||\mathbf{x}||^2 + ||\mathbf{y}||^2 + 2|(\mathbf{x} \cdot \mathbf{y})| \leq \big(||\mathbf{x}|| + ||\mathbf{y}||\big)^2,$$

so that $||\mathbf{x} - \mathbf{y}|| \leq ||\mathbf{x}|| + ||\mathbf{y}||$. If we now replace $\mathbf{x}$ and $\mathbf{y}$ by $\mathbf{x} - \mathbf{y}$ and $\mathbf{z} - \mathbf{y}$, respectively, we obtain the following result.

Theorem 4.2.4. the triangle inequality *For* $\mathbf{x}, \mathbf{y}$ *and* $\mathbf{z}$ *in* $\mathbb{R}^3$,

$$||\mathbf{x} - \mathbf{z}|| \leq ||\mathbf{x} - \mathbf{y}|| + ||\mathbf{y} - \mathbf{z}||.$$

This inequality expresses the fact that the length of any side of a triangle in $\mathbb{R}^3$ with vertices $\mathbf{x}, \mathbf{y}$ and $\mathbf{z}$ is no greater than the sum of the lengths of the other two sides (see Theorem 3.1.2).

We end this section by characterizing the scalar product in terms of the important notion of a linear map. A function $f : \mathbb{R}^3 \to \mathbb{R}$ is *linear* if for all vectors $\mathbf{u}$ and $\mathbf{v}$, and for all scalars λ,

$$f(\mathbf{u} + \mathbf{v}) = f(\mathbf{u}) + f(\mathbf{v}), \quad f(\lambda \mathbf{u}) = \lambda f(\mathbf{u}) \qquad (4.2.3)$$

(that is, f 'preserves' linear combinations). The scalar product $\mathbf{x} \cdot \mathbf{y}$ is linear in

x (for a fixed **y**), and linear in **y** (for a fixed **x**). In fact, the scalar product gives *all* linear maps in the following sense.

Theorem 4.2.5 *The most general linear map* $f : \mathbb{R}^3 \to \mathbb{R}$ *is of the form* $\mathbf{x} \mapsto \mathbf{x} \cdot \mathbf{a}$, *for some* **a** *in* $\mathbb{R}^3$.

Proof Suppose that $f : \mathbb{R}^3 \to \mathbb{R}$ is linear, and let $\mathbf{a} = (a_1, a_2, a_3)$, where $a_1 = f(\mathbf{i})$, $a_2 = f(\mathbf{j})$, $a_3 = f(\mathbf{k})$. Then

$$f(\mathbf{x}) = f(x_1\mathbf{i} + x_2\mathbf{j} + x_3\mathbf{k}) = x_1 f(\mathbf{i}) + x_2 f(\mathbf{j}) + x_3 f(\mathbf{k}) = \mathbf{x} \cdot \mathbf{a}.$$

$\square$

Exercise 4.2

1. Let θ be the acute angle formed by two diagonals of a cube. Show that $\cos\theta = 1/3$, and hence find θ.
2. Let Q be a quadilateral whose sides have lengths ℓ_1, ℓ_2, ℓ_3 and ℓ_4 taken in this order around the quadilateral. Show that the diagonals of Q are orthogonal if and only if $\ell_1^2 + \ell_3^2 = \ell_2^2 + \ell_4^2$. [*Hint*: let the vertices of Q be **a**, **b**, **c** and **d**.] Deduce that the diagonals of a rectangle are orthogonal if and only if the rectangle is a square, and that the diagonals of a parallelogram are orthogonal if and only if the parallelogram is a rhombus.

4.3 The vector product

We shall now study the *vector product* of two vectors. This product is itself a vector, and as we shall see, it is an extremely important and useful tool. Nevertheless, it is not ideal (it is not associative, or commutative), and it seems worthwhile to pause and see that we cannot do much better. We know that $\mathbb{R}$ is a field, and that we can extend addition and multiplication from $\mathbb{R}$ to $\mathbb{C}$ in such a way that $\mathbb{C}$ is a field. Now we may regard $\mathbb{C}$ as the horizontal coordinate plane in $\mathbb{R}^3$, and as addition on $\mathbb{C}$ extends to addition on $\mathbb{R}^3$, it is natural to ask whether we can extend multiplication from $\mathbb{C}$ to $\mathbb{R}^3$ so that $\mathbb{R}^3$ becomes a field. Unfortunately we cannot, and we shall now see why (and in Chapter 6 we prove a similar result for all dimensions).

In this discussion it is convenient to represent the vector **x** in $\mathbb{R}^3$ by the pair (z, x_3), where $z = x_1 + ix_2$. We shall now suppose that there is a multiplication, say $*$, on $\mathbb{R}^3$ which, with the given addition of vectors, makes $\mathbb{R}^3$ into a field. We want this multiplication to embrace the usual product of complex numbers, and the scalar multiple of vectors, so we require that $(z, 0)*(w, 0) = (zw, 0)$ and $(\lambda, 0)*(z, t) = (\lambda z, \lambda t)$ whenever λ is real. As $\mathbb{R}^3$ is closed under $*$, we can write $(i, 0)*(0, 1) = (w, s)$, where $w \in \mathbb{C}$ and $s \in \mathbb{R}$, and it follows from this that

$$(i, 0)*(w, s) = (i, 0)*(w, 0) + (i, 0)*[(s, 0)*(0, 1)]$$
$$= (iw, 0) + (s, 0)*[(i, 0)*(0, 1)]$$
$$= (iw, 0) + (s, 0)*(w, s)$$
$$= (iw + sw, s^2).$$

Next, by the associative law,

$$[(i, 0)*(i, 0)]*(0, 1) = (i, 0)*[(i, 0)*(0, 1)].$$

The left-hand side of this is $(-1, 0)*(0, 1)$, which is $(0, -1)$; the right-hand side is $(i, 0)*(w, s)$, which is $(iw + sw, s^2)$. As $-1 \neq s^2$, we see that no such multiplication $*$ can exist. □

This argument shows that we cannot extend multiplication from $\mathbb{C}$ to $\mathbb{R}^3$ in a way that it is associative, distributive and commutative. Thus we will have to be satisfied with a vector product that only has some of these properties. In fact, the vector product will not be commutative or associative; interesting examples of non-associative operations are rare, and the vector product is one of these. We now define the vector product.

Suppose that the two segments $[\mathbf{0}, \mathbf{x}]$ and $[\mathbf{0}, \mathbf{y}]$ in $\mathbb{R}^3$ do not lie on the same line. The vector product of $\mathbf{x}$ and $\mathbf{y}$ will be a vector $\mathbf{n}$ such that $[\mathbf{0}, \mathbf{n}]$ is orthogonal to $[\mathbf{0}, \mathbf{x}]$ and $[\mathbf{0}, \mathbf{y}]$. In terms of coordinates, this means that

$$n_1 x_1 + n_2 x_2 + n_3 x_3 = 0, \quad n_1 y_1 + n_2 y_2 + n_3 y_3 = 0,$$

and the general solution of these equations is

$$\mathbf{n} = \lambda(x_2 y_3 - x_3 y_2)\mathbf{i} + \lambda(x_3 y_1 - x_1 y_3)\mathbf{j} + \lambda(x_1 y_2 - x_2 y_1)\mathbf{k},$$

for any real λ. This motivates the following definition.

Definition 4.3.1 The *vector product* of the vectors $\mathbf{x}$ and $\mathbf{y}$ is

$$\mathbf{x} \times \mathbf{y} = (x_2 y_3 - x_3 y_2)\mathbf{i} + (x_3 y_1 - x_1 y_3)\mathbf{j} + (x_1 y_2 - x_2 y_1)\mathbf{k}.$$

□

This definition produces the cyclic identities

$$\mathbf{i} \times \mathbf{j} = \mathbf{k}, \quad \mathbf{j} \times \mathbf{k} = \mathbf{i}, \quad \mathbf{k} \times \mathbf{i} = \mathbf{j}, \qquad (4.3.1)$$

and

$$\mathbf{j} \times \mathbf{i} = -\mathbf{k}, \quad \mathbf{k} \times \mathbf{j} = -\mathbf{i}, \quad \mathbf{i} \times \mathbf{k} = -\mathbf{j}.$$

As $\mathbf{i} \times (\mathbf{i} \times \mathbf{j}) \neq \mathbf{0} = (\mathbf{i} \times \mathbf{i}) \times \mathbf{j}$, these identities confirm that the vector product is *not* commutative or associative.

4.3 The vector product

The following properties of the vector product are immediate:

(1) $\mathbf{x} \times \mathbf{y}$ is orthogonal to $\mathbf{x}$ and to $\mathbf{y}$;
(2) $(\lambda\mathbf{x} + \mu\mathbf{y}) \times \mathbf{z} = \lambda(\mathbf{x} \times \mathbf{z}) + \mu(\mathbf{y} \times \mathbf{z})$;
(3) $\mathbf{x} \times \mathbf{y} = -\mathbf{y} \times \mathbf{x}$;
(4) $\mathbf{x} \times \mathbf{y} = \mathbf{0}$ if and only if there are scalars λ and μ, not both zero, such that $\lambda\mathbf{x} = \mu\mathbf{y}$. In particular, $\mathbf{x} \times \mathbf{x} = \mathbf{0}$.

As (4) is so important we give a proof. Suppose that $\lambda\mathbf{x} = \mu\mathbf{y}$, where say, $\mu \neq 0$. Then $\mathbf{y} = (\lambda/\mu)\mathbf{x}$ and hence $\mathbf{x} \times \mathbf{y} = \mathbf{0}$. Conversely, suppose that $\mathbf{x} \times \mathbf{y} = \mathbf{0}$; then $x_i y_j = x_j y_i$ for $i, j = 1, 2, 3$, so that for each j, $x_j \mathbf{y} = y_j \mathbf{x}$. If $\mathbf{x} = \mathbf{0}$, we take $\lambda = 1$ and $\mu = 0$. If $\mathbf{x} \neq \mathbf{0}$, then $x_k \neq 0$ for some k and we take $\lambda = y_k$ and $\mu = x_k$. □

The length of the vector $\mathbf{x} \times \mathbf{y}$ is easily calculated. Indeed,

$$||\mathbf{x} \times \mathbf{y}||^2 = (x_2 y_3 - x_3 y_2)^2 + (x_3 y_1 - x_1 y_3)^2 + (x_1 y_2 - x_2 y_1)^2$$
$$= ||\mathbf{x}||^2 ||\mathbf{y}||^2 - (\mathbf{x} \cdot \mathbf{y})^2 \qquad (4.3.2)$$

which we prefer to write as

$$||\mathbf{x} \times \mathbf{y}||^2 + (\mathbf{x} \cdot \mathbf{y})^2 = ||\mathbf{x}||^2 ||\mathbf{y}||^2. \qquad (4.3.3)$$

This shows that if the angle between $[\mathbf{0}, \mathbf{x}]$ and $[\mathbf{0}, \mathbf{y}]$ is θ, then

$$||\mathbf{x} \times \mathbf{y}|| = ||\mathbf{x}|| . ||\mathbf{y}|| |\sin\theta|.$$

The vectors $\mathbf{x}$ and $\mathbf{y}$ determine a parallelogram $\mathcal{P}$ whose vertices are $\mathbf{0}$, $\mathbf{x}$, $\mathbf{y}$ and $\mathbf{x} + \mathbf{y}$. If we regard the segment $[\mathbf{0}, \mathbf{x}]$ as the base of $\mathcal{P}$, then its height is $||\mathbf{y}|| |\sin\theta|$, and its area is $||\mathbf{x} \times \mathbf{y}||$. When $\mathbf{x}$ and $\mathbf{y}$ lie on a line L through the origin, all of the vertices of $\mathcal{P}$ are on L and the area of $\mathcal{P}$ is zero; this is another way of expressing (4) above.

Finally, we comment on the 'direction' of $\mathbf{x} \times \mathbf{y}$. Given two segments $[\mathbf{0}, \mathbf{x}]$ and $[\mathbf{0}, \mathbf{y}]$ that are not on the same line, the vector product $\mathbf{x} \times \mathbf{y}$ is a vector that is orthogonal to the plane Π that contains them. Which way does $\mathbf{x} \times \mathbf{y}$ 'point'? The answer is given by the so-called 'right-handed corkscrew rule': if we rotate the segment $[\mathbf{0}, \mathbf{x}]$ to the segment $[\mathbf{0}, \mathbf{y}]$, the rotation being in the plane Π and sweeping out *the smaller of the two angles between* $[\mathbf{0}, \mathbf{x}]$ *and* $[\mathbf{0}, \mathbf{y}]$, then $[\mathbf{0}, \mathbf{x} \times \mathbf{y}]$ lies in the direction that a 'right-handed corkscrew' would travel if it were rotated in the same way. This is illustrated by the vector products in (4.3.1); however, it is not mathematics, and we shall make this idea precise in Section 4.6.

Exercise 4.3

1. Suppose that $\mathbf{a} \neq \mathbf{0}$. Show that $\mathbf{x} = \mathbf{y}$ if and only if $\mathbf{a}\cdot\mathbf{x} = \mathbf{a}\cdot\mathbf{y}$ and $\mathbf{a} \times \mathbf{x} = \mathbf{a} \times \mathbf{y}$.
2. Use the vector product to find the area of the triangle with vertices $(1, 2, 0)$, $(2, 5, 2)$ and $(4, -1, 2)$.
3. Prove that $(\mathbf{a} \times \mathbf{b})\cdot(\mathbf{c} \times \mathbf{d}) = (\mathbf{a}\cdot\mathbf{c})(\mathbf{b}\cdot\mathbf{d}) - (\mathbf{a}\cdot\mathbf{d})(\mathbf{b}\cdot\mathbf{c})$.
4. Suppose that $\mathbf{a}$, $\mathbf{b}$ and $\mathbf{c}$ do not lie on any straight line. Show that the normal to the plane that contains them lies in the direction $(\mathbf{a} \times \mathbf{b}) + (\mathbf{b} \times \mathbf{c}) + (\mathbf{c} \times \mathbf{a})$. Consider the special case $\mathbf{c} = \mathbf{0}$.
5. Suppose that $\mathbf{a} \times \mathbf{x} \neq \mathbf{0}$. Define vectors $\mathbf{x}_0, \mathbf{x}_1, \ldots$ by $\mathbf{x}_0 = \mathbf{x}$ and $\mathbf{x}_{n+1} = \mathbf{a} \times \mathbf{x}_n$. As $||\mathbf{x}_n|| \leq ||\mathbf{a}||^n ||\mathbf{x}||$, it is clear that $\mathbf{x}_n \to \mathbf{0}$ if $||\mathbf{a}|| < 1$. What happens as $n \to \infty$ if $||\mathbf{a}|| = 1$, or if $||\mathbf{a}|| > 1$?

4.4 The scalar triple product

There is a natural way to combine three vectors to obtain a scalar.

Definition 4.4.1 The *scalar triple product* $[\mathbf{x}, \mathbf{y}, \mathbf{z}]$ of the vectors $\mathbf{x}, \mathbf{y}$ and $\mathbf{z}$ is defined by $[\mathbf{x}, \mathbf{y}, \mathbf{z}] = \mathbf{x}\cdot(\mathbf{y} \times \mathbf{z})$. □

If $\mathbf{y} = \mathbf{z}$, then $\mathbf{y} \times \mathbf{z} = \mathbf{0}$. If $\mathbf{x} = \mathbf{y}$, or $\mathbf{x} = \mathbf{z}$, then $\mathbf{x} \perp \mathbf{y} \times \mathbf{z}$; thus $[\mathbf{x}, \mathbf{y}, \mathbf{z}] = 0$ unless the three vectors are distinct. We now use this to show that $[\mathbf{x}, \mathbf{y}, \mathbf{z}]$ is invariant under cyclic permutations of $\mathbf{x}, \mathbf{y}$ and $\mathbf{z}$, and changes by a factor -1 under transpositions.

Theorem 4.4.2 *For any vectors $\mathbf{x}, \mathbf{y}$ and $\mathbf{z}$, we have*

$$[\mathbf{x}, \mathbf{y}, \mathbf{z}] = [\mathbf{z}, \mathbf{x}, \mathbf{y}] = [\mathbf{y}, \mathbf{z}, \mathbf{x}]; \tag{4.4.1}$$

$$[\mathbf{x}, \mathbf{y}, \mathbf{z}] = -[\mathbf{x}, \mathbf{z}, \mathbf{y}] = -[\mathbf{y}, \mathbf{x}, \mathbf{z}] = -[\mathbf{z}, \mathbf{y}, \mathbf{x}]. \tag{4.4.2}$$

Proof It suffices to prove (4.4.2) as (4.4.1) follows by repeated applications of (4.4.2) (equivalently, a 3-cycle is the product of two transpositions). As the scalar triple product is linear in each vector, and zero if two of the vectors are the same, we have

$$0 = [\mathbf{x} + \mathbf{y}, \mathbf{x} + \mathbf{y}, \mathbf{z}] = [\mathbf{x}, \mathbf{y}, \mathbf{z}] + [\mathbf{y}, \mathbf{x}, \mathbf{z}];$$

thus $[\mathbf{x}, \mathbf{y}, \mathbf{z}] = -[\mathbf{y}, \mathbf{x}, \mathbf{z}]$. The rest of the proof is similar. □

It is easy to see that $|[\mathbf{x}, \mathbf{y}, \mathbf{z}]|$ is the volume of the parallelepiped $\mathcal{P}'$ formed from the three segments $[\mathbf{0}, \mathbf{x}], [\mathbf{0}, \mathbf{y}]$ and $[\mathbf{0}, \mathbf{z}]$. Indeed, the volume of $\mathcal{P}'$ is the area of the base, namely $||\mathbf{x} \times \mathbf{y}||$, multiplied by the height h of $\mathcal{P}$. As h is the

4.4 The scalar triple product

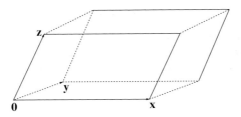

Figure 4.4.1

length of the projection of $\mathbf{z}$ onto the unit vector $(\mathbf{x} \times \mathbf{y}/\|\mathbf{x} \times \mathbf{y}\|)$, this gives the required result (see Figure 4.4.1).

The next result can be predicted from the interpretation of a scalar triple product as the volume of a parallelepiped.

Theorem 4.4.3 *The segments* $[\mathbf{0}, \mathbf{x}]$, $[\mathbf{0}, \mathbf{y}]$ *and* $[\mathbf{0}, \mathbf{z}]$ *are coplanar if and only if* $[\mathbf{x}, \mathbf{y}, \mathbf{z}] = 0$.

Proof We may assume that the three segments lie on different lines through the origin (as the result is trivial otherwise). Then the three segments are coplanar if and only if the segment $[\mathbf{0}, \mathbf{x}]$ is orthogonal to the normal to the plane Π that contains the segments $[\mathbf{0}, \mathbf{y}]$ and $[\mathbf{0}, \mathbf{z}]$. As this normal is in the direction $\mathbf{y} \times \mathbf{z}$, the result follows. □

The last result in this section shows how to find the coordinates of a point $\mathbf{x}$ with reference to any given set of (*not necessarily orthogonal*) coordinates axes along the directions of $\mathbf{a}$, $\mathbf{b}$ and $\mathbf{c}$. This result points the way to the important idea of a *basis* in an abstract vector space, and it says that any such triple $\mathbf{a}$, $\mathbf{b}$ and $\mathbf{c}$ is a basis of $\mathbb{R}^3$.

Theorem 4.4.4 *Suppose that* $[\mathbf{0}, \mathbf{a}]$, $[\mathbf{0}, \mathbf{b}]$ *and* $[\mathbf{0}, \mathbf{c}]$ *are not coplanar. Then any* $\mathbf{x}$ *in* $\mathbb{R}^3$ *can be written as the linear combination*

$$\mathbf{x} = \frac{[\mathbf{x}, \mathbf{b}, \mathbf{c}]}{[\mathbf{a}, \mathbf{b}, \mathbf{c}]}\mathbf{a} + \frac{[\mathbf{a}, \mathbf{x}, \mathbf{c}]}{[\mathbf{a}, \mathbf{b}, \mathbf{c}]}\mathbf{b} + \frac{[\mathbf{a}, \mathbf{b}, \mathbf{x}]}{[\mathbf{a}, \mathbf{b}, \mathbf{c}]}\mathbf{c} \qquad (4.4.3)$$

of the three vectors $\mathbf{a}$, $\mathbf{b}$ *and* $\mathbf{c}$.

Proof As $[\mathbf{0}, \mathbf{a}]$, $[\mathbf{0}, \mathbf{b}]$ and $[\mathbf{0}, \mathbf{c}]$ are not coplanar, $[\mathbf{a}, \mathbf{b}, \mathbf{c}] \neq 0$, and we can write $\mathbf{x} = \lambda \mathbf{a} + \mu \mathbf{b} + \nu \mathbf{c}$ for some scalars λ, μ and ν. As $\mathbf{b} \times \mathbf{c}$ is orthogonal to both $\mathbf{b}$ and $\mathbf{c}$, we see that

$$[\mathbf{x}, \mathbf{b}, \mathbf{c}] = [\lambda \mathbf{a} + \mu \mathbf{b} + \nu \mathbf{c}, \mathbf{b}, \mathbf{c}] = \lambda[\mathbf{a}, \mathbf{b}, \mathbf{c}],$$

and similarly for μ and ν. □

Exercise 4.4

1. Show that $(\mathbf{a} + \mathbf{b}) \cdot \big((\mathbf{b} + \mathbf{c}) \times (\mathbf{c} + \mathbf{a})\big) = 2[\mathbf{a}, \mathbf{b}, \mathbf{c}]$.
2. Show that the four vectors $(0, 0, 0)$, $(-2, 4, s)$, $(1, 6, 2)$ and $(t, 7, 0)$ are coplanar if and only if $6st = 8t + 7s + 28$.
3. Given non-coplanar segments $[\mathbf{0}, \mathbf{a}]$, $[\mathbf{0}, \mathbf{b}]$ and $[\mathbf{0}, \mathbf{c}]$, we say that the vectors $\mathbf{a}'$, $\mathbf{b}'$ and $\mathbf{c}'$ are *reciprocal vectors* to $\mathbf{a}$, $\mathbf{b}$ and $\mathbf{c}$ if

$$\mathbf{a}' = \frac{1}{[\mathbf{a}, \mathbf{b}, \mathbf{c}]} \mathbf{b} \times \mathbf{c}, \quad \mathbf{b}' = \frac{1}{[\mathbf{a}, \mathbf{b}, \mathbf{c}]} \mathbf{c} \times \mathbf{a}, \quad \mathbf{c}' = \frac{1}{[\mathbf{a}, \mathbf{b}, \mathbf{c}]} \mathbf{a} \times \mathbf{b}.$$

Prove that

(i) $[\mathbf{0}, \mathbf{a}']$, $[\mathbf{0}, \mathbf{b}']$ and $[\mathbf{0}, \mathbf{c}']$ are non-coplanar, and
(ii) $\mathbf{a}'' = \mathbf{a}$, $\mathbf{b}'' = \mathbf{b}$, $\mathbf{c}'' = \mathbf{c}$.

4.5 The vector triple product

The scalar triple product produces a scalar from three vectors; the vector triple product produces a vector.

Definition 4.5.1 The *vector triple product* of the three vectors $\mathbf{x}$, $\mathbf{y}$ and $\mathbf{z}$ is $\mathbf{x} \times (\mathbf{y} \times \mathbf{z})$. □

In general, $\mathbf{y}$ and $\mathbf{z}$ determine a plane Π through the origin with normal in the direction $\mathbf{y} \times \mathbf{z}$. As $\mathbf{x} \times (\mathbf{y} \times \mathbf{z})$ is perpendicular to this normal, it lies in Π, and it follows that $\mathbf{x} \times (\mathbf{y} \times \mathbf{z})$ is a linear combination of $\mathbf{y}$ and $\mathbf{z}$, say $\mathbf{x} \times (\mathbf{y} \times \mathbf{z}) = \alpha \mathbf{y} + \beta \mathbf{z}$. As the left-hand side is a linear function of $\mathbf{x}$, so is the right-hand side, and this means that α and β are linear scalar-valued functions of $\mathbf{x}$. According to Theorem 4.2.5, there are vectors $\mathbf{a}$ and $\mathbf{b}$ such that $\mathbf{x} \times (\mathbf{y} \times \mathbf{z}) = (\mathbf{x} \cdot \mathbf{a})\mathbf{y} - (\mathbf{x} \cdot \mathbf{b})\mathbf{z}$. This motivates the following formula for the vector triple product.

Theorem 4.5.2 *For all vectors $\mathbf{x}$, $\mathbf{y}$ and $\mathbf{z}$,*

$$\mathbf{x} \times (\mathbf{y} \times \mathbf{z}) = (\mathbf{x} \cdot \mathbf{z})\mathbf{y} - (\mathbf{x} \cdot \mathbf{y})\mathbf{z}. \tag{4.5.1}$$

Proof If we expand both sides of (4.5.1) in terms of the x_i, y_j and z_k we see that the two expressions are identical. □

This proof is elementary, but tedious in the extreme, so we give another proof that is based on the important notions of linearity and continuity.

A second proof We may suppose that $\mathbf{y}$ and $\mathbf{z}$ lie along different lines through the origin, for otherwise (4.5.1) is trivially true. Then any $\mathbf{x}$ can be written as a linear combination of $\mathbf{y}$, $\mathbf{z}$ and $\mathbf{y} \times \mathbf{z}$. As both sides of (4.5.1) are linear in $\mathbf{x}$, it is only necessary to verify (4.5.1) when **x** is each of these three vectors. When **x** = **y** × **z** all terms in (4.5.1) are zero, so it is true. The two cases **x** = **y** and **x** = **z** are the same (apart from a factor -1 on each side of the equation), so we have reduced our task to giving a proof of the identity

$$\mathbf{y} \times (\mathbf{y} \times \mathbf{z}) = (\mathbf{y} \cdot \mathbf{z})\mathbf{y} - (\mathbf{y} \cdot \mathbf{y})\mathbf{z} \quad (4.5.2)$$

for all **y** and **z**.

Now (4.5.2) is true when **y** and **z** lie along the same line through the origin. Thus, as both sides of (4.5.2) are linear in **z**, it only remains to prove this when **z** is orthogonal to **y**. Moreover, it is clearly sufficient to prove this when **y** and **z** are unit vectors; thus we have to show that if $||\mathbf{y}|| = ||\mathbf{z}|| = 1$, and $\mathbf{y} \perp \mathbf{z}$, then $\mathbf{y} \times (\mathbf{y} \times \mathbf{z}) = -\mathbf{z}$. Now $\mathbf{y} \times (\mathbf{y} \times \mathbf{z})$ is of length one and also a scalar multiple of **z** (because **y**, **z** and $\mathbf{y} \times \mathbf{z}$ are mutually orthogonal unit vectors). Thus it is $\pm \mathbf{z}$, so if we let $\mu = \mathbf{z} \cdot [\mathbf{y} \times (\mathbf{y} \times \mathbf{z})]$, then $\mu = \pm 1$. However, μ is clearly a continuous function of the coefficients of **y** and **z**, so it must be independent of **y** and **z**. As $\mu = -1$ when $\mathbf{y} = \mathbf{i}$ and $\mathbf{z} = \mathbf{j}$, we see that $\mathbf{y} \times (\mathbf{y} \times \mathbf{z}) = -\mathbf{z}$ and the proof is complete. □

Exercise 4.5

1. Prove that $(\mathbf{a} \times \mathbf{b}) \times \mathbf{c} = \mathbf{a} \times (\mathbf{b} \times \mathbf{c})$ if and only if $(\mathbf{a} \times \mathbf{c}) \times \mathbf{b} = \mathbf{0}$.
2. Prove the following vector identities:

   $\mathbf{a} \times (\mathbf{b} \times \mathbf{c}) + \mathbf{b} \times (\mathbf{c} \times \mathbf{a}) + \mathbf{c} \times (\mathbf{a} \times \mathbf{b}) = \mathbf{0}$;

   $(\mathbf{a} \times \mathbf{b}) \times (\mathbf{c} \times \mathbf{d}) = [\mathbf{a}, \mathbf{b}, \mathbf{d}]\mathbf{c} - [\mathbf{a}, \mathbf{b}, \mathbf{c}]\mathbf{d}$;

   $(\mathbf{a} \times \mathbf{b}) \cdot \big((\mathbf{c} \times \mathbf{d}) \times (\mathbf{e} \times \mathbf{f})\big) = [\mathbf{a}, \mathbf{b}, \mathbf{d}][\mathbf{c}, \mathbf{e}, \mathbf{f}] - [\mathbf{a}, \mathbf{b}, \mathbf{c}][\mathbf{d}, \mathbf{e}, \mathbf{f}]$;

   $(\mathbf{b} \times \mathbf{c}) \cdot (\mathbf{a} \times \mathbf{d}) + (\mathbf{c} \times \mathbf{a}) \cdot (\mathbf{b} \times \mathbf{d}) + (\mathbf{a} \times \mathbf{b}) \cdot (\mathbf{c} \times \mathbf{d}) = 0$.

3. Give a geometric proof of (4.5.2) (you may use the 'corkscrew rule').

4.6 Orientation and determinants

Our first task is to convert the 'corkscrew rule' (Section 4.3) into a mathematical result. Consider two non-zero vectors **u** and **v**, *taken in this order*, in the plane given by $x_3 = 0$. Let θ be the angle between the segments $[\mathbf{0}, \mathbf{u}]$ and $[\mathbf{0}, \mathbf{v}]$, measured in the anti-clockwise direction. We wish to formalize the idea that the vectors **u** and **v** are *positively orientated* if $0 < \theta < \pi$, and they are *negatively orientated* if $\pi < \theta < 2\pi$. We can identify each vector $(x_1, x_2, 0)$ with the complex number $x_1 + ix_2$, and if we write $\mathbf{u} = u_1 + iu_2 = |\mathbf{u}|e^{i\alpha}$, and

$\mathbf{v} = v_1 + iv_2 = |\mathbf{v}|e^{i\beta}$, we easily see that

$$u_1 v_2 - u_2 v_1 = |\mathbf{u}|\,|\mathbf{v}|\sin(\beta - \alpha) = |\mathbf{u}|\,|\mathbf{v}|\sin\theta.$$

It is convenient to let

$$\Delta(\mathbf{u}, \mathbf{v}) = u_1 v_2 - u_2 v_1;$$

then $\mathbf{u}$ and $\mathbf{v}$ are positively or negatively orientated according as $\Delta(\mathbf{u}, \mathbf{v})$ is positive or negative. Note that the orientation of $\mathbf{u}$ and $\mathbf{v}$ is now determined *algebraically*, and without reference to the 'clockwise direction', corkscrews, or even angles. Moreover, as $\mathbf{u} = (u_1, u_2, 0)$, and similarly for $\mathbf{v}$, we see that

$$\mathbf{u} \times \mathbf{v} = \Delta(\mathbf{u}, \mathbf{v})\mathbf{k}.$$

Thus $\mathbf{u}$ and $\mathbf{v}$ in the plane $x_3 = 0$ are positively or negatively orientated according as $\mathbf{u} \times \mathbf{v}$ is a positive or negative multiple of $\mathbf{k}$. Thus we have the following simple definition of the orientation of two vectors in $\mathbb{C}$.

Definition 4.6.1 Two vectors $\mathbf{u}$ and $\mathbf{v}$ lying in the complex plane in $\mathbb{R}^3$ are *positively orientated* if $\Delta(\mathbf{u}, \mathbf{v}) > 0$, and are *negatively orientated* if $\Delta(\mathbf{u}, \mathbf{v}) < 0$.

The ordered pair $\mathbf{i}$ and $\mathbf{j}$ are positively orientated; the pair $\mathbf{j}$ and $\mathbf{i}$ are negatively orientated. More generally, as $\Delta(\mathbf{v}, \mathbf{u}) = -\Delta(\mathbf{u}, \mathbf{v})$, $\mathbf{u}$ and $\mathbf{v}$ are positively orientated if and only if $\mathbf{v}$ and $\mathbf{u}$ are negatively orientated. Also, $\mathbf{u}$ and $\mathbf{v}$ are positively orientated if and only if $\mathbf{u} \times \mathbf{v}$ is a *positive* multiple of $\mathbf{k}$. These statements are derived formally from Definition 4.6.1.

The expression $\Delta(\mathbf{u}, \mathbf{v})$ suggests that it will be useful to study the general 2×2 *determinant*

$$\begin{vmatrix} a & b \\ c & d \end{vmatrix} = ad - bc$$

whose entries a, b, c and d are real or complex numbers. If we let $\epsilon_{12} = 1$, $\epsilon_{21} = -1$ and $\epsilon_{11} = \epsilon_{22} = 0$, then

$$\Delta(\mathbf{u}, \mathbf{v}) = \begin{vmatrix} u_1 & u_2 \\ v_1 & v_2 \end{vmatrix} = \sum_{i,j=1}^{2} \epsilon_{ij} u_i v_j.$$

The generalization of 2×2 determinants to $n \times n$ determinants depends on generalizing the symbols ϵ_{ij} from two to n suffices; this uses the theory of permutations and will be done later. However, we need 3×3 determinants now. First, we define ϵ_{ijk}, where $\{i, j, k\} \subset \{1, 2, 3\}$, by

$$\epsilon_{123} = \epsilon_{312} = \epsilon_{231} = 1, \quad \epsilon_{321} = \epsilon_{132} = \epsilon_{213} = -1,$$

4.6 Orientation and determinants

and then let $\epsilon_{ijk} = 0$ when i, j and k are not distinct. To understand why these terms are relevant we recall that the sign $\varepsilon(\rho)$ of a permutation ρ of $\{1, 2, 3\}$ is $(-1)^k$, where ρ can be written as the product of k transpositions. Thus the ϵ_{ijk} are connected to permutations by the relation

$$\epsilon_{ijk} = \varepsilon(\rho), \quad \rho = \begin{pmatrix} 1 & 2 & 3 \\ i & j & k \end{pmatrix},$$

with $\epsilon_{ijk} = 0$ if the map ρ is not invertible. We are now ready to discuss 3×3 determinants.

Definition 4.6.2 A 3×3 *determinant* is a 3×3 array of (real or complex) numbers whose value is given by

$$\begin{vmatrix} x_1 & x_2 & x_3 \\ y_1 & y_2 & y_3 \\ z_1 & z_2 & z_3 \end{vmatrix} = \sum_{i,j,k=1}^{3} \epsilon_{ijk} x_i y_j z_k \quad (4.6.1)$$

$$= \sum_{\rho} \varepsilon(\rho) x_{\rho(1)} y_{\rho(2)} z_{\rho(3)},$$

where the last sum is over all permutations ρ of $\{1, 2, 3\}$. □

It is possible to express the components of a vector product in a convenient form in terms of the ϵ_{ijk} for (by inspection) the i-th component of a product is given by

$$(\mathbf{x} \times \mathbf{y})_i = \sum_{j,k=1}^{3} \epsilon_{ijk} x_j y_k \quad (4.6.2)$$

(note that only the two terms make a contribution to this sum because $\epsilon_{ijk} = 0$ unless i, j and k are distinct). The expression (4.6.2) leads directly to a similar expression for the scalar triple product, namely

$$[\mathbf{x}, \mathbf{y}, \mathbf{z}] = \sum_{i=1}^{3} x_i (\mathbf{y} \times \mathbf{z})_i = \sum_{i,j,k=1}^{3} \epsilon_{ijk} x_i y_j z_k, \quad (4.6.3)$$

and this is the same sum as in (4.6.1).

Following the case of two variables, we now define $\Delta(\mathbf{u}, \mathbf{v}, \mathbf{w})$, for vectors $\mathbf{u}$, $\mathbf{v}$ and $\mathbf{w}$ in $\mathbb{R}^3$, to be the determinant obtained by taking the rows of the array to be the components of the vector; explicitly

$$\Delta(\mathbf{u}, \mathbf{v}, \mathbf{w}) = \begin{vmatrix} u_1 & u_2 & u_3 \\ v_1 & v_2 & v_3 \\ w_1 & w_2 & w_3 \end{vmatrix}. \quad (4.6.4)$$

Next, we make the following definition.

Definition 4.6.3 The three vectors **u**, **v** and **w** in $\mathbb{R}^3$ are said to be *positively orientated* if $\Delta(\mathbf{u}, \mathbf{v}, \mathbf{w}) > 0$, and *negatively orientated* if $\Delta(\mathbf{u}, \mathbf{v}, \mathbf{w}) < 0$. □

Note that **i**, **j** and **k** are positively orientated, and **j**, **i** and **k** are negatively orientated. More generally, we have the following result which explains the close connection between vector products and orientation.

Theorem 4.6.4 *For any three vectors* **u**, **v** *and* **w** *in* $\mathbb{R}^3$,
$$\Delta(\mathbf{u}, \mathbf{v}, \mathbf{w}) = [\mathbf{u}, \mathbf{v}, \mathbf{w}] = \mathbf{u} \cdot (\mathbf{v} \times \mathbf{w}).$$

This follows immediately from (4.6.3), (4.6.1) and (4.6.4). Moreover, Theorem 4.4.3 implies that if [**0**, **x**], [**0**, **y**] and [**0**, **z**] are not coplanar then the vectors **x**, **y** and **z** are either positively orientated or negatively orientated. Theorem 4.4.2 leads to the following result which shows that our choice of $\mathbf{x} \times \mathbf{y}$ was made so that **x**, **y** and $\mathbf{x} \times \mathbf{y}$ (in this order) are always *positively orientated*; this is the formal statement of the 'corkscrew rule'.

Corollary 4.6.5 *Suppose that* **x** *and* **y** *are not scalar multiples of each other. Then the vectors* **x**, **y** *and* $\mathbf{x} \times \mathbf{y}$ *are positively orientated.*

The proof is simply that
$$[\mathbf{x}, \mathbf{y}, \mathbf{x} \times \mathbf{y}] = [\mathbf{x} \times \mathbf{y}, \mathbf{x}, \mathbf{y}] = ||\mathbf{x} \times \mathbf{y}||^2 > 0.$$
□

We now list some properties of 3×3 determinants. Briefly, these properties enable one to evaluate determinants with only a few calculations; however, as these calculations can now be performed by machines (which are faster and more accurate than humans), we shall not spend long on this matter. The general aim in manipulating determinants is to alter the entries, without changing the value of the determinant, so as to obtain as many zero entries as possible for, obviously, this will simplify the calculations. There are five basic rules for manipulating determinants, and these are as follows. A determinant

(1) is unaltered if we interchange rows and columns: that is,
$$\begin{vmatrix} x_1 & x_2 & x_3 \\ y_1 & y_2 & y_3 \\ z_1 & z_2 & z_3 \end{vmatrix} = \begin{vmatrix} x_1 & y_1 & z_1 \\ x_2 & y_2 & z_2 \\ x_3 & y_3 & z_3 \end{vmatrix};$$

(2) is a linear function of each column (and of each row);
(3) is zero if two columns (or two rows) are identical;
(4) is unaltered if we add to any given column any linear combination of the other columns (and similarly for rows);

(5) changes sign when we interchange any two columns (or rows). In addition,
(6) a 3 × 3 determinant can be expressed as a linear combination of 2 × 2 determinants, namely

$$\begin{vmatrix} a_{11} & a_{12} & a_{13} \\ a_{21} & a_{22} & a_{23} \\ a_{31} & a_{32} & a_{33} \end{vmatrix} = a_{11} \begin{vmatrix} a_{22} & a_{23} \\ a_{32} & a_{33} \end{vmatrix} - a_{12} \begin{vmatrix} a_{21} & a_{23} \\ a_{31} & a_{33} \end{vmatrix} + a_{13} \begin{vmatrix} a_{21} & a_{22} \\ a_{31} & a_{32} \end{vmatrix}.$$

The rules (1)–(6) can be proved (in a trivial but tedious way) by writing out all determinants in full. Some of the rules can also be proved from the known properties of the scalar triple product for, given any determinant D, we can always find vectors $\mathbf{u}$, $\mathbf{v}$ and $\mathbf{w}$ such that $D = \Delta(\mathbf{u}, \mathbf{v}, \mathbf{w})$. However, as there is no generalization of the vector product to all dimensions (see Chapter 6), these properties will eventually have to be proved for $n \times n$ determinants from a definition similar to (4.6.1). Finally, we mention that (6) is the first step of an inductive definition of the $n \times n$ determinant in terms of the $(n-1) \times (n-1)$ determinant. We shall not stop to verify these rules, and we end with two examples in which we evaluate determinants using (1)–(6). Of course, the reader may feel (as the author does) that it would be simpler to evaluate the determinant directly from the definition, but our purpose here is to illustrate the use of the rules.

Example 4.6.6 Using (6) above we see that $\begin{vmatrix} 1 & 3 & 5 \\ 2 & 6 & 8 \\ 0 & 1 & 4 \end{vmatrix} = 2$. However, the reader may wish to verify the following, which depends on the linearity as a function of the rows. Let $\mathbf{a} = (1, 3, 5)$, $\mathbf{b} = (1, 3, 4)$ and $\mathbf{c} = (0, 1, 4)$, so that the rows of D are $\mathbf{a}$, $2\mathbf{b}$ and $\mathbf{c}$. First,

$$D = [\mathbf{a}, 2\mathbf{b}, \mathbf{c}] = 2[\mathbf{a}, \mathbf{b}, \mathbf{c}] = 2[\mathbf{a} - \mathbf{b}, \mathbf{b}, \mathbf{c}]$$
$$= 2[\mathbf{a} - \mathbf{b}, \mathbf{b} - 4(\mathbf{a} - \mathbf{b}), \mathbf{c} - 4(\mathbf{a} - \mathbf{b})].$$

We write this last scalar triple product as $[\mathbf{u}, \mathbf{v}, \mathbf{w}]$, and then

$$D = 2[\mathbf{u}, \mathbf{v} - 3\mathbf{w}, \mathbf{w}] = 2 \begin{vmatrix} 0 & 0 & 1 \\ 1 & 0 & 0 \\ 0 & 1 & 0 \end{vmatrix} = 2.$$

$\square$

Example 4.6.7 We leave the reader to verify the following steps:

$$\begin{vmatrix} 1 & 3 & 4 \\ 2 & 0 & 1 \\ 1 & 6 & 5 \end{vmatrix} = \begin{vmatrix} 1 & 3 & 4 \\ 2 & 0 & 1 \\ 0 & 3 & 1 \end{vmatrix} = \begin{vmatrix} 1 & 0 & 3 \\ 2 & 0 & 1 \\ 0 & 3 & 1 \end{vmatrix} = \begin{vmatrix} 1 & 0 & 0 \\ 2 & 0 & -5 \\ 0 & 3 & 1 \end{vmatrix} = \begin{vmatrix} 0 & -5 \\ 3 & 1 \end{vmatrix} = 15.$$

Exercise 4.6

1. Show that the determinants

$$\begin{vmatrix} 1 & 10 & 4 \\ 8 & 82 & 30 \\ 6 & 62 & 23 \end{vmatrix}, \quad \begin{vmatrix} 100 & 20 & 13 \\ 6 & 1 & 2 \\ 80 & 20 & 5 \end{vmatrix}, \quad \begin{vmatrix} 999 & 998 & 997 \\ 996 & 995 & 994 \\ 993 & 992 & 991 \end{vmatrix}$$

are 2, -380 and 0, respectively.

2. Evaluate the determinants

$$\begin{vmatrix} 1 & 1 & 1 \\ x & a & b \\ x^2 & a^2 & b^2 \end{vmatrix}, \quad \begin{vmatrix} x & a & b \\ x^2 & a^2 & b^2 \\ x^3 & a^3 & b^3 \end{vmatrix},$$

and factorize both answers.

3. Show that for any vectors $\mathbf{a}, \mathbf{b}, \mathbf{c}, \mathbf{u}, \mathbf{v}, \mathbf{w}$ in $\mathbb{R}^3$,

$$[\mathbf{a}, \mathbf{b}, \mathbf{c}][\mathbf{u}, \mathbf{v}, \mathbf{w}] = \begin{vmatrix} \mathbf{a}\cdot\mathbf{u} & \mathbf{a}\cdot\mathbf{v} & \mathbf{a}\cdot\mathbf{w} \\ \mathbf{b}\cdot\mathbf{u} & \mathbf{b}\cdot\mathbf{v} & \mathbf{b}\cdot\mathbf{w} \\ \mathbf{c}\cdot\mathbf{u} & \mathbf{c}\cdot\mathbf{v} & \mathbf{c}\cdot\mathbf{w} \end{vmatrix}.$$

4.7 Applications to geometry

Vectors give us a convenient way of describing geometry in $\mathbb{R}^3$, and this section consists of various applications to the geometry of lines, planes, triangles and tetrahedra. The discussion will be brief, and the reader is asked to supply the missing details.

(A) The geometry of lines

The vector equation of the line L through $\mathbf{x}_0$ *and* $\mathbf{x}_0 + \mathbf{a}$ *is*

$$(\mathbf{x} - \mathbf{x}_0) \times \mathbf{a} = \mathbf{0},$$

because $\mathbf{x} \in L$ if and only if $\mathbf{x} - \mathbf{x}_0 = t\mathbf{a}$ for some real t. If we take $\mathbf{x}_1 = \mathbf{x}_0 + \mathbf{a}$, we see that *the line through* $\mathbf{x}_0$ *and* $\mathbf{x}_1$ *has equation*

$$(\mathbf{x} - \mathbf{x}_0) \times (\mathbf{x}_1 - \mathbf{x}_0) = \mathbf{0}.$$

The distance from $\mathbf{y}$ *to the line through* $\mathbf{x}_0$ *and parallel to* $[\mathbf{0}, \mathbf{a}]$ *is*

$$||(\mathbf{y} - \mathbf{x}_0) \times \mathbf{a}||/||\mathbf{a}||.$$

By taking $\mathbf{y} = \mathbf{x}_1$, we see that *the distance between the parallel lines given by* $(\mathbf{x} - \mathbf{x}_0) \times \mathbf{a} = \mathbf{0}$ *and* $(\mathbf{x} - \mathbf{x}_1) \times \mathbf{a} = \mathbf{0}$ *is* $||(\mathbf{x}_1 - \mathbf{x}_0) \times \mathbf{a}||/||\mathbf{a}||$.

The non-parallel lines given by $(\mathbf{x} - \mathbf{x}_0) \times \mathbf{a} = \mathbf{0}$ and $(\mathbf{x} - \mathbf{x}_1) \times \mathbf{b} = \mathbf{0}$ meet if and only if $[\mathbf{x}_0, \mathbf{a}, \mathbf{b}] = [\mathbf{x}_1, \mathbf{a}, \mathbf{b}]$, for this is so if and only if $[\mathbf{0}, \mathbf{a}]$, $[\mathbf{0}, \mathbf{b}]$ and $[\mathbf{0}, \mathbf{x}_0 - \mathbf{x}_1]$ are coplanar (Theorem 4.4.3). Notice that as the lines are not parallel, $\mathbf{a} \times \mathbf{b} \neq \mathbf{0}$. Now suppose that these lines meet at $\mathbf{y}$. Then there are real s and t such that $\mathbf{x}_0 + t\mathbf{a} = \mathbf{y} = \mathbf{x}_1 + s\mathbf{b}$. Now write $\mathbf{z} = \mathbf{x}_1 - \mathbf{x}_0$; then

$$t(\mathbf{a} \times \mathbf{b}) = (t\mathbf{a}) \times \mathbf{b} = (\mathbf{z} + s\mathbf{b}) \times \mathbf{b} = \mathbf{z} \times \mathbf{b},$$

so that $t = [\mathbf{a} \times \mathbf{b}, \mathbf{z}, \mathbf{b}]/\|\mathbf{a} \times \mathbf{b}\|^2$. Thus *if the lines meet then they intersect at the point*

$$\mathbf{x}_0 + \left(\frac{[\mathbf{x}_1 - \mathbf{x}_0, \mathbf{b}, \mathbf{a} \times \mathbf{b}]}{\|\mathbf{a} \times \mathbf{b}\|^2} \right) \mathbf{a}.$$

(B) The geometry of planes

The equation of the plane Π *which contains* $\mathbf{a}$, *and which has normal in the direction* $\mathbf{n}$, *is* $\mathbf{x} \cdot \mathbf{n} = \mathbf{a} \cdot \mathbf{n}$ (because $\mathbf{x} \in \Pi$ if and only if $\mathbf{x} - \mathbf{a} \perp \mathbf{n}$).
The equation of the plane Σ *through the three non-collinear points* $\mathbf{a}$, $\mathbf{b}$ *and* $\mathbf{c}$ *is*

$$[\mathbf{x}, \mathbf{b}, \mathbf{c}] + [\mathbf{a}, \mathbf{x}, \mathbf{c}] + [\mathbf{a}, \mathbf{b}, \mathbf{x}] = [\mathbf{a}, \mathbf{b}, \mathbf{c}].$$

Indeed, this equation is linear in the coordinates x_j and so defines a plane. As this plane obviously contains $\mathbf{a}$, $\mathbf{b}$ and $\mathbf{c}$ (Theorem 4.4.3), it is the equation of Σ.

The distance of $\mathbf{y}$ *from the plane* Π *given by* $\mathbf{x} \cdot \mathbf{n} = d$, *where* $\|\mathbf{n}\| = 1$, *is* $|t|$, where t is such that $\mathbf{y} + t\mathbf{n} \in \Pi$. This last condition is $(\mathbf{y} + t\mathbf{n}) \cdot \mathbf{n} = d$, so the required distance is $|d - (\mathbf{y} \cdot \mathbf{n})|$.

The equation of the line L *of intersection of the non-parallel planes* $\mathbf{x} \cdot \mathbf{a} = d_1$ *and* $\mathbf{x} \cdot \mathbf{b} = d_2$ *is* $\mathbf{x} \times (\mathbf{a} \times \mathbf{b}) = d_2\mathbf{a} - d_1\mathbf{b}$. As the planes are not parallel, $\mathbf{a} \times \mathbf{b} \neq \mathbf{0}$, and as L is orthogonal to the normals of both planes it is in the direction $\mathbf{a} \times \mathbf{b}$. It follows that L is given by an equation of the form $\mathbf{x} \times (\mathbf{a} \times \mathbf{b}) = \mathbf{c}$. This equation is $(\mathbf{x} \cdot \mathbf{b})\mathbf{a} - (\mathbf{x} \cdot \mathbf{a})\mathbf{b} = \mathbf{c}$, and taking $\mathbf{x}$ on L we see that $\mathbf{c} = d_2\mathbf{a} - d_1\mathbf{b}$.

The lines L_0 and L_1 are said to be *skew* if they do not lie in any plane. We shall show that *skew lines lie in a pair of parallel planes, and so have a common normal*. Suppose that the skew lines L_0 and L_1 are given by

$$(\mathbf{x} - \mathbf{x}_0) \times \mathbf{a} = \mathbf{0}, \quad (\mathbf{x} - \mathbf{x}_1) \times \mathbf{b} = \mathbf{0}. \tag{4.7.1}$$

As L_0 and L_1 are not parallel, $\mathbf{a} \times \mathbf{b} \neq \mathbf{0}$. Now consider the two planes Π_0 and Π_1 given by

$$(\mathbf{x} - \mathbf{x}_0)\cdot(\mathbf{a} \times \mathbf{b}) = 0, \quad (\mathbf{x} - \mathbf{x}_1)\cdot(\mathbf{a} \times \mathbf{b}) = 0, \tag{4.7.2}$$

respectively. These planes are parallel for they have a common normal, namely $\mathbf{a} \times \mathbf{b}$. Moreover, L_0 lies in Π_0 because any point of L_0 is of the form $\mathbf{x}_0 + t\mathbf{a}$ for some real t and, similarly, L_1 lies in Π_1.

The shortest distance between the skew lines given by (4.7.1) *is*

$$\frac{|(\mathbf{x}_0 - \mathbf{x}_1)\cdot(\mathbf{a} \times \mathbf{b})|}{\|\mathbf{a} \times \mathbf{b}\|}. \tag{4.7.3}$$

Indeed, the shortest distance between the lines in (4.7.1) is also the shortest distance between the planes in (4.7.2). As $\mathbf{x}_0 \in \Pi_0$ and $\mathbf{x}_1 \in \Pi_1$, this distance is the projection of $\mathbf{x}_0 - \mathbf{x}_1$ onto the common unit normal of the planes, and this gives the stated formula.

Suppose that the three planes $\mathbf{x}\cdot\mathbf{a} = \lambda$, $\mathbf{x}\cdot\mathbf{b} = \mu$ and $\mathbf{x}\cdot\mathbf{c} = \nu$ intersect in a single point; then this point is (by inspection)

$$\frac{\lambda(\mathbf{b} \times \mathbf{c}) + \mu(\mathbf{c} \times \mathbf{a}) + \nu(\mathbf{a} \times \mathbf{b})}{[\mathbf{a}, \mathbf{b}, \mathbf{c}]}.$$

As the planes meet in only a single point, their normals, in the directions $\mathbf{a}$, $\mathbf{b}$ and $\mathbf{c}$, are not coplanar; thus $[\mathbf{a}, \mathbf{b}, \mathbf{c}] \neq 0$ (Theorem 4.4.3). This implies that for any $\mathbf{x}$,

$$\mathbf{x} = \left(\frac{\mathbf{x}\cdot\mathbf{a}}{[\mathbf{a}, \mathbf{b}, \mathbf{c}]}\right)(\mathbf{b} \times \mathbf{c}) + \left(\frac{\mathbf{x}\cdot\mathbf{b}}{[\mathbf{a}, \mathbf{b}, \mathbf{c}]}\right)(\mathbf{c} \times \mathbf{a}) + \left(\frac{\mathbf{x}\cdot\mathbf{c}}{[\mathbf{a}, \mathbf{b}, \mathbf{c}]}\right)(\mathbf{a} \times \mathbf{b}),$$

because both sides of this equation have the same scalar product with each of $\mathbf{a}$, $\mathbf{b}$ and $\mathbf{c}$. If $\mathbf{x}$ is the point of intersection, then $\mathbf{x}\cdot\mathbf{a} = \lambda$, and so on, and the result follows.

(C) The geometry of triangles

A *median* of a triangle is the segment joining a vertex v of the triangle to the midpoint of the opposite side s. The *altitude* from v is the segment from v to the line L that contains s that is orthogonal to L. *The altitudes of a triangle are concurrent; the medians of a triangle are concurrent; the angle bisectors of a triangle are concurrent.*

We only consider the medians. Let the vertices of the triangle be $\mathbf{a}$, $\mathbf{b}$ and $\mathbf{c}$. The general point on the median joining $\mathbf{a}$ to $\mathbf{d}$, where $\mathbf{d} = \frac{1}{2}(\mathbf{b}+\mathbf{c})$ is $t\mathbf{a} + (1-t)\mathbf{d}$, where $0 \leq t \leq 1$. When $t = 1/3$ this point is $(\mathbf{a}+\mathbf{b}+\mathbf{c})/3$. By symmetry, this point also lies on the other medians.

4.7 Applications to geometry

The sine rule: suppose that a triangle T has angles α, β and γ opposite sides of lengths a, b and c, respectively. Then

$$\frac{a}{\sin \alpha} = \frac{b}{\sin \beta} = \frac{c}{\sin \gamma}.$$

Let the vertices be **a**, **b** and **c**, and let $\mathbf{p} = \mathbf{a} - \mathbf{b}$, $\mathbf{q} = \mathbf{b} - \mathbf{c}$, $\mathbf{r} = \mathbf{c} - \mathbf{a}$. As $\mathbf{p} + \mathbf{q} + \mathbf{r} = \mathbf{0}$, we see that $\mathbf{p} \times \mathbf{q} = \mathbf{q} \times \mathbf{r} = \mathbf{r} \times \mathbf{p}$, and this gives the sine rule.

(D) The geometry of tetrahedra

A *tetrahedron* T is formed by attaching four triangles together along their edges so as to form a surface which has four triangular faces, six edges and four vertices. Any tetrahedron has three pairs of 'opposite' edges (that is, pairs of edges with no common end-points), and *the segments that join the midpoints of opposite edges are concurrent at their midpoints*. Further, *these segments are mutually orthogonal if and only if each pair of opposite edges of T have the same length*. Let the vertices of T be at **a**, **b**, **c**, and **d**. One pair of opposite edges has midpoints $\frac{1}{2}(\mathbf{a} + \mathbf{b})$ and $\frac{1}{2}(\mathbf{c} + \mathbf{d})$, and the midpoint of the segment joining these points is $\mathbf{p} = \frac{1}{4}(\mathbf{a} + \mathbf{b} + \mathbf{c} + \mathbf{d})$. By symmetry, **p** must be the midpoint of each of the segments that join the midpoints of each pair of opposite edges of T. Next, these three segments passing through **p** are parallel to the three vectors $\frac{1}{2}(\mathbf{a} + \mathbf{b}) - \mathbf{p}$, $\frac{1}{2}(\mathbf{a} + \mathbf{c}) - \mathbf{p}$, $\frac{1}{2}(\mathbf{a} + \mathbf{d}) - \mathbf{p}$; thus the segments are mutually orthogonal if and only if these three vectors are mutually orthogonal. Now the first two vectors are

$$[(\mathbf{a} - \mathbf{d}) + (\mathbf{b} - \mathbf{c})]/4, \quad [(\mathbf{a} - \mathbf{d}) - (\mathbf{b} - \mathbf{c})]/4,$$

and these are orthogonal if and only if $||\mathbf{a} - \mathbf{d}|| = ||\mathbf{b} - \mathbf{c}||$ because, in general, $\mathbf{u} - \mathbf{v} \perp \mathbf{u} + \mathbf{v}$ if and only if $||\mathbf{u}|| = ||\mathbf{v}||$.

Exercise 4.7

1. Show that the distance between the point $(1, 1, 1)$ and the line through $(2, 0, 3)$ and $(-1, 0, 1)$ is $\sqrt{29/13}$.
2. Show that the distance between the point $(-3, 0, 1)$ and the line given by $(1, 0, 2) + t(1, 1, 2)$, where $t \in \mathbb{R}$, is $7/\sqrt{3}$.
3. Show that the distance between the two lines in the direction $(1, 2, 1)$ that pass through the points $(4, 2, -1)$ and $(3, 1, 0)$, respectively is $7/\sqrt{21}$.
4. Show that the distance between the two skew lines given in parametric form by $(1, 2, 3) + t(2, 0, 1)$ and $(0, 0, 1) + t(1, 0, 1)$ is 2.

5. Find the equation of the plane through the points $(0, 1, 2)$, $(-4, 3, 1)$ and $(10, 0, 7)$.
6. Find the intersection of the three planes given by $\mathbf{x}\cdot\mathbf{a} = 1$, $\mathbf{x}\cdot\mathbf{b} = 2$ and $\mathbf{x}\cdot\mathbf{c} = 3$, where $\mathbf{a} = (3, 1, 1)$, $\mathbf{b} = (2, 0, 8)$ and $\mathbf{c} = (1, 0, 2)$.
7. Show that the minimum distance between the origin and the plane given by $2x_1 + 5x_2 - x_3 = 1$ is $1/\sqrt{30}$, and that this is attained at the point $(2/30, 5/30, -1/30)$ on the plane.
8. Find the vectorial equation of the line of intersection of the planes $3x_1 + 2x_2 + x_3 = 3$ and $x_1 + x_2 + x_3 = 4$.
9. Consider the cube with vertices at the points (r, s, t), where each of r, s and t is 0 or 1. What is the surface area of the tetrahedron whose vertices are at the points $\mathbf{0}$ and the centres of the three faces of the cube that do not contain $\mathbf{0}$?
10. Use vectors to show that the diagonals of a parallelogram bisect each other, and that the diagonals are orthogonal if and only if the parallelogram is a rhombus. Show that the midpoints of the sides of any quadrilateral form the vertices of a parallelogram.
11. Let T be the tetrahedron with vertices $\mathbf{0}$, $a\mathbf{i}$, $b\mathbf{j}$ and $c\mathbf{k}$, and let the faces opposite these vertices have areas A_0, A_a, A_b, and A_c, respectively. Show that $A_0^2 = A_a^2 + A_b^2 + A_c^2$ (Pythagoras' theorem for a tetrahedron).
12. Show that the minimum distance between a pair of opposite edges of a regular tetrahedron T with edge length ℓ is $\ell/\sqrt{2}$.

4.8 Vector equations

In this section we see how to solve each of the equations

$$\lambda \mathbf{x} + \mu(\mathbf{x}\cdot\mathbf{a})\mathbf{c} = \mathbf{b}, \qquad (4.8.1)$$
$$\lambda \mathbf{x} + \mu(\mathbf{x} \times \mathbf{a}) = \mathbf{b}, \qquad (4.8.2)$$

where λ, μ, $\mathbf{a}$, $\mathbf{b}$ and $\mathbf{c}$ are given scalars and non-zero vectors. If we write either of these equations in terms of coordinates, we obtain three linear equation in the three components of $\mathbf{x}$. The set of solutions (of each of these equations) is therefore the intersection of three planes, and so is either empty, a point, a line, a plane or (possibly) $\mathbb{R}^3$. This tells us what to expect, but our objective is to solve these equations by vector methods. The cases when $\mu = 0$, or $\lambda = 0$, are either trivial or have been considered earlier, so we shall assume that $\lambda \neq 0$ and $\mu \neq 0$. If we write $\mathbf{a}' = (\mu/\lambda)\mathbf{a}$, and $\mathbf{b}' = \lambda^{-1}\mathbf{b}$, the equations are converted into similar equations with $\lambda = \mu = 1$. Thus we may assume that $\lambda = \mu = 1$. We consider each in turn.

Because $\mathbf{x}\cdot\mathbf{a}$ is a scalar, any solution of the equation $\mathbf{x} + (\mathbf{x}\cdot\mathbf{a})\mathbf{c} = \mathbf{b}$ must be of the form $\mathbf{x} = \mathbf{b} + t\mathbf{c}$, where t is real. Thus all solutions (if any) of this equation lie on the line L through $\mathbf{b}$ in the direction $\mathbf{c}$. If we now check to see whether or not $\mathbf{x} = \mathbf{b} + t\mathbf{c}$ is a solution, we find that (i) there is a unique solution if $1 + \mathbf{a}\cdot\mathbf{c} \neq 0$; (ii) there is no solution if $1 + \mathbf{a}\cdot\mathbf{c} = 0$ and $\mathbf{a}\cdot\mathbf{b} \neq \mathbf{0}$; (iii) every point on L is a solution if $1 + \mathbf{a}\cdot\mathbf{c} = 0$ and $\mathbf{a}\cdot\mathbf{b} = \mathbf{0}$.

Now consider the equation $\mathbf{x} + (\mathbf{x} \times \mathbf{a}) = \mathbf{b}$. If $\mathbf{y}$ is the difference of any two solutions, then $\mathbf{y} + (\mathbf{y} \times \mathbf{a}) = \mathbf{0}$, and so $\mathbf{y} = \mathbf{0}$ (because $\mathbf{y} \perp \mathbf{y} \times \mathbf{a}$). This shows that the given equation has at most one solution. If $\mathbf{a} \times \mathbf{b} = \mathbf{0}$, then $\mathbf{b}$ is a solution. If $\mathbf{a} \times \mathbf{b} \neq \mathbf{0}$, then every $\mathbf{x}$ can be expressed in the form $\mathbf{x} = x_1\mathbf{a} + x_2\mathbf{b} + x_3(\mathbf{a} \times \mathbf{b})$, and it is then a simple matter to check that

$$\mathbf{x} = \left(\frac{1}{1 + ||\mathbf{a}||^2}\right)((\mathbf{a}\cdot\mathbf{b})\mathbf{a} + \mathbf{b} + (\mathbf{a} \times \mathbf{b}))$$

is the unique solution to the equation.

Exercise 4.8

1. Solve the equations $\mathbf{x} + (\mathbf{x}\cdot\mathbf{i})\mathbf{i} = \mathbf{j}$, and $\mathbf{x} + (\mathbf{x} \times \mathbf{i}) = \mathbf{j}$.
2. Solve the equations $\mathbf{x} + (\mathbf{x}\cdot\mathbf{a})\mathbf{a} = \mathbf{a}$, and $\mathbf{x} + (\mathbf{x} \times \mathbf{a}) = \mathbf{a}$.
3. Show that the solution of the simultaneous equations $\mathbf{x} + (\mathbf{c} \times \mathbf{y}) = \mathbf{a}$ and $\mathbf{y} + (\mathbf{c} \times \mathbf{x}) = \mathbf{b}$ is given by

$$\mathbf{x} = [(\mathbf{a}\cdot\mathbf{c})\mathbf{c} + \mathbf{a} + \mathbf{b} \times \mathbf{c}]/(1 + ||\mathbf{c}||^2),$$
$$\mathbf{y} = [(\mathbf{b}\cdot\mathbf{c})\mathbf{c} + \mathbf{b} + \mathbf{a} \times \mathbf{c}]/(1 + ||\mathbf{c}||^2).$$

4. Solve the simultaneous vector equations $\mathbf{x} + \mathbf{y} = \mathbf{a}$ and $\mathbf{x} \times \mathbf{y} = \mathbf{b}$.

5
Spherical geometry

5.1 Spherical distance

This chapter is devoted to *spherical geometry*; that is, to geometry on the surface of the sphere

$$S = \{\mathbf{x} \in \mathbb{R}^3 : ||\mathbf{x}|| = 1\}.$$

Later in this chapter we shall use spherical trigonometry to derive Euler's formula for polyhedral surfaces, and this will lead (eventually) to a discussion of the symmetry groups of regular polyhedra.

We take for granted the fact that one can measure the length of a smooth curve, and the area of a (reasonably simple) set, on S. A *great circle*, of length 2π, is the intersection of S and a plane that passes through the origin. Every other circle on S is a plane section of S of length less than 2π. If σ is an arc of a great circle, then the length of σ is the angle (in radians) subtended by σ at 0. Given any two points $\mathbf{a}$ and $\mathbf{b}$ on S there is a unique great circle, say C, that contains them (and C lies in the plane through $\mathbf{0}$, $\mathbf{a}$ and $\mathbf{b}$). Also, $\mathbf{a}$ and $\mathbf{b}$ divide C into two arcs which have different lengths unless $\mathbf{b} = -\mathbf{a}$.

Definition 5.1.1 Let $\mathbf{a}$ and $\mathbf{b}$ be two points on S. Then the *spherical distance* $\delta(\mathbf{a}, \mathbf{b})$ between $\mathbf{a}$ and $\mathbf{b}$ is the length of the shorter of the two arcs of the (unique) great circle through $\mathbf{a}$ and $\mathbf{b}$. Clearly,

$$\delta(\mathbf{a}, \mathbf{b}) = \cos^{-1}(\mathbf{a} \cdot \mathbf{b}), \qquad (5.1.1)$$

where $\cos^{-1}(\mathbf{a} \cdot \mathbf{b})$ is chosen in the range $[0, \pi]$. □

As an application, consider the Earth to be a perfect sphere of radius R whose centre lies at the origin $\mathbf{0}$ in $\mathbb{R}^3$. We may suppose that the $\mathbf{i}$-axis meets the surface of the earth at the point with zero latitude and longitude, and that the positive $\mathbf{k}$-axis passes through the north pole. Thus the point on the Earth's surface with latitude α (positive in the northern hemisphere, and negative in the southern hemisphere), and longitude β is given by the vector

$$R\bigl(\cos\alpha \cos\beta\, \mathbf{i} + \cos\alpha \sin\beta\, \mathbf{j} + \sin\alpha\, \mathbf{k}\bigr)$$

(the reader should draw a diagram). Suppose, now that $\mathbf{x}_1$ and $\mathbf{x}_2$ are two points on the Earth's surface, and write these as

$$\mathbf{x}_1 = R\bigl(\cos\alpha_1 \cos\beta_1\, \mathbf{i} + \cos\alpha_1 \sin\beta_1\, \mathbf{j} + \sin\alpha_1\, \mathbf{k}\bigr),$$
$$\mathbf{x}_2 = R\bigl(\cos\alpha_2 \cos\beta_2\, \mathbf{i} + \cos\alpha_2 \sin\beta_2\, \mathbf{j} + \sin\alpha_2\, \mathbf{k}\bigr).$$

Then, from (5.1.1),

$$\delta(\mathbf{x}_1, \mathbf{x}_2) = R \cos^{-1}\Bigl[\cos\alpha_1 \cos\alpha_2 \cos(\beta_1 - \beta_2) + \sin\alpha_1 \sin\alpha_2\Bigr]. \quad (5.1.2)$$

This formula gives us the distance (measured on the surface of the Earth) between the two points with latitude α_i and longitude β_i, $i = 1, 2$.

Exercise 5.1

1. Verify that any point with latitude α is a spherical distance $R(\pi/2 - \alpha)$ from the north pole.
2. Assume that the Earth is a sphere of radius 4000 miles. Show that the spherical distance between London (latitude 51° north, longitude 0°) and Sydney (latitude 34° south, longitude 151° east) is approximately 10 500 miles.
3. Suppose that an aircraft flies on the shortest route from London (latitude 51° north, longitude 0°) to Los Angeles (latitude 34° north, longitude 151° east). How close does the aircraft get to the north pole?
4. Let $\mathbf{x}$ and $\mathbf{y}$ be two points on the sphere. Show that the normal to the plane determined by the great circle through $\mathbf{x}$ and $\mathbf{y}$ intersects the sphere at the points $\pm \mathbf{z}$, where $\mathbf{z} = (\mathbf{x} \times \mathbf{y})/\|\mathbf{x} \times \mathbf{y}\|$. Suppose that $\mathbf{w}$ lies on the same side of the plane as $\mathbf{z}$. Show that $\cos\delta(\mathbf{w}, \mathbf{z}) = [\mathbf{w}, \mathbf{x}, \mathbf{y}]$.

5.2 Spherical trigonometry

We begin our discussion of spherical trigonometry with the the spherical version of Pythagoras' theorem. Consider a triangle on $\mathcal{S}$, by which we mean three points $\mathbf{a}$, $\mathbf{b}$ and $\mathbf{c}$ of $\mathcal{S}$ which do not lie on a great circle, where the sides of the triangle are the arcs of great circles that join these points in pairs. We assume that the angle in the triangle at $\mathbf{c}$ is $\pi/2$, and we may position this triangle so

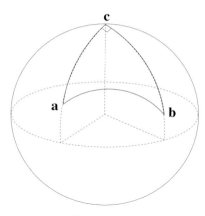

Figure 5.2.1

that $\mathbf{c} = \mathbf{k}$, $\mathbf{a}$ is in the $(\mathbf{i}, \mathbf{k})$-plane, and that $\mathbf{b}$ is in the $(\mathbf{j}, \mathbf{k})$-plane (if necessary we may interchange the labels on $\mathbf{a}$ and $\mathbf{b}$). With this, we have

$$\mathbf{c} = \mathbf{k}, \quad \mathbf{a} = \cos\alpha_1 \, \mathbf{i} + \sin\alpha_1 \, \mathbf{k}, \quad \mathbf{b} = \cos\alpha_2 \, \mathbf{j} + \sin\alpha_2 \, \mathbf{k} \qquad (5.2.1)$$

for some α_1 and α_2 (see Figure 5.2.1) so that $(\mathbf{a} \cdot \mathbf{b}) = (\mathbf{a} \cdot \mathbf{c})(\mathbf{b} \cdot \mathbf{c})$. If we now apply (5.1.1) we immediately obtain the following result.

Pythagoras' theorem *Let $\mathbf{a}$, $\mathbf{b}$ and $\mathbf{c}$ be the vertices of a spherical triangle on S, with the sides making an angle of $\pi/2$ at $\mathbf{c}$. Then*

$$\cos\delta(\mathbf{a}, \mathbf{b}) = \cos\delta(\mathbf{a}, \mathbf{c}) \cos\delta(\mathbf{b}, \mathbf{c}). \qquad (5.2.2)$$

Consider (informally) an 'infinitesimal' right-angled triangle. As this triangle is nearly flat we would expect (5.2.2) to look roughly like the usual Euclidean version $a^2 + b^2 = c^2$ of Pythagoras' theorem. This is indeed the case because for small θ, $\cos\theta$ is approximately $1 - \theta^2/2$.

Next, we consider a spherical triangle (as illustrated in Figure 5.2.1) with vertices $\mathbf{a}$, $\mathbf{b}$ and $\mathbf{c}$. As is customary in Euclidean geometry, we let the angles at $\mathbf{a}$, $\mathbf{b}$ and $\mathbf{c}$ be α, β and γ, respectively, and we denote the lengths of the sides by a, b and c; thus $a = \delta(\mathbf{b}, \mathbf{c})$, $b = \delta(\mathbf{c}, \mathbf{a})$ and $c = \delta(\mathbf{a}, \mathbf{b})$. We now prove the following identity; note that the right-hand side here involves *two sides* and *one angle*.

Theorem 5.2.1 *In any spherical triangle with vertices $\mathbf{a}$, $\mathbf{b}$ and $\mathbf{c}$ we have*

$$[\mathbf{a}, \mathbf{b}, \mathbf{c}] = \sin a \, \sin b \, \sin \gamma. \qquad (5.2.3)$$

Proof We may choose our axes so that $\mathbf{c} = \mathbf{k}$, and that $\mathbf{a}$ lies in the $(\mathbf{i}, \mathbf{k})$-plane. As $\delta(\mathbf{a}, \mathbf{c}) = b$, we see that $\mathbf{a} = (\sin b, 0, \cos b)$. Similarly, $\mathbf{b}$ has latitude

$\pi/2 - a$ and longitude γ; thus

$$\mathbf{b} = (\sin a \, \cos \gamma, \, \sin a \, \sin \gamma, \, \cos a),$$

and (5.2.3) follows directly from these expressions. □

We end this section with the sine rule, and the cosine rule, for spherical geometry.

The sine and cosine rules *In a spherical triangle with vertices* **a**, **b** *and* **c**, *we have the sine rule:*

$$\frac{\sin \alpha}{\sin a} = \frac{\sin \beta}{\sin b} = \frac{\sin \gamma}{\sin c},$$

and the cosine rule: $\cos c = \cos a \, \cos b + \sin a \, \sin b \, \cos \gamma$.

Proof The sine rule follows immediately from (5.2.3) and the fact that $[\mathbf{a}, \mathbf{b}, \mathbf{c}] = [\mathbf{c}, \mathbf{a}, \mathbf{b}] = [\mathbf{b}, \mathbf{c}, \mathbf{a}]$. To prove the cosine rule we may choose **a**, **b** and **c** as in the proof of Theorem 5.2.1. Then

$$\cos c = \cos \delta(\mathbf{a}, \mathbf{b}) = \mathbf{a} \cdot \mathbf{b} = \cos a \, \cos b + \sin a \, \sin b \, \cos \gamma,$$

as required. □

Exercise 5.2

1. Derive Pythagoras' Theorem from the cosine rule.
2. Show that if an equilateral spherical triangle has sides of length a and interior angles α, then $\cos(a/2) \sin(\alpha/2) = 1/2$. Deduce that $\alpha > \pi/3$ (so that the angle sum of the triangle exceeds π).
3. Calculate the perimeter of a spherical triangle all of whose angles are $\pi/2$.

5.3 Area on the sphere

Let us now find the formula for the area of a spherical triangle on $\mathcal{S}$. We denote the spherical area (that is, the area on $\mathcal{S}$) of a set E by $\mu(E)$, and as the surface area of a sphere of radius r is $4\pi r^2$, we see that $\mu(\mathcal{S}) = 4\pi$. Two (distinct) great circles meet at diametrically opposite points, and they divide the sphere into four regions, called *lunes*. The angle of a lune is the angle (in the lune) at which the circles meet, and it is clear that *the area of a lune of angle α is 2α*, for it is obviously proportional to α, and equal to 4π when $\alpha = 2\pi$.

Unlike Euclidean geometry, the area of a spherical triangle is completely determined by its angles (there are no similarity maps in spherical geometry), and the following formula was first found by A. Girard in 1625.

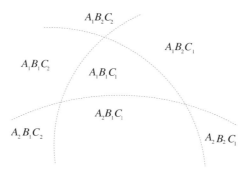

Figure 5.3.1

Theorem 5.3.1 *Let T be a spherical triangle with angles α, β and γ. Then $\mu(T) = \alpha + \beta + \gamma - \pi$.*

Proof The triangle T is formed from sides that lie on three great circles which we denote by $\mathcal{A}$, $\mathcal{B}$ and $\mathcal{C}$. The great circle $\mathcal{A}$ subdivides the sphere into two hemispheres which we denote by A_1 and A_2. We define B_1, B_2, C_1 and C_2 similarly, and these may be chosen so that $T = A_1 \cap B_1 \cap C_1$, with T having angles α, β and γ at its vertices on $\mathcal{B} \cap \mathcal{C}$, $\mathcal{C} \cap \mathcal{A}$ and $\mathcal{A} \cap \mathcal{B}$, respectively. Now $\mathcal{A}$, $\mathcal{B}$ and $\mathcal{C}$ divide the sphere into the eight triangles $A_i \cap B_j \cap C_k$, where $i, j, k = 1, 2$, and which, for brevity, we write as $A_i B_j C_k$. The triangle T ($= A_1 B_1 C_1$)) and its six neighbours are illustrated (symbolically) in Figure 5.3.1; the only triangle that does not appear in this illustration is $A_2 B_2 C_2$.

Now (for example) $A_1 B_1 C_1$ and $A_2 B_1 C_1$ together comprise a lune of angle α determined by the great circles $\mathcal{B}$ and $\mathcal{C}$. Thus we find that

$$\mu(A_1 B_1 C_1) + \mu(A_2 B_1 C_1) = 2\alpha;$$
$$\mu(A_1 B_1 C_1) + \mu(A_1 B_2 C_1) = 2\beta;$$
$$\mu(A_1 B_1 C_1) + \mu(A_1 B_1 C_2) = 2\gamma;$$
$$\mu(A_2 B_2 C_2) + \mu(A_1 B_2 C_2) = 2\alpha;$$
$$\mu(A_2 B_2 C_2) + \mu(A_2 B_1 C_2) = 2\beta;$$
$$\mu(A_2 B_2 C_2) + \mu(A_2 B_2 C_1) = 2\gamma.$$

Adding each side of these six equations, and noting that

$$\mu(A_1 B_1 C_1) = \mu(A_2 B_2 C_2), \quad \sum_{i,j,k=1}^{2} \mu(A_i B_j C_k) = 4\pi,$$

we see that $\mu(A_1 B_1 C_1) + \pi = \alpha + \beta + \gamma$ as required. □

The formula for the area of a triangle extends readily to the area of a spherical polygon.

Theorem 5.3.2 *Let P be a polygon on the sphere (with each of its n sides being an arc of a great circle), and let the interior angles of the polygon be $\theta_1, \ldots, \theta_n$. Then the area $\mu(P)$ of the polygon is given by*

$$\mu(P) = \theta_1 + \cdots + \theta_n - (n-2)\pi. \qquad (5.3.1)$$

The proof of (5.3.1) in the case of a *convex polygon* on the sphere is easy. Suppose that a polygon P has the property that there is some point $\mathbf{x}$ in P that can be joined to each vertex $\mathbf{v}_j$ by an arc of a great circle, with these arcs being non-intersecting (except at $\mathbf{x}$) and lying in P. Then these arcs divide P into n triangles, and applying Theorem 5.3.1 to each, and summing the results, we obtain (5.3.1). The proof of (5.3.1) for a non-convex polygon will be given in the next section. □

Exercise 5.3

1. Calculate the area of a spherical triangle all of whose angles are $\pi/2$, and also the area of a spherical triangle all of whose angles are $3\pi/2$.
2. For which values of θ is it possible to construct an equilateral spherical triangle with each angle equal to θ?
3. Prove the famous result of Archimedes that the area of the part of S that lies between the two parallel planes given, say, by $x_3 = a$ and $x_3 = b$, is the same as the area of the part of the circumscribing cylinder (given by $x_1^2 + x_2^2 = 1$) that lies between these two planes. Hence find the area of the 'polar cap' $\{\mathbf{x} \in S : \delta(\mathbf{x}, \mathbf{k}) < r\}$.

5.4 Euler's formula

A spherical triangle T is a region of S bounded by three arcs σ_1, σ_2 and σ_3 of great circles. The arcs σ_j are the *edges* of T, and the three points $\sigma_i \cap \sigma_j$ are the *vertices* of T. A *triangulation* of the sphere S is a partitioning of S into a finite number of non-overlapping spherical triangles T_j such that the intersection of any two of the T_j is either empty, or a common edge, or a common vertex, of the two triangles. The *edges* of the triangulation are the edges of all of the T_j; the *vertices* of the triangulation are all of the vertices of the T_j.

The simplest example of a triangulation on the sphere (say, the surface of the Earth) is found by drawing the equator and n lines of longitude. In this case the triangulation contains $2n$ triangles, $n + 2$ vertices (n on the equator, and one at each pole), and $3n$ edges. If we denote the numbers of triangles (which we now call *faces*), edges and vertices by F, E and V, respectively, we find that $F - E + V = 2$; thus the expression $F - E + V$ *does not depend on the choice of* n. It is even more remarkable that the formula $F - E + V = 2$ holds for *all* triangulations of the sphere. This famous result is due to the Swiss mathematician Leonard Euler (1707–83) who was one of the most productive mathematicians of all time.

Euler's theorem *Suppose that a triangulation of S has F triangles, E edges and V vertices. Then $F - E + V = 2$.*

We shall give Legendre's beautiful proof that is based on the area of a spherical triangle.

Proof The area of a spherical triangle Δ with angles θ_1, θ_2 and θ_3 is $\theta_1 + \theta_2 + \theta_3 - \pi$. Suppose that there are F triangles, E edges and V vertices in a triangulation of the sphere. Then, summing over all angles in all triangles, the total angle sum is $2\pi V$ (for all of the angles occur at a vertex without overlap, and the angle sum at any one of these V vertices is exactly 2π). Also, the sum of the areas of the triangles is the area of the sphere; thus $2\pi V - F\pi = 4\pi$, or $2V = F + 4$. Now (by counting the edges of each triangle, and noting that this counts each edge twice), we obtain $3F = 2E$; thus

$$F - E + V = F - 3F/2 + (F + 4)/2 = 2.$$

□

There are various important extensions of Euler's theorem which we shall now discuss briefly and informally. A *spherical polygon* is a region bounded by a finite number of arcs of great circles in such a way that the arcs form a 'closed' curve on S that divides S into exactly two regions. A spherical polygon P is *convex* if any two points in P can be joined by an arc (of a great circle) that lies entirely within P. Now partition S into a finite number of non-overlapping convex spherical polygons (which now need not be spherical triangles), and suppose that there are F polygons (or faces), E edges (each edge counted only once), and V vertices (each vertex counted only once); then, again, $F - E + V = 2$. To prove this consider one face of the triangulation, and suppose that this is bounded by a polygonal closed curve that comprises m edges and m vertices, say v_j. We take any point $\mathbf{x}$ in the face and join this to each v_j by a segment of a great circle, thus producing a triangulation of this face that includes, among its edges and vertices, all of the original edges and vertices of the face in the original partition. The contribution of this face (without its boundary) to the count $F - E + V$ from the new triangulation is 1, for it contributes m triangles, m edges and one vertex (at $\mathbf{x}$). As its contribution to $F - E + V$ in the original partition of S is also 1 (it contributes one to F and zero to E and V), it makes no difference to the count of $F - E + V$ whether we subdivide the face into triangles or not. If we carry out this subdivision for all faces, and then use Euler's theorem, we find that $F - E + V = 2$.

Next, suppose that we partition the sphere into convex polygons and that we then deform the partition continuously in such a way that the numbers F, E and V do not change during the deformation; then (obviously) after the deformation we still have $F - E + V = 2$. In particular, this will still be true when the edges are not necessarily arcs of great circles (or even arcs of any circles) providing that the given 'curvilinear partition' can be 'deformed' into a partition of the prescribed type in such a way that the numbers of faces, edges and vertices remains constant during the 'deformation'. Of course, all this is obvious; the difficulty (if one is to be rigorous) lies in the deformations.

Finally, we can extend this idea and now allow deformations of the sphere itself (together with any given partition of it). For example, as we can deform the sphere to a cube, the formula $F - E + V = 2$ will still hold for a cube. Indeed, if we partition the cube in the natural way into its six square faces, we see that $F = 6$, $E = 12$ and $V = 8$, so $F - E + V = 2$. If we wish, we can now divide each square face into two triangles by drawing one diagonal across each face, and then regard these diagonals together with the 'natural' edges of the cube as the edges of a new triangulation. This triangulation has twelve triangles, eighteen edges and eight vertices and again, $12 - 18 + 8 = 2$. Collectively, these ideas are described by saying that Euler's formula is a *topological invariant*, but we shall not discuss this any further.

Roughly speaking, a *polyhedron* is a 'closed' surface (that is, a surface with no bounding edges) made up by 'joining' polygons together along edges (of the same length), with the resulting surface being capable of being deformed into a sphere. We can also define a *convex polyhedron* to be the (non-empty) intersection of a finite number of half-spaces (a *half-space* is one 'side' of a plane); for example, a cube is easily seen to be the intersection of exactly six half-spaces. A convex polyhedron can be easily deformed into a sphere (by choosing a sphere which contains the polyhedron, and whose centre lies inside the polyhedron, and then projecting the polyhedron radially from the centre onto the sphere); thus, if a convex polyhedron has F faces, E edges and V vertices, then $F - E + V = 2$.

We complete this informal discussion by considering Euler's formula for a plane polygon. Suppose that a closed polygonal curve C (lying in the plane) divides the plane into two regions. Exactly one of these regions will be bounded, and we denote this by P. Suppose now that we triangulate P in the manner considered above. It seems clear that we can deform the polygon P until it lies on the 'southern hemisphere' of S, in such a way that the boundary curve of P lies in the 'equator'. Then there will be, say, m vertices and m edges lying on the equator (for each edge ends at a vertex which is the starting point of the next edge). Let us now include the 'northern hemisphere' N, say, as one of the polygons on the sphere, with the m edges and m vertices being the edges and vertices of N regarded as a the polygon. Then we have constructed a partitioning of S and, for this partition, $F - E + V = 2$. As we have simply added one face to the original picture, it follows that for the triangulation of the plane polygon P, we must have $F - E + V = 1$.

We end this section by completing the proof of Theorem 5.3.2.

The proof of Theorem 5.3.2 Let P be any polygon on the sphere. We extend each side of P to the great circle that contains it, and in this way we subdivide the sphere into a finite number of convex polygons. As each of these polygons either lies within P or outside of P, this shows that *we can subdivide any spherical polygon into a finite number of convex polygons*, and hence also *into a finite number of spherical triangles*.

Now take any polygon P on S, and divide it into triangles T_j as described above. Suppose that the triangulation of P into these triangles T_j has F triangles, E edges and V vertices; then by Euler's formula (as described in the previous paragraph), $F - E + V = 1$. In order to describe the next step of the proof, let us denote the original polygonal curve by C, and its 'interior' (which we have just divided into triangles) by P_0. Suppose now that E_0 of the edges of this new triangulation lie on some side of C, so that $E - E_0$ edges lie in P_0. As each of the F triangles T_j has three sides, we see that $3F = E_0 + 2(E - E_0)$; thus $3F + E_0 = 2E$. We now compute areas. We suppose that the original polygon has n vertices with internal angles $\theta_1, \ldots, \theta_n$ at these vertices. These vertices lie on C. Now E_0 of the V vertices in the new triangulation lie on C, and of these, n occur as original vertices of P (on C), while the remaining $E_0 - n$ occur as 'new' vertices lying on C (and interior to an original side of P). The angle sum at each of these $E_0 - n$ vertices is π. The remaining $V - E_0$ vertices are in P_0, and the angle sum at each of these vertices is 2π. Thus

$$\mu(P) = \sum_{j=1}^{F} \mu(T_j)$$
$$= (\theta_1 + \cdots + \theta_n) + (E_0 - n)\pi + (V - E_0)2\pi - \pi F$$
$$= (\theta_1 + \cdots + \theta_n) - n\pi + (2V - F - E_0)\pi$$
$$= (\theta_1 + \cdots + \theta_n) - (n - 2)\pi,$$

because $F - E + V = 1$ and $3F + E_0 = 2E$. □

Exercise 5.4

1. Verify Euler's formula for a 'pyramid' that has an n-gon as a base.
2. What is Euler's formula for a plane polygon from which two polygonal 'holes' have been removed?

5.5 Regular polyhedra

A *regular polyhedron* is a polyhdron that is made by joining together a finite number of congruent, regular polygons, each with p sides, and with exactly q polygons meeting at each vertex, and we say that this a regular polyhedron of type (p, q). The *faces* of the polyhedron are the regular p-gons, the *edges* are the segments where two faces are joined, and the *vertices* are the points at the ends of the edges. Such solids, which have a high degree of symmetry, are also known as *Platonic solids* (after Plato who associated them with earth, water, fire, air and the cosmos).

It is clear that if we are to construct a regular polyhedron of type (p, q), then we must be able to places q of the polygons together, in the plane and without overlapping, each having one vertex at the origin. Indeed, the sum of their angles at the origin must be strictly less than 2π, so that p and q must satisfy $q(\pi - 2\pi/p) < 2\pi$. Now this inequality is equivalent to

$$(p-2)(q-2) < 4 \tag{5.5.1}$$

and the only solutions (p, q) of this that are consistent with the obvious geometric constraints $p \geq 3$ and $q \geq 3$ are $(3, 3)$, $(3, 4)$, $(3, 5)$, $(4, 3)$ and $(5, 3)$. In particular, there are at most five such polyhedra (up to scaling).

Now consider a regular polyhedron of type (p, q) and suppose that this has F faces, E edges and V. Then, by Euler's formula, $F - E + V = 2$. As each edge is the edge of exactly two faces, and each face has p edges, we see that $2E = pF$ (we simply count each 'side' of each edge in two different ways). Similarly, as each edge has two 'ends', and each vertex is the endpoint of q 'ends', we see that $2E = qV$. We now have three simultaneous equations, namely $F - E + V = 2$, $pF = 2E$ and $qV = 2E$, and if we solve these for F, E and V in terms of p and q, we find that

$$F = \frac{4q}{2p + 2q - pq}, \quad E = \frac{2pq}{2p + 2q - pq}, \quad V = \frac{4p}{2p + 2q - pq}, \tag{5.5.2}$$

Notice, from (5.5.1), that $2p + 2q - pq > 0$. If we now substitute the five distinct possibilities for (p, q) in the formulae in (5.5.2) we obtain the following

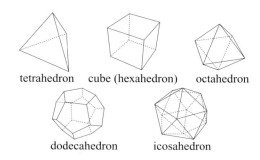

Figure 5.5.1

complete list of regular polyhedra:

polyhedron	faces	edges	vertices
tetrahedron	4	6	4
cube	6	12	8
octahedron	8	12	6
icosahedron	20	30	12
dodecahedron	12	30	20

Of course, this does not show that these solids do actually exist but they do (see Figure 5.5.1), and we shall accept this for the moment.

Finally, we remark that to obtain the Platonic solids it is not enough to merely assume that the faces of the polyhedron are congruent regular polygons. For example, we can attach five equilateral triangles together to form a pyramid with a regular pentagonal base and then join two of these pyramids together across their base; the resulting polyhedron has congruent faces that are equilateral triangles, but it is not a Platonic solid.

Exercise 5.5

1. Establish the formulae (5.5.2). These arise from three simultaneous equations; those who know about matrices should write these equations in matrix form and solve them by finding the inverse matrix. What is the determinant of the matrix?
2. The deficiency of a polyhedron is $2\pi V - \Theta$, where Θ is the sum of all the interior angles of all of its faces. Show that each of the five Platonic solids has deficiency 4π.
3. Show that the mid-points of the faces of a cube form the vertices of a regular octahedron.

4. Let A, B, C and D be the vertices of a regular tetrahedron. Show that the midpoints of the sides AB, BC, CD and DA are coplanar, and form the vertices of a square.

5.6 General polyhedra

In this section we shall examine several results that are closely related to Euler's formula. We have seen that the condition $F - E + V = 2$ is a *necessary* condition for the existence of a polyhedron (that can be deformed into a sphere) with F faces, E edges and V vertices. However, this is not the only necessary condition. Suppose that we 'cut' each edge of the polyhedron at its midpoint and then count the segments that are formed by this cutting. As each edge produces two segments, this count is clearly $2E$. On the other hand, every vertex has at least three segments ending there, and each segment ends at only one vertex; thus the count is at least $3V$. We conclude that $2E \geq 3V$. In a similar way, imagine that we have 'separated' the polyhedron into its faces, and that we now count the edges of all of these faces. Clearly this count is $2E$. On the other hand, each face has at least three sides so the count must be at least $3F$. We deduce that $2E \geq 3F$. Thus

$$F - E + V = 2, \quad 2E \geq 3V, \quad 2E \geq 3F. \qquad (5.6.1)$$

are all necessary conditions for the existence of a polyhedron (that can be deformed into a sphere), with F faces, E edges and V vertices. These show that Euler's relation $F - E + V = 2$ is not by itself sufficient for the existence of a polyhedron; for example, there is no polyhedron with $F = 4$, $E = 7$ and $V = 5$ because for these values, $2E < 3V$.

Theorem 5.6.1 *There exists a convex polyhedron with F faces, E edges and V vertices if and only if* (5.6.1) *holds.*

Proof We know that (5.6.1) are necessary conditions for the existence of a polyhedron. To prove their sufficiency; we take positive integers F, E and V satisfying (5.6.1), and we divide the proof into two cases, namely (i) $V \geq F$ and (ii) $V < F$.

(i) Suppose that a convex polyhedron has f faces, e edges and v vertices, and that it has at least one vertex of valency three (that is, exactly three faces meet at the vertex). Then we can truncate the polyhedron at that vertex (that is, we can 'slice' the vertex from the polyhedron) and the resulting polyhedron has $f + 1$ faces, $e + 3$ edges and $v + 2$ vertices. As the resulting polyhedron also has vertices of valency three, we can repeat this operation as often as we wish, say t times, and so obtain a convex polyhedron with $f + t$ faces, $e + 3t$ edges and $v + 2t$ vertices. If we apply this to the polyhedron that is a pyramid P_m whose base is an m-gon (this has $m + 1$ faces, $2m$ edges and $m + 1$ vertices), the resulting polyhedron will have $m + 1 + t$ faces, $2m + 3t$ edges and $m + 1 + 2t$ vertices; thus it suffices to show that the equations

$$m + 1 + t = F,$$
$$2m + 3t = E,$$
$$m + 1 + 2t = V$$

have a solution in integers m and t with $m \geq 3$ and $t \geq 0$. These equations are consistent (because $F - E + V = 2$), and the solution is $t = V - F$ and $m = 2F - V - 1$. Finally, $t \geq 0$ by (i), and $m = 2F - V - 1 = (2E - 3V) + 3 \geq 3$.

(ii) Suppose that a convex polyhedron has f faces, e edges and v vertices, and that one of the faces is triangular. Then we can glue a tetrahedron onto that face, and if the tetrahedron is sufficiently 'flat' the resulting polyhedron will be convex, with a triangular face, and it will have $f + 2$ faces, $e + 3$ edges and $v + 1$ vertices. If we do this t times we obtain a convex polyhedron with $f + 2t$ faces, $e + 3t$ edges and $v + t$ vertices. If the original polyhedron is the polyhedron P_m used above, the resulting polyhedron will have $m + 1 + 2t$ faces, $2m + 3t$ edges and $m + 1 + t$ vertices, and we now have to solve the equations

$$m + 1 + 2t = F,$$
$$2m + 3t = E,$$
$$m + 1 + t = V,$$

again with $t \geq 0$ and $m \geq 3$. These equations are consistent, with solution $t = F - V$ and $m = 2V - F - 1$. By (ii), $t > 0$, and $m = 2V - F - 1 = (2E - 3F) + 3 \geq 3$. $\square$

Our next result was known to Euler. Consider a polyhedron with F faces, E edges and V vertices. To examine the geometry in greater detail, let F_n be the number of faces with exactly n edges (so F_3 is the number of triangular faces, and so on), and let V_m be the number of vertices which have exactly m edges ending at that vertex (equivalently, exactly m faces containing that vertex). Clearly,

$$F = F_3 + F_4 + \cdots, \quad V = V_3 + V_4 + \cdots. \qquad (5.6.2)$$

Theorem 5.6.2

(1) *For any polyhedron, $3F_3 + 2F_4 + F_5 \geq 12$. In particular, among all of the faces of a polyhedron, there are at least four faces each of which has at most five edges.*

(2) *For any polyhedron, $F_3 + V_3 \geq 8$. In particular, any polyhedron has either a triangular face, or a vertex with exactly three edges meeting at that vertex (or both).*

(3) *The integers $F_3 + F_5 + F_7 + \cdots$ and $V_3 + V_5 + V_7 + \cdots$ are even.*

Remark There is equality in (1) and (2) in the case of the tetrahedron and the cube, so both inequalities are best possible. Note that (1) implies that one cannot make a polyhedron (however large and complicated) out of a 'random' collection of, say, hexagons. This is not obvious!

Proof As every edge bounds exactly two faces, we have

$$2E = 3F_3 + 4F_4 + 5F_5 + \cdots,$$

and as every edge contains exactly two vertices,

$$2E = 3V_3 + 4V_4 + 5V_5 + \cdots.$$

This proves (3). To prove (2), we note that (summing over $k = 3, 4 \ldots$)

$$\sum (4-k)F_k + \sum (4-k)V_k = 4\sum F_k + 4\sum V_k - \sum kF_k - \sum kV_k$$
$$= 4F + 4V - 2E - 2E$$
$$= 8.$$

This shows that

$$F_3 + V_3 = 8 + \sum_{k \geq 5}(k-4)(F_k + V_k) \geq 8, \qquad (5.6.3)$$

which is stronger than (2). Finally, Euler's relation in the form

$$6\sum F_k + 6\sum V_k = 12 + \sum kF_k + 2\sum kV_k$$

leads to the identity

$$3F_3 + 2F_4 + F_5 = 12 + \sum_{k \geq 6}(k-6)F_k + \sum_{k \geq 3}(2k-6)V_k \geq 12,$$

and this gives (1). □

We end this chapter with a result, due to Descartes, that is equivalent to Euler's formula. The *deficiency of a vertex v* of a polyhedron P is $2\pi - \sum_j \theta_j$, where the θ_j are the angles at v of the faces that meet at v. The *total deficiency* of P is the sum of the deficiencies of each of its vertices. For example, if P is a cube, then the deficiency of each vertex is $\pi/2$, and so the total deficiency of a cube is 4π. This is no accident.

Descartes' theorem *The total deficiency of any polyhedron is 4π.*

Proof We suppose that a polyhedron has F faces, E edges and V vertices. We have seen that $F = \sum_{m \geq 3} F_m$ and $2E = \sum_{m \geq 3} m F_m$. Thus

$$2(F - E + V) = 2V - (2E - 2F)$$
$$= 2V - \sum_{m \geq 3}(m - 2)F_m.$$

As the sum of the interior angle of an m-gon is $(m-2)\pi$, the total deficiency of the polyhedron is D, where

$$D = 2\pi V - \sum_m (m-2) F_m \pi = 2(F - E + V)\pi.$$

It follows that $D = 4\pi$ if and only if $F - E + V = 2$; thus Descartes' theorem is equivalent to Euler's theorem. □

Exercise 5.6

1. Prove (in the notation of the text) that $3V_3 + 2V_4 + V_5 \geq 12$.
2. We have seen that $F_3 + V_3 \geq 8$. Suppose that $F_3 + V_3 = 8$. Use (5.6.3) to show that F_3 and V_3 are even, so that the possible values of (F_3, V_3) are $(0, 8)$, $(2, 6)$, $(4, 4)$, $(6, 2)$ and $(8, 0)$. Show that each of these pairs is attained by some polyhedron.
3. A *deltahedron* is a polyhdedron (that can be deformed into a sphere) whose faces are congruent equilateral triangles. Show that every vertex of a convex deltahedron has valency at most five. Deduce that for any convex deltahedron, (F, E, V) must be of the form $(2k, 3k, k+2)$, where $k = 2, 3, \ldots, 10$. In fact, convex deltahedra exist for all of these k except 9. There exists such a polyhedron with eighteen triangular 'faces', but some of these 'faces' are coplanar.

6
Quaternions and isometries

6.1 Isometries of Euclidean space

Our first objective is to understand isometries of $\mathbb{R}^3$.

Definition 6.1.1 A map $f : \mathbb{R}^3 \to \mathbb{R}^3$ is an *isometry* if it preserves distances; that is, if for all $\mathbf{x}$ and $\mathbf{y}$, $||f(\mathbf{x}) - f(\mathbf{y})|| = ||\mathbf{x} - \mathbf{y}||$.

Each reflection across a plane is an isometry, and we shall see later that every isometry is a composition of reflections. Consider the plane Π given by $\mathbf{x}\cdot\mathbf{n} = d$, where $||\mathbf{n}|| = 1$, and let R be the reflection across Π. As $\mathbf{n}$ is the normal to Π, we see that $R(\mathbf{x}) = \mathbf{x} + 2t\mathbf{n}$, where t is chosen so that the midpoint, $\mathbf{x} + t\mathbf{n}$, of $\mathbf{x}$ and $R(\mathbf{x})$ lies on Π. This last condition gives $d = \mathbf{x}\cdot\mathbf{n} + t$, so that

$$R(\mathbf{x}) = \mathbf{x} + 2(d - \mathbf{x}\cdot\mathbf{n})\mathbf{n}. \tag{6.1.1}$$

It is geometrically obvious that (a) $R(\mathbf{x}) = \mathbf{x}$ if and only if $\mathbf{x} \in \Pi$, (b) for all $\mathbf{x}$, $R(R(\mathbf{x})) = \mathbf{x}$, and (c) R is an isometry. These properties can be verified algebraically from (6.1.1), as can the next result.

Theorem 6.1.2 *A reflection across a plane Π is a linear map if and only if $0 \in \Pi$.*

Now consider two parallel planes, say $\mathbf{x}\cdot\mathbf{n} = d_1$ and $\mathbf{x}\cdot\mathbf{n} = d_2$, and let R_1 and R_2 denote the reflections in these planes. It is easy to see that $R_1 R_2(\mathbf{x}) = \mathbf{x} + 2(d_1 - d_2)\mathbf{n}$, so that *the composition of two reflections in parallel planes is a translation* (and conversely). Clearly, the translation is by $2d\mathbf{n}$, where $\mathbf{n}$ is the common normal of the two planes, and d is their distance apart.

Next, we consider the composition of reflections R_j in two distinct intersecting planes Π_1 and Π_2. The planes intersect in a line L, and each R_j fixes every point of L. In addition, $R_j(\Pi) = \Pi$ for every plane Π that is orthogonal to L. The action of $R_2 R_1$ on Π is a reflection across the line $\Pi \cap \Pi_1$ followed by a reflection across the line $\Pi \cap \Pi_2$; thus we see that $R_2 R_1$ *is a rotation of* $\mathbb{R}^3$ *about the axis* L *of an angle equal to twice the angle between the planes* Π_j. This leads us to the following definition of a rotation.

Definition 6.1.3 A *rotation* of $\mathbb{R}^3$ is the composition of reflections across two distinct non-parallel planes. The line of intersection of the planes is the *axis* of the rotation.

Notice that the fact that every rotation has an axis is part of our definition of a rotation. Later we will give an alternative (but equivalent) definition of a rotation in terms of matrices, but then we will need to *prove* that every rotation has an axis. As a rotation of $\mathbb{R}^2$ has one fixed point and no axis, the proof cannot be entirely trivial (in fact, it depends on the fact that every real cubic polynomial has a real root). Each rotation of an odd-dimensional space has an axis, but a rotation of an even-dimensional space need not have an axis (because, as we shall see later, every real polynomial of odd degree has a real root, whereas a real polynomial of even degree need not have any real roots). As each reflection is an isometry, so is any composition of reflections. The converse is also true, and in the following stronger form.

Theorem 6.1.4 *Every isometry of* $\mathbb{R}^3$ *is the composition of at most four reflections. In particular, every isometry is a bijection of* $\mathbb{R}^3$ *onto itself. Every isometry that fixes* $\mathbf{0}$ *is the composition of at most three reflections in planes that contain* $\mathbf{0}$.

A reflection R across a plane is a permutation of $\mathbb{R}^3$ whose square is the identity map. Thus, in some sense, Theorem 6.1.4 is analogous to the result that a permutation of a finite set can be expressed as a product of transpositions. The common feature of these two results (and other results too) is that a given map f is being expressed as a composition of simpler maps f_j for which $f_j^2 = I$. Our proof of Theorem 6.1.4 is based on the following three simple results.

Lemma 6.1.5 *Suppose that* f *is an isometry with* $f(\mathbf{0}) = \mathbf{0}$. *Then for all* $\mathbf{x}$ *and* $\mathbf{y}$, $||f(\mathbf{x})|| = ||\mathbf{x}||$ *and* $f(\mathbf{x}) \cdot f(\mathbf{y}) = \mathbf{x} \cdot \mathbf{y}$ *(that is, f preserves norms and scalar products).*

Proof First, $||f(\mathbf{x})|| = ||f(\mathbf{x}) - f(\mathbf{0})|| = ||\mathbf{x} - \mathbf{0}|| = ||\mathbf{x}||$, so that f preserves norms. Next,

$$||f(\mathbf{x})||^2 + ||f(\mathbf{y})||^2 - 2(f(\mathbf{x}) \cdot f(\mathbf{y})) = ||f(\mathbf{x}) - f(\mathbf{y})||^2$$
$$= ||\mathbf{x} - \mathbf{y}||^2$$
$$= ||\mathbf{x}||^2 + ||\mathbf{y}||^2 - 2(\mathbf{x} \cdot \mathbf{y}),$$

so that f also preserves scalar products. □

Lemma 6.1.6 *If an isometry f fixes $\mathbf{0}$, $\mathbf{i}$, $\mathbf{j}$ and $\mathbf{k}$ then $f = I$.*

Proof Let $f(\mathbf{x}) = \mathbf{y}$. Then $||\mathbf{y} - \mathbf{i}|| = ||f(\mathbf{x}) - f(\mathbf{i})|| = ||\mathbf{x} - \mathbf{i}||$, and as $||\mathbf{y}|| = ||\mathbf{x}||$, we have $\mathbf{y} \cdot \mathbf{i} = \mathbf{x} \cdot \mathbf{i}$. The same holds for $\mathbf{j}$ and $\mathbf{k}$, so that $\mathbf{y} - \mathbf{x}$ is orthogonal to $\mathbf{i}$, $\mathbf{j}$ and $\mathbf{k}$. Thus $\mathbf{y} = \mathbf{x}$. □

Lemma 6.1.7 *Suppose that $||\mathbf{a}|| = ||\mathbf{b}|| \neq 0$. Then there is a reflection R across a plane Π through $\mathbf{0}$ such that $R(\mathbf{a}) = \mathbf{b}$ and $R(\mathbf{b}) = \mathbf{a}$.*

Proof We may suppose that $\mathbf{a} \neq \mathbf{b}$, since if $\mathbf{a} = \mathbf{b}$ we can take any plane Π through $\mathbf{0}$ and $\mathbf{a}$. Our geometric intuition tells us that Π should be $\mathbf{x} \cdot \mathbf{n} = 0$, where $\mathbf{n} = (\mathbf{a} - \mathbf{b})/(||\mathbf{a} - \mathbf{b}||)$, so let R be the reflection in this plane. Then $R(\mathbf{x}) = \mathbf{x} - 2(\mathbf{x} \cdot \mathbf{n})\mathbf{n}$, and a computation shows that $R(\mathbf{a}) = \mathbf{b}$. Note the use of the identity $2\mathbf{a} \cdot (\mathbf{a} - \mathbf{b}) = ||\mathbf{a} - \mathbf{b}||^2$ in this computation. □

The proof of Theorem 6.1.4 Let f be any isometry of $\mathbb{R}^3$. If $f(\mathbf{0}) \neq \mathbf{0}$ then then there is a reflection R_1 that interchanges $\mathbf{0}$ and $f(\mathbf{0})$. If f fixes $\mathbf{0}$, we let $R_1 = I$. Thus $R_1 f$ is an isometry that fixes $\mathbf{0}$, and we let $f_1 = R_1 f$. As $||f_1(\mathbf{k})|| = ||\mathbf{k}||$ there is a reflection R_2 in some plane through the origin that interchanges $f_1(\mathbf{k})$ and $\mathbf{k}$; thus $R_2 f_1$ is an isometry that fixes $\mathbf{0}$ and $\mathbf{k}$. Let $f_2 = R_2 f_1$, and note that by Lemma 6.1.5, $f(\mathbf{i})$ and $f(\mathbf{j})$ are orthogonal unit vectors in the plane $\mathbf{x} \cdot \mathbf{k} = 0$. There is a reflection R_3 in some vertical plane through the origin that maps $f(\mathbf{j})$ to $\mathbf{j}$ (and fixes $\mathbf{0}$ and $\mathbf{k}$) so that now, $R_3 f_2$ fixes $\mathbf{0}$, $\mathbf{j}$ and $\mathbf{k}$. Let $f_3 = R_3 f_2$; then f_3 maps $\mathbf{i}$ to $\pm \mathbf{i}$. If f_3 fixes $\mathbf{i}$, let $R_4 = I$. If not, let R_4 be the reflection in $\mathbf{x} \cdot \mathbf{i} = 0$. Then Lemma 6.1.6 implies that $R_4 f_3 = I$, so that $f = R_1 R_2 R_3 R_4$. □

A slight modification of the proof just given yields the next result.

Theorem 6.1.8 *The most general isometry f is of the form $f(\mathbf{x}) = A(\mathbf{x}) + f(\mathbf{0})$, where A is a linear map.*

Proof Suppose that f is an isometry. Then $A(\mathbf{x}) = f(\mathbf{x}) - f(\mathbf{0})$ is an isometry that fixes $\mathbf{0}$, and the proof of Theorem 6.1.4 shows that g is a composition of (at most three) reflections in planes through $\mathbf{0}$. As each such reflection is a linear map, so is A. □

It is clear that the composition of two isometries is an isometry, and the composition of functions is always associative. Trivially, the identity map is an isometry. If we write an isometry f as a composition, say $R_1 \cdots R_p$, of reflections R_j, then $f^{-1} = R_p \cdots R_1$ and this is an isometry. This proves the next result.

Theorem 6.1.9 *The set of isometries of $\mathbb{R}^3$ is a group with respect to composition.*

Although it may seem obvious that the inverse of an isometry f is an isometry, one has to prove (somewhere) that f^{-1} exists, and also that it is defined on $\mathbb{R}^3$.

We have seen that every isometry can be expressed as the composition of at most four reflections, and we want to give an example of an isometry that cannot be expressed as a composition of fewer than four reflections. We need some preliminary remarks. The reflection R given in (6.1.1) can be written as $R = T\tilde{R}$, where $\tilde{R}$ is the reflection in the plane $\mathbf{x} \cdot \mathbf{n} = 0$, and $T(\mathbf{x}) = \mathbf{x} + 2d\mathbf{n}$. Thus every reflection is a reflection in a plane through $\mathbf{0}$ followed by a translation. Suppose we have (with the obvious notation) $R_1(\mathbf{x}) = \tilde{R}_1(\mathbf{x}) + \mathbf{a}_1$, and similarly for R_2. Then, as $\tilde{R}_1$ is linear, we have

$$R_1 R_2(\mathbf{x}) = \tilde{R}_1\big(\tilde{R}_2(\mathbf{x}) + \mathbf{a}_2\big) + \mathbf{a}_1 = \tilde{R}_1 \tilde{R}_2(\mathbf{x}) + \tilde{R}_1(\mathbf{a}_2) + \mathbf{a}_1.$$

A similar argument applies to any composition of reflections; thus *given reflections $R_1, \ldots, R_n$ there are reflections $\tilde{R}_1, \ldots, \tilde{R}_n$ across planes through the origin, and a vector $\mathbf{b}$, such that $R_1 \cdots R_n(\mathbf{x}) = \tilde{R}_1 \cdots \tilde{R}_n(\mathbf{x}) + \mathbf{b}$.* We can now give our example.

Example 6.1.10 Let f be the isometry defined as the rotation of angle π about the $\mathbf{k}$-axis followed by the translation $\mathbf{x} \mapsto \mathbf{x} + \mathbf{k}$; thus f is a 'screw motion' which is given explicitly by $(x_1, x_2, x_3) \mapsto (-x_1, -x_2, x_3 + 1)$. Let A_1 and A_2 be the reflections in the planes $\mathbf{x} \cdot \mathbf{i} = 0$ and $\mathbf{x} \cdot \mathbf{j} = 0$, respectively; then $f(\mathbf{x}) = A_1 A_2(\mathbf{x}) + \mathbf{k}$. Now suppose that f can be written as a composition of p reflections and, as above, write $f(\mathbf{x}) = \tilde{R}_1 \cdots \tilde{R}_p(\mathbf{x}) + \mathbf{b}$, where each $\tilde{R}_j$ is a reflection in a plane through $\mathbf{0}$. As $\mathbf{b} = f(\mathbf{0}) = \mathbf{k}$, we find that $A_2 A_1 = \tilde{R}_1 \cdots \tilde{R}_p$. We shall prove below that if Q is any reflection in a plane though the origin then, for all vectors $\mathbf{x}, \mathbf{y}$ and $\mathbf{z}$, $[Q(\mathbf{x}), Q(\mathbf{y}), Q(\mathbf{z})] = -[\mathbf{x}, \mathbf{y}, \mathbf{z}]$ (this is because Q reverses the orientation of the vectors). Repeated applications of this rule applied to $A_1 A_2$ and to $\tilde{R}_1 \cdots \tilde{R}_p$ give $(-1)^2 = (-1)^p$ so that p is even. It is clear that f is not the composition of exactly two reflections (for then f would be a translation, or would have a line of fixed points); thus f cannot be expressed as the composition of fewer than four reflections. □

The next result (which we have used in Example 6.1.10) shows how scalar and vector products behave under a reflection in a plane through the origin. In particular, such a reflection reverses the orientation of any three vectors, and this fact was used in Example 6.1.10.

6.1 Isometries of Euclidean space

Theorem 6.1.11 *Let R be the reflection in a plane Π that contains the origin. Then for all vectors $\mathbf{x}$, $\mathbf{y}$ and $\mathbf{z}$,*

$$R(\mathbf{x}) \cdot R(\mathbf{y}) = \mathbf{x} \cdot \mathbf{y}, \tag{6.1.2}$$

$$R(\mathbf{x}) \times R(\mathbf{y}) = -R(\mathbf{x} \times \mathbf{y}), \tag{6.1.3}$$

$$[R(\mathbf{x}), R(\mathbf{y}), R(\mathbf{z})] = -[\mathbf{x}, \mathbf{y}, \mathbf{z}]. \tag{6.1.4}$$

Proof We may assume that Π is given by $\mathbf{x} \cdot \mathbf{n} = 0$, where $||\mathbf{n}|| = 1$, so that $R(\mathbf{x}) = \mathbf{x} - 2(\mathbf{x} \cdot \mathbf{n})\mathbf{n}$, and (6.1.2) follows immediately. Next, (6.1.3) follows from two applications of the formula (4.5.1) for the vector triple product, for

$$\begin{aligned} R(\mathbf{x}) \times R(\mathbf{y}) &= (\mathbf{x} \times \mathbf{y}) - 2(\mathbf{x} \cdot \mathbf{n})(\mathbf{n} \times \mathbf{y}) - 2(\mathbf{y} \cdot \mathbf{n})(\mathbf{x} \times \mathbf{n}) \\ &= (\mathbf{x} \times \mathbf{y}) - 2\mathbf{n} \times \left[(\mathbf{x} \cdot \mathbf{n})\mathbf{y} - (\mathbf{y} \cdot \mathbf{n})\mathbf{x}\right] \\ &= (\mathbf{x} \times \mathbf{y}) - 2\mathbf{n} \times \left[\mathbf{n} \times (\mathbf{y} \times \mathbf{x})\right] \\ &= (\mathbf{x} \times \mathbf{y}) - 2\left[(\mathbf{n} \cdot (\mathbf{y} \times \mathbf{x}))\mathbf{n} - (\mathbf{y} \times \mathbf{x})\right] \\ &= -R(\mathbf{x} \times \mathbf{y}). \end{aligned}$$

This proves (6.1.3). Finally, (6.1.4) holds because

$$\begin{aligned} [R(\mathbf{x}), R(\mathbf{y}), R(\mathbf{z})] &= R(\mathbf{x}) \cdot (R(\mathbf{y}) \times R(\mathbf{z})) \\ &= -R(\mathbf{x}) \cdot (R(\mathbf{y} \times \mathbf{z})) \\ &= -\mathbf{x} \cdot (\mathbf{y} \times \mathbf{z}) \\ &= -[\mathbf{x}, \mathbf{y}, \mathbf{z}]. \end{aligned}$$

□

It is now clear how scalar and vector products behave under rotations that fix the origin; *if A is a rotation about the origin, then*

$$A(\mathbf{x}) \cdot A(\mathbf{y}) = \mathbf{x} \cdot \mathbf{y}, \quad A(\mathbf{x}) \times A(\mathbf{y}) = A(\mathbf{x} \times \mathbf{y}), \tag{6.1.5}$$

and

$$[A(\mathbf{x}), A(\mathbf{y}), A(\mathbf{z})] = [\mathbf{x}, \mathbf{y}, \mathbf{z}].$$

Note that rotations about the origin preserve the orientation of vectors, and this leads to our final result (which is proved without reference to matrices).

Theorem 6.1.12 *The set of rotations of $\mathbb{R}^3$ that fix $\mathbf{0}$ is a group with respect to composition.*

Proof We need only show that the composition of two rotations is again a rotation, for the rest of the requirements for a group are easily verified. Let

$R_3 = R_2 R_1$, where R_1 and R_2 are rotations that fix $\mathbf{0}$. As we have remarked above, each rotation, and therefore R_3 also, preserves the orientation of vectors, and this shows that if we express R_3 as a composition of, say, p reflections in planes through the origin, then p is even. However, by Theorem 6.1.4, R_3 can be expressed as a composition of at most three reflections; thus we must be able to express R_3 as a composition of exactly two reflections, and hence R_3 is a rotation. □

Let f be an isometry, say $f(\mathbf{x}) = \mathbf{x}_0 + A(\mathbf{x})$, where A is the identity map I, or a composition of k reflections in planes through the origin, where k is one, two or three. We say that A is a *direct* isometry if $A = I$ or $k = 2$; otherwise we say that f is *indirect*. Direct isometries preserve orientation; indirect isometries reverse orientation.

Definition 6.1.13 A *screw-motion* in $\mathbb{R}^3$ is a rotation about some line in $\mathbb{R}^3$, followed by a translation in the direction of that that line.

We include the possibilities that the rotation is trivial (so the screw-motion is a translation) and that the translation is trivial (so that the screw-motion is a rotation). It is clear that every screw-motion is a direct isometry of $\mathbb{R}^3$; it is less clear that the converse is true.

Theorem 6.1.14 *Every direct isometry of $\mathbb{R}^3$ is a screw-motion.*

Proof As every translation is a screw-motion, we may confine our attention to a direct isometry f that is of the form $f(\mathbf{x}) = \mathbf{x}_0 + A(\mathbf{x})$, where A is a rotation (so $k = 2$). We choose a vector, say $\mathbf{a}$, lying along the axis of A. If $\mathbf{x}_0$ is a scalar multiple of $\mathbf{a}$, there is nothing to prove, but, in general, this will not be so.

A rotation of the same angle as A, but whose axis is a line in the direction $\mathbf{a}$, and passing through a point $\mathbf{w}$, is given by $\mathbf{x} \mapsto \mathbf{w} + A(\mathbf{x} - \mathbf{w})$; thus we need to be able to show that f can be written in the form $f(\mathbf{x}) = \mathbf{w} + A(\mathbf{x} - \mathbf{w}) + \lambda \mathbf{a}$ for some vector $\mathbf{w}$ and some real λ. Equivalently, because A is linear (so that $A(\mathbf{x} - \mathbf{w}) = A(\mathbf{x}) - A(\mathbf{w})$), we need to be able to find $\mathbf{w}$ and λ such that $\mathbf{x}_0 = \mathbf{w} - A(\mathbf{w}) + \lambda \mathbf{a}$. Now write $\mathbf{x}_0$ in the form $\mathbf{x}_1 + \mu \mathbf{a}$, where $\mathbf{x}_1 \perp \mathbf{a}$; then we need $\mathbf{w}$ and λ to satisfy $\mathbf{w} - A(\mathbf{w}) + \lambda \mathbf{a} = \mathbf{x}_1 + \mu \mathbf{a}$. We take $\lambda = \mu$, and then seek a solution $\mathbf{w}$ of $\mathbf{w} - A(\mathbf{w}) = \mathbf{x}_1$. If W is the plane given by $\mathbf{x} \cdot \mathbf{a} = 0$, it is apparent (by elementary geometry, or by using complex numbers) that there is some $\mathbf{w}$ in W such that $\mathbf{w} - A(\mathbf{w}) = \mathbf{x}_1$ and the proof is complete. □

Finally, we consider the following natural problem. Given rotations R_1 and R_2, how can we find the axis and angle of rotation of the rotation $R_2 R_1$? One way is to let Π_0 be the plane containing the axes (which we may assume are distinct) of R_1 and R_2, and let α_0 be the reflection across Π_0. Then choose planes

Π_1 and Π_2 (both containing **0**) so that if α_1 and α_2 denote reflections across these planes, then $R_2 = \alpha_2\alpha_0$ and $R_1 = \alpha_0\alpha_1$. Then $R_2 R_1 = \alpha_2\alpha_0\alpha_0\alpha_1 = \alpha_2\alpha_1$. Further, if we examine the intersection of these planes with the unit sphere, then the calculations of angles and distances involved are a matter of spherical (rather than Euclidean) geometry. We also give an algebraic solution to this problem in terms of quaternions (in Section 6.3), and in terms of matrices (in Chapter 11).

Exercise 6.1

1. Find the formula for the reflection in the plane $x_2 = 0$ followed by the reflection in the plane $x_3 = 0$. Find the fixed points of this transformation, and identify it geometrically.
2. Given unit vectors **a** and **b**, let $R_\mathbf{a}$ be the reflection across the plane $\mathbf{x}\cdot\mathbf{a}$, and similarly for $R_\mathbf{b}$. Show that $R_\mathbf{a} R_\mathbf{b} = R_\mathbf{b} R_\mathbf{a}$ if and only if $\mathbf{a} \perp \mathbf{b}$.
3. Let $R_\mathbf{a}$ and $R_\mathbf{b}$ be the reflections in $\mathbf{x}\cdot\mathbf{a} = 0$ and $\mathbf{x}\cdot\mathbf{b} = 0$, respectively, where $||\mathbf{a}|| = 1$ and $||\mathbf{b}|| = 1$. Find a formula for the composition $R_\mathbf{a}(R_\mathbf{b}(\mathbf{x}))$. This composition of reflections is a rotation whose axis is the line of intersection of the two planes. Explain why this axis is the set of scalar multiples of $\mathbf{a} \times \mathbf{b}$, and verify analytically that both $R_\mathbf{a}$ and $R_\mathbf{b}$ fix every scalar multiple of $\mathbf{a} \times \mathbf{b}$.
4. Let L_1, L_2 and L_3 be three lines in the complex plane $\mathbb{C}$, each containing the origin, and let R_j be the reflection (of $\mathbb{C}$ into itself) across L_j. Show that $R_1 R_2 R_3$ is a reflection across some line in $\mathbb{C}$. Show, however, that if the L_j do not have a common point then $R_1 R_2 R_3$ need not be a reflection.
5. Suppose that $R_1, \ldots, R_p, S_1, \ldots, S_q$ are all reflections across planes that contain **0**. Show that if $R_1 \cdots R_p = S_1 \cdots S_q$ then $(-1)^p = (-1)^q$. Compare this result with the definition of the signature of a permutation.

6.2 Quaternions

In 1843 Hamilton introduced quaternions as a way to generalize the algebra of complex numbers to higher dimensions. We shall use them to represent reflections and rotations in $\mathbb{R}^3$ algebraically. There are various equivalent ways of describing quaternions but fundamentally they are points in Euclidean four-dimensional space $\mathbb{R}^4$. Before we consider the various representations of quaternions it is convenient to introduce the idea of a Cartesian product of two sets. Given any two sets A and B, the Cartesian product set, which we denote by $A \times B$, is the set of all ordered pairs (a, b), where $a \in A$ and $b \in B$. In as far as there is a natural identification of the objects

$$(w, x, y, z), \quad ((w, x), (y, z)), \quad (w, (x, y, z)),$$

in $\mathbb{R}^4$, $\mathbb{R}^2 \times \mathbb{R}^2$ and $\mathbb{R} \times \mathbb{R}^3$, respectively, we can identify each of these spaces with the others. Formally, a complex number is an ordered pair of real numbers so that $\mathbb{C} = \mathbb{R}^2$ (this really is equality here, and no identification is necessary), so we can add $\mathbb{C} \times \mathbb{C}$ to this list. Notice that a point in $\mathbb{R} \times \mathbb{R}^3$ is an ordered pair $(a, \mathbf{x})$, where $a \in \mathbb{R}$ and $\mathbf{x} \in \mathbb{R}^3$, and we can write this as $a + x_1\mathbf{i} + x_2\mathbf{j} + x_3\mathbf{k}$.

Definition 6.2.1 A *quaternion* is an expression $a + b\mathbf{i} + c\mathbf{j} + d\mathbf{k}$, where a, b, c and d are real numbers. The set of quaternions is denoted by $\mathbb{H}$.

We now define an addition and a multiplication on the set $\mathbb{H}$ of quaternions, and it is sufficient to do this (and to carry out any subsequent calculations) in any of the models given above since a statement about one model can be converted on demand into a statement about another model. The addition of quaternions is just the natural vector addition in $\mathbb{R}^4$, namely

$$(x_1, x_2, x_3, x_4) + (y_1, y_2, y_3, y_4) = (x_1 + y_1, x_2 + y_2, x_3 + y_3, x_4 + y_4),$$

and clearly $\mathbb{R}^4$ is an abelian group with respect to addition. In terms of the representation in $\mathbb{R} \times \mathbb{R}^3$ this becomes $(a, \mathbf{x}) + (b, \mathbf{y}) = (a + b, \mathbf{x} + \mathbf{y})$, and in terms of $\mathbb{C} \times \mathbb{C}$ it is $(z_1, z_2) + (w_1, w_2) = (z_1 + w_1, z_2 + w_2)$.

The multiplication of quaternions is rather special. We recall how in Section 3.1 we motivated the multiplication of complex numbers by considering the formal product

$$(x + iy)(u + iv) = ux + i(xv + yu) + yvi^2,$$

and then imposing the condition that $i^2 = -1$ to return us to something of the form $p + iq$. We shall now adopt a similar strategy for quaternions. We write the general quaternion as $a + b\mathbf{i} + c\mathbf{j} + d\mathbf{k}$ and then, using the distributive laws, we take a formal product of two such expressions. We now need to specify what we mean by the products of any two of $\mathbf{i}$, $\mathbf{j}$ and $\mathbf{k}$, and these are specified as follows:

(1) $\mathbf{i}^2 = \mathbf{j}^2 = \mathbf{k}^2 = -1$, and
(2) $\mathbf{ij} = \mathbf{k} = -\mathbf{ji}, \quad \mathbf{jk} = \mathbf{i} = -\mathbf{kj}, \quad \mathbf{ki} = \mathbf{j} = -\mathbf{ik}.$

Notice that the product of one of these vectors with itself is a *scalar*; otherwise the product is a vector. Note also how (2) mimics the vector product of these vectors, and (1) mimics the formula $i^2 = -1$. We shall also assume that any real number commutes with each of the vectors $\mathbf{i}$, $\mathbf{j}$ and $\mathbf{k}$; for example, $3\mathbf{j} = \mathbf{j}3$. Of course, we could rewrite this whole discussion formally in terms of four-tuples of real numbers, but to do so would not contribute to an understanding of the underlying ideas. Indeed, the following example is probably more useful:

$$\begin{align}(2+3\mathbf{i}-4\mathbf{k})(\mathbf{i}-2\mathbf{j}) &= 2(\mathbf{i}-2\mathbf{j})+3\mathbf{i}(\mathbf{i}-2\mathbf{j})-4\mathbf{k}(\mathbf{i}-2\mathbf{j})\\ &= 2\mathbf{i}-4\mathbf{j}+3\mathbf{i}^2-6\mathbf{ij}-4\mathbf{ki}+8\mathbf{kj}\\ &= 2\mathbf{i}-4\mathbf{j}-3-6\mathbf{k}-4\mathbf{j}-8\mathbf{i}\\ &= -3-6\mathbf{i}-8\mathbf{j}-6\mathbf{k}.\end{align}$$

We note that the set of quaternions is closed under multiplication, but that multiplication is *not* commutative (for example, $\mathbf{ij} \neq \mathbf{ji}$); the reader should constantly be aware of this fact. It is an elementary (though tedious) matter to check that the associative law holds for multiplication, and that the distributive laws also hold. If we rewrite the definition of multiplication in terms of quaternions in the form $(a, \mathbf{x})$ we see a striking link with the vector algebra on $\mathbb{R}^3$; this is given in the next result.

Theorem 6.2.2 *The product of quaternions is given by the formula*

$$(a, \mathbf{x})(b, \mathbf{y}) = \big(ab - \mathbf{x}\cdot\mathbf{y}, a\mathbf{y}+b\mathbf{x}+(\mathbf{x}\times\mathbf{y})\big), \qquad (6.2.1)$$

where $\mathbf{x}\cdot\mathbf{y}$ and $\mathbf{x}\times\mathbf{y}$ are the scalar and vectors products, respectively.

The proof is by pure computation and we omit the details, but the reader should certainly verify this when $a = b = 0$. Note that multiplication by the 'real number' $(a, \mathbf{0})$ gives $(a, \mathbf{0})(b, \mathbf{y}) = (ab, a\mathbf{y}) = (b, \mathbf{y})(a, \mathbf{0})$. This shows that the quaternion $(1, \mathbf{0})$ acts as the identity element with respect to multiplication. Theorem 6.2.2 is of special interest when the two quaternions are pure quaternions, which we now define (these play the role of the purely imaginary numbers in $\mathbb{C}$).

Definition 6.2.3 A *pure quaternion* is a quaternion of the form $(0, \mathbf{x})$ where $\mathbf{x} \in \mathbb{R}^3$. The set of pure quaternions is denoted by $\mathbb{H}_0$.

When applied to pure quaternions, Theorem 6.2.2 yields the important formula

$$(0, \mathbf{x})(0, \mathbf{y}) = \big(-\mathbf{x}\cdot\mathbf{y}, \mathbf{x}\times\mathbf{y}\big) \qquad (6.2.2)$$

in which the quaternion product is expressed only in terms of the scalar and vector products. This shows that if q is the pure quaternion $(0, \mathbf{x})$, and $||\mathbf{x}|| = 1$, then $q^2 = (-1, \mathbf{0})$, and this extends the identities $\mathbf{i}^2 = \mathbf{j}^2 = \mathbf{k}^2 = -1$ to *all* pure quaternions of unit length. It also shows that some quaternions, for example -1, have *infinitely many square roots*.

Let us return to the notation given in Definition 6.1.1 so that we now write $a + \mathbf{x}$ instead of $(a, \mathbf{x})$. The conjugate of a quaternion is defined by analogy with the conjugate of a complex number.

Definition 6.2.4 The *conjugate* of the quaternion $a + \mathbf{x}$ is $a - \mathbf{x}$. We denote the conjugate of a quaternion q by $\bar{q}$.

Theorem 6.2.5 *For quaternions p and q, we have $\overline{(pq)} = \bar{q}\bar{p}$.*

This follows directly from Theorem 6.2.2, but note that the order of terms here is important. Finally, we note an important identity: if $q = a + \mathbf{x}$ then, from (6.2.1),

$$q\bar{q} = (a + \mathbf{x})(a - \mathbf{x}) = a^2 + ||\mathbf{x}||^2.$$

Of course,

$$(a^2 + ||\mathbf{x}||^2)^{1/2} = \sqrt{a^2 + x_1^2 + x_2^2 + x_3^2},$$

and as this is the distance between the origin and the point $a + \mathbf{x}$ viewed as a point of $\mathbb{R}^4$, we write this as $||a + \mathbf{x}||$. This is the *norm* of the quaternion, and it obeys the important rule that for any quaternions p and q,

$$||pq|| = ||p|| \, ||q||.$$

The proof is easy for, using the associative law and Theorem 6.2.5,

$$||pq||^2 = (pq)\overline{(pq)} = pq\bar{q}\bar{p} = p\,||q||^2\,\bar{p} = ||q||^2 \, p\bar{p} = ||q||^2 \, ||p||^2 \quad (6.2.3)$$

as required. $\square$

Finally, we have just seen that if q is not the zero quaternion, and if $q' = ||q||^{-2}\bar{q}$, then $qq' = 1 = q'q$. Thus each non-zero quaternion has a multiplicative inverse, and we have proved the following result.

Theorem 6.2.6 *The set of non-zero quaternions is a non-abelian group with respect to the multiplication defined above.*

In conclusion, the set $\mathbb{H}$ of quaternions, with the addition and multiplication defined above, satisfies all of the axioms for a field *except* that multiplication is not commutative. Accordingly, the quaternions are sometimes referred to as a *skew field*.

Exercise 6.2

1. Simplify the sequence of quaternions $\mathbf{i}, \mathbf{ij}, \mathbf{ijk}, \mathbf{ijki}, \mathbf{ijkij}, \ldots$.
2. Let $q = a + \mathbf{x}$. Express a and $\mathbf{x}$ in terms of q and $\bar{q}$.
3. Let q be a pure quaternion. Compute q^2, and hence show that $q^{-1} = -||q||^{-2}q$.

4. Show that $\{1, -1, \mathbf{i}, -\mathbf{i}, \mathbf{j}, -\mathbf{j}, \mathbf{k}, -\mathbf{k}\}$ is a group under multiplication.
5. Prove (6.2.1).
6. Prove Theorem 6.2.5.
7. Show that a quaternion q is a pure quaternion if and only if q^2 is real and not positive.
8. Let p and q be pure quaternions. Show that (in the obvious sense) pq is the vector product $p \times q$ if and only if $p \perp q$ (in $\mathbb{R}^3$). In the same sense, show that $pq = qp$ if and only if $p \times q = 0$.

6.3 Reflections and rotations

The formula (6.1.1) for a reflection across a plane is simple; however the formula for a rotation, obtained by combining two reflections, is not. In this section we shall see how quaternions provide an alternative algebraic way to express reflections and rotations. Quaternions are points of $\mathbb{R}^4$, but we shall pay special attention to the space $\mathbb{H}_0$ of pure quaternions which we identify with $\mathbb{R}^3$.

Consider the map $\theta : \mathbb{H} \to \mathbb{H}$ given by $\theta(y) = -qyq^{-1}$, where q is a non-zero pure quaternion. It is clear that θ is a linear map; that is, $\theta(\lambda_1 y_1 + \lambda_2 y_2) = \lambda_1 \theta(y_1) + \lambda_2 \theta(y_2)$ and, trivially, $\theta(0) = 0$ and $\theta(q) = -q$. These facts suggest that θ might be related to the reflection across the plane with normal q, and we shall now show that this is so.

Theorem 6.3.2 *The map $\theta : \mathbb{H} \to \mathbb{H}$ given by $\theta(y) = -qyq^{-1}$, where q is a non-zero pure quaternion, maps the set $\mathbb{H}_0$ of pure quaternions into itself. Further, when $\mathbb{H}_0$ is identified with $\mathbb{R}^3$, and q with $\mathbf{q}$, θ is the reflection across the plane $\mathbf{x} \cdot \mathbf{q} = 0$. If $\mathbf{q}$ is a pure quaternion of unit length, then $\theta(y) = qyq$.*

Proof We begin by looking at the fixed points of θ in $\mathbb{H}_0$. A general quaternion y is fixed by θ if and only if $-qy = yq$, and if we write $q = (0, \mathbf{q})$ and $y = (0, \mathbf{y})$, and use (6.2.2), we see that a pure quaternion $\mathbf{y}$ is fixed by θ if and only if $\mathbf{y} \cdot \mathbf{q} = 0$. Now let Π be the plane in $\mathbb{R}^3$ given by $\mathbf{x} \cdot \mathbf{q} = 0$; then θ fixes each point of Π. Now take any pure quaternion $\mathbf{y}$. We can write $\mathbf{y} = \mathbf{p} + \lambda \mathbf{q}$, where λ is real and $\mathbf{p} \in \Pi$. Thus, as θ is linear,

$$\theta(\mathbf{y}) = \theta(\mathbf{p}) + \lambda \theta(\mathbf{q}) = \mathbf{p} - \lambda \mathbf{q}.$$

This shows that $\theta(\mathbf{y})$ is a pure quaternion, and also that it is the reflection of $\mathbf{y}$ in Π. Finally, if $\mathbf{q}$ is a pure quaternion of unit length, then $q^2 = -1$ so that $q^{-1} = -q$ and $\theta(y) = qyq$. □

Before moving on to discuss rotations, we give a simple example.

Example 6.3.3 The reflection across the plane $\mathbf{x}\cdot\mathbf{k}=0$ is given by $(x_1, x_2, x_3) \mapsto (x_1, x_2, -x_3)$ or, in terms of quaternions, $\mathbf{x} \mapsto \mathbf{kxk}$. We also note (in preparation for what follows) that as $\mathbf{k}^{-1} = -\mathbf{k}$, the map $\mathbf{x} \mapsto \mathbf{kxk}^{-1}$ is given by $(x_1, x_2, x_3) \mapsto (-x_1, -x_2, x_3)$, and this is a rotation of angle π about the $\mathbf{k}$-axis. □

We now use quaternions to describe rotations that fix the origin. Consider the rotation R obtained by the reflection across the plane with normal unit $\mathbf{p}$ followed by the reflection across the plane with unit normal $\mathbf{q}$. According to Theorem 6.3.2,

$$R(\mathbf{x}) = q(p\mathbf{x}p)q = (qp)\mathbf{x}(pq), \tag{6.3.1}$$

where $p = (0, \mathbf{p})$ and $q = (0, \mathbf{q})$. This observation yields the following result.

Theorem 6.3.4 *Let $r = (\cos \frac{1}{2}\theta, \sin \frac{1}{2}\theta\, \mathbf{n})$, where $\mathbf{n}$ is a unit vector. Then r is a unit quaternion, and the map $\mathbf{x} \mapsto r\mathbf{x}r^{-1} = r\mathbf{x}\bar{r}$ is a clockwise rotation by an angle θ about the axis in the direction $\mathbf{n}$.*

Proof Choose unit vectors $\mathbf{p}$ and $\mathbf{q}$ such that $\mathbf{p}\cdot\mathbf{q} = \cos \frac{1}{2}\theta$ and $\mathbf{p} \times \mathbf{q} = \sin \frac{1}{2}\theta\, \mathbf{n}$. Then the rotation described (geometrically) in the theorem is reflection in $\mathbf{x}\cdot\mathbf{p} = 0$ followed by reflection in $\mathbf{x}\cdot\mathbf{q} = 0$; thus it is given by R in (6.3.1). Now a simple calculation using (6.2.2) shows that $qp = -r$ and $pq = -\bar{r}$. Thus $R(\mathbf{x}) = (-r)\mathbf{x}(-\bar{r}) = r\mathbf{x}\bar{r}$. Finally, as $||r|| = 1$ we see that $\bar{r} = r^{-1}$. □

Theorem 6.3.4 provides a way to find the geometric description of the composition of two given rotations. Let R_r and R_s be the rotations associated with the quaternions

$$r = \cos \tfrac{1}{2}\theta + \sin \tfrac{1}{2}\theta\, \mathbf{n}, \quad s = \cos \tfrac{1}{2}\varphi + \sin \tfrac{1}{2}\varphi\, \mathbf{m}.$$

Then the composition $R_s R_r$ is the map $\mathbf{x} \mapsto s(r\mathbf{x}\bar{r})\bar{s} = (sr)\mathbf{x}(\overline{sr})$, and as sr is also a unit quaternion (which is necessarily of the form $\cos \frac{1}{2}\psi + \sin \frac{1}{2}\psi\, \mathbf{h}$ for some unit vector $\mathbf{h}$), we see that $R_s R_r = R_{sr}$. By computing sr as a quaternion product, we can find the axis and angle of rotation of the composition $R_s R_r$. We urge the reader to experiment with some simple examples to see this idea in action.

Exercise 6.3

1. Let r be a unit quaternion ($||r|| = 1$). Show that we can write $r = \cos\theta + \sin\theta\, \mathbf{p}$, where $\mathbf{p}$ is a unit pure quaternion.

6.3 Reflections and rotations

2. Use Theorem 6.2.2 to show that if p and q are pure quaternions ($p = \mathbf{p}$ and $q = \mathbf{q}$), and if $\mathbf{p} \perp \mathbf{q}$, then $-qpq^{-1} = p$.
3. Use quaternions to find the image of $\mathbf{x}$ under a rotation about the $\mathbf{k}$-axis of angle $\pi/6$. Now verify your result by elementary geometry.
4. Let Π be the plane given by $\mathbf{x} \cdot \mathbf{n} = 0$, where $\mathbf{n} = (1, -1, 0)$, and let R be the reflection across Π. Write $\mathbf{x} = (a, b, c)$ and $\mathbf{y} = R(\mathbf{x})$. Verify, both using elementary geometry and quaternion algebra, that $\mathbf{y} = (b, a, c)$.

7
Vector spaces

7.1 Vector spaces

The essential feature of vectors is that if we add them, or take scalar multiples of them, we obtain another vector; thus *a linear combination of vectors is again a vector*. The general theory of vector spaces is based on this simple idea, and in this chapter we study abstract spaces with this property. The fact that the general theory of vector spaces is so widely applicable makes the use of boldface type for vectors an awkward convention, so we now abandon it. It should be clear from the discussion which symbols are vectors and which are scalars (and if we think it might not be, we shall write $\lambda \in \mathbb{R}$, $x \in V$, and so on or, sometimes, add V or $\mathbb{F}$ as a suffix).

A *vector space* consists of a set V of *vectors* and a field $\mathbb{F}$ of *scalars*. In this book, the field $\mathbb{F}$ of scalars will always be $\mathbb{R}$ or $\mathbb{C}$, and when the argument holds for both of these we shall use $\mathbb{F}$. The vectors form an abelian group with respect to an operation which is always called 'addition' and denoted by $+$. The scalars allow us to take a 'scalar multiple' of a vector; thus to each scalar λ and each vector v we can associate a vector λv. As a consequence of these assumptions, given scalars $\lambda_1, \ldots, \lambda_n$, and vectors $v_1, \ldots, v_n$, we can form the *linear combination* $\lambda_1 v_1 + \cdots + \lambda_n v_n$ and this too is a vector. We stress that some statements may be true for one choice of $\mathbb{F}$ and false for the other; for example, in the vector space $\mathbb{C}$, the vectors 1 and i are scalar multiples of each other if $\mathbb{F} = \mathbb{C}$ (the vector i is the scalar multiple i of the vector 1), but not if $\mathbb{F} = \mathbb{R}$ (for there is no real scalar λ such that $1 = \lambda i$, or $i = \lambda 1$). In order to develop a worthwhile theory, we need to include some (natural) assumptions that will ensure that our algebraic manipulations run smoothly, and these are given in the following formal definition of a vector space.

7.1 Vector spaces

Definition 7.1.1 A *vector space* V *over the field* $\mathbb{F}$ is a set V of *vectors*, a field $\mathbb{F}$ of *scalars*, an addition $u + v$ of vectors u and v, and a scalar multiplication λv of a scalar λ with a vector v, which together satisfy the following properties:

(1) V is an abelian group with respect to the operation $+$;
(2) if $\lambda \in \mathbb{F}$ and $v \in V$, then $\lambda v \in V$;
(3) for all λ, μ, v and w, $(\lambda + \mu)v = \lambda v + \mu v$, and $\lambda(v + w) = \lambda v + \lambda w$.
(4) for all λ, μ and v, $\lambda(\mu v) = (\lambda \mu)v$;
(5) for all v, $1v = v$, where 1 is the multiplicative identity in $\mathbb{F}$.

We say that V is a *real vector space* if $\mathbb{F} = \mathbb{R}$, and a *complex vector space* if $\mathbb{F} = \mathbb{C}$.

We stress again that the underlying idea is that it is possible to form linear combinations of vectors, and that the associated 'arithmetic' satisfies all of the familiar rules. The simplest examples of vector spaces are the spaces of n-tuples of real (or complex) numbers.

Example 7.1.2 The set of ordered n-tuples $(x_1, \ldots, x_n)$ of real numbers is denoted by $\mathbb{R}^n$. The rules for addition and scalar multiplication are

$$(x_1, \ldots, x_n) + (y_1, \ldots, y_n) = (x_1 + y_1, \ldots, x_n + y_n),$$
$$\lambda(x_1, \ldots, x_n) = (\lambda x_1, \ldots, \lambda x_n),$$

where λx_j is the usual product in $\mathbb{R}$, and $\mathbb{R}^n$ is a real vector space. Although we shall not use it for a while, this is an appropriate point to define the scalar product in $\mathbb{R}^n$. Given x and y in $\mathbb{R}^n$, their scalar product is (by analogy with $\mathbb{R}^3$) given by

$$x \cdot y = x_1 y_1 + \cdots + x_n y_n.$$

Similarly, the set $\mathbb{C}^n$ of ordered n-tuples $(z_1, \ldots, z_n)$ of complex numbers is a complex vector space with the same rules as above (except that λ is now a complex number). However, in this case, the scalar product is

$$z \cdot w = z_1 \bar{w}_1 + \cdots + z_n \bar{w}_n$$

(and not $\sum z_i w_i$ because we want to insist that $z \cdot z \geq 0$ with equality if and only if z is the zero vector).

We shall also need to use column vectors of real and complex numbers; for example

$$\begin{pmatrix} x \\ y \end{pmatrix}, \quad \begin{pmatrix} 2 + 3i \\ 5 - i \end{pmatrix}.$$

The space of real n-tuples written as column vectors is denoted by $\mathbb{R}^{n,t}$, where the symbol 't' here stands for the *transpose* (which converts rows into columns, and columns into rows, in the obvious way). The addition and scalar multiplication in $\mathbb{R}^{n,t}$ are the obvious ones and $\mathbb{R}^{n,t}$ is also a real vector space. Of course, $\mathbb{R}^{n,t}$ differs from $\mathbb{R}^n$ only in the way that we write the vectors, but the reason for making this distinction will emerge later. In the same way the space $\mathbb{C}^{n,t}$ of columns of n complex numbers is a complex vector space. □

The following examples illustrate the wide variety of vector spaces that exist and, consequently, the wide applicability of the results that we shall prove shortly.

Example 7.1.3 The set of all real-valued functions on a set X is a real vector space. Addition and scalar multiplication are defined so that $f + g$ is the function $x \mapsto f(x) + g(x)$, and λf is the function $x \mapsto \lambda f(x)$. Similarly, the space of complex valued functions on X is a complex vector space. We are *not* assuming here that the set X has any algebraic structure. □

The class of vector spaces constructed in Example 7.1.3 contains a huge number of interesting special cases; for example, the space of all real polynomials is a real vector space, and the complex polynomials form a complex vector space. The space of all real (or complex) polynomials of degree at most k is also a vector space, but the set of polynomials of exact degree k is not, for the sum of two polynomials of degree k need not be of degree k. In these examples the zero vector is the *zero polynomial*; this is the constant *function* with value 0 and it is *not* a number. We shall discuss vector spaces of polynomials in greater detail later. The set of all *trigonometric polynomials*

$$a_0 + a_1 \cos t + \cdots + a_n \cos nt + b_1 \sin t + \cdots + b_n \sin nt,$$

where the a_j and b_j are real numbers, is a vector space over $\mathbb{R}$. The set V of solutions of the differential equation $\ddot{x} + \mu^2 x = 0$ is a vector space over $\mathbb{R}$. We know that each solution is of the form $x(t) = a \cos \mu t + b \sin \mu t$, but it is important to realize that V is a vector space because a linear combination of solutions is again a solution, and we can see this *without* solving the equation. This is true of all *linear* differential equations, and this is where the terminology comes from. Finally, a real sequence is a map $f : \{1, 2, \ldots\} \to \mathbb{R}$, the sequence being $f(1), f(2), \ldots$, and the set of real sequences is a vector space. The same is true of complex sequences, and also of finite sequences, which brings us back to Example 7.1.2.

We end this section by proving a simple, and not unexpected, result about general vector spaces which will be used frequently (and without explicit mention). In general, we denote the additive and multiplicative identities in the field $\mathbb{F}$ by 0 and 1, respectively, and the zero vector in V also by 0 (it should be obvious in any context whether 0 means the zero vector or the zero scalar). However, in order to to make the following lemma completely clear, we shall temporarily use $0_\mathbb{F}$ and $1_\mathbb{F}$ for these elements in $\mathbb{F}$, and 0_V for the zero vector. This lemma is fundamental, and its proof is tedious, but we shall soon move on to more interesting things. Note that $-1_\mathbb{F}$ is the additive inverse of the multiplicative identity $1_\mathbb{F}$ in the field $\mathbb{F}$.

Lemma 7.1.4 *Let V be a vector space over $\mathbb{F}$, and suppose that $\lambda \in \mathbb{F}$ and $v \in V$. Then*

(1) $\lambda v = 0_V$ *if and only if* $\lambda = 0_\mathbb{F}$ *or* $v = 0_V$;
(2) $(-1_\mathbb{F})v$ *is the additive inverse of* v.

Proof First, for any scalar λ, and any vector v,

$$0_\mathbb{F} v = (0_\mathbb{F} + 0_\mathbb{F})v = 0_\mathbb{F} v + 0_\mathbb{F} v,$$
$$\lambda 0_V = \lambda(0_V + 0_V) = \lambda 0_V + \lambda 0_V.$$

By Lemma 1.2.5, $0_\mathbb{F} v = 0_V = \lambda 0_V$. Now suppose that $\lambda v = 0_V$. If $\lambda \neq 0_\mathbb{F}$, then λ^{-1} exists in $\mathbb{F}$ so (by what we have just proved), $0_V = \lambda^{-1} 0_V = \lambda^{-1}(\lambda v) = (\lambda^{-1}\lambda)v = 1_\mathbb{F} v = v$. Thus if $\lambda v = 0_V$ then $\lambda = 0_\mathbb{F}$ or $v = 0_V$. Finally,

$$v + (-1_\mathbb{F})v = 1_\mathbb{F} v + (-1_\mathbb{F})v = \big(1_\mathbb{F} + (-1_\mathbb{F})\big)v = 0_\mathbb{F} v = 0_V,$$

so that $(-1_\mathbb{F})v$ is the (unique) additive inverse of v. □

Exercise 7.1

1. Let Π_1 and Π_2 be the planes in $\mathbb{R}^3$ given by $3x + 2y - 5z = 0$ and $3x + 2y - 5z = 1$, respectively. Show that Π_1 is a vector space, but that Π_2 is not.
2. Show that $\mathbb{C}^n$ is a vector space over $\mathbb{R}$.
3. Verify that the set of the even complex polynomials p (that is, the p with $p(-z) = p(z)$) is a complex vector space. Is the set of all odd complex polynomials p (with $p(-z) = -p(z)$) a complex vector space?
4. Solve the differential equation $d^2x/dt^2 = x$, and show that the set of solutions is a real vector space. Solve the equation $(dx/dt)^2 = x$. Is this set of solutions a vector space? Which of these differential equations is linear?
5. Let V be a vector space over $\mathbb{F}$ and let $v_1, \ldots, v_m$ be any vectors in V. Show that the set V_0 of all linear combinations of $v_1, \ldots, v_m$ is a vector space over $\mathbb{F}$; it is the vector space *generated by* $v_1, \ldots, v_m$.

7.2 Dimension

We shall now define what we mean by the dimension of a vector space, and we shall confirm that $\mathbb{R}^2$ has dimension two, $\mathbb{R}^3$ has dimension three, and so on. We begin with a definition and a theorem which we need in order to define dimension.

Definition 7.2.1 The set of vectors $v_1, \ldots, v_n$ of vectors in a vector space V is a *basis* of V if, for every v in V, there exist unique scalars λ_j such that $v = \sum_j \lambda_j v_j$.

We stress that this definition asserts both the *existence* of the scalars λ_j and their *uniqueness*. The next result is fundamental.

Theorem 7.2.2 *Let V be a vector space and suppose that $v_1, \ldots, v_n$ and $w_1, \ldots, w_m$ are two bases of V. Then $m = n$.*

Theorem 7.2.2 enables us to say what we mean by the dimension of a vector space. To each non-trivial vector space V with a finite basis we can assign a positive integer, namely the number of elements in *any* basis of V, and this is the dimension of V.

Definition 7.2.3 We say that a vector space V has *finite dimension*, or is *finite dimensional*, if it has a finite basis, and the *dimension* dim(V) of V is then the number of elements in any basis of V. If $V = \{0\}$ we put dim(V) = 0. If V has no finite basis, then V is *infinite dimensional*.

The proof of Theorem 7.2.2 As $v_1, \ldots, v_n$ is a basis of V, each w_k can be expressed as a linear combination of the v_j; thus for each k there are scalars $\lambda_{1k}, \ldots, \lambda_{nk}$ such that

$$w_k = \sum_{j=1}^{n} \lambda_{jk} v_j, \qquad k = 1, \ldots, m.$$

Likewise, for each j there are scalars $\mu_{1j}, \ldots, \mu_{mj}$ such that

$$v_j = \sum_{i=1}^{m} \mu_{ij} w_i, \qquad j = 1, \ldots, n.$$

These give

$$w_k = \sum_{j=1}^{n} \sum_{i=1}^{m} \lambda_{jk} \mu_{ij} w_i, \qquad k = 1, \ldots, m.$$

As the w_j form a basis, we may equate coefficients of w_k on each side of this equation (recall the uniqueness in Definition 7.2.1) and obtain

$$1 = \sum_{j=1}^{n} \lambda_{jk}\mu_{kj}, \qquad k = 1, \ldots, m.$$

Summing both sides of this equation over $k = 1, \ldots, m$ now gives

$$m = \sum_{k=1}^{m}\left(\sum_{j=1}^{n} \lambda_{jk}\mu_{kj}\right).$$

However, the same argument with v_j and w_k interchanged gives

$$n = \sum_{s=1}^{n}\left(\sum_{t=1}^{m} \mu_{ts}\lambda_{st}\right)$$

and as these last two double sums are the same (they differ only in the order of summation), we see that $n = m$. □

It is now time to introduce some useful terminology which separates the two essential features of a basis.

Definition 7.2.4 The set of vectors $v_1, \ldots, v_n$ *span* V if every v in V can be expressed as a linear combination of $v_1, \ldots, v_n$ in *at least one way*; that is, if *there exist* scalars λ_j such that $v = \sum_j \lambda_j v_j$.

It is possible for $v_1, \ldots, v_n$ to span V and have $\sum_j \lambda_j v_j = \sum_j \mu_j v_j$ with $(\lambda_1, \ldots, \lambda_n) \neq (\mu_1, \ldots, \mu_n)$. For example, if v_1 and v_2 span V, then so do v_1, v_2, v_3, where $v_3 = v_1 + v_2$, and $1v_1 + 1v_2 + 0v_3 = 0v_1 + 0v_2 + 1v_3$.

Definition 7.2.5 The vectors $v_1, \ldots, v_n$ are *linearly independent* if every v in V can be expressed as a combination of $v_1, \ldots, v_n$ in *at most one way*; that is, if $\sum_j \lambda_j v_j = \sum_j \mu_j v_j$, then $\lambda_j = \mu_j$ for each j.

If the vectors $v_1, \ldots, v_n$ are linearly independent, there may be some v that cannot be expressed as $\sum_j v_j$; for example **i, j** are linearly independent, but no linear combination of them is **k**.

Clearly, $v_1, \ldots, v_n$ is a basis of V if and only if $v_1, \ldots, v_n$ are linearly independent and span V. There is an alternative (equivalent) approach to linear independence which is often taken as the definition of linear independence. The vectors $v_1, \ldots, v_m$ are linearly independent if

$$\lambda_1 v_1 + \cdots + \lambda_m v_m = \mu_1 v_1 + \cdots + \mu_m v_m$$

implies that $\lambda_1 = \mu_1, \ldots, \lambda_m = \mu_m$; in other words, if we can equate the coefficients of $v_1, \ldots, v_n$ in any such equation. In particular, *the vectors $v_1, \ldots, v_n$*

are linearly independent if and only if $\rho_1 v_1 + \cdots + \rho_m v_m = 0$ implies that $\rho_1 = \cdots = \rho_m = 0$. This is probably the easiest way to check the linear independence of vectors.

To show that $\mathbb{R}^2$ has dimension two it suffices to show that $e_1 = (1, 0)$ and $e_2 = (0, 1)$ is a basis of $\mathbb{R}^2$. As $(x, y) = xe_1 + ye_2$ we see that e_1, e_2 span $\mathbb{R}^2$. Next, if some vector v can be expressed as $ae_1 + be_2 = v = ce_1 + de_2$, then $(a, b) = (c, d)$ so that $a = c$ and $b = d$; thus e_1 and e_2 are linearly independent. This shows that $\mathbb{R}^2$ has a basis consisting of two vectors and so $\dim(\mathbb{R}^2) = 2$. This argument extends immediately to $\mathbb{R}^n$, and shows that the vectors

$$e_1 = (1, 0, \ldots, 0), \; e_2 = (0, 1, 0, \ldots, 0), \ldots, \; e_n = (0, \ldots, 0, 1) \quad (7.2.1)$$

form a basis of $\mathbb{R}^n$; thus $\dim(\mathbb{R}^n) = n$. We call $e_1, \ldots, e_n$ the *standard basis* of $\mathbb{R}^n$. An entirely similar statement holds for $\mathbb{C}^n$, and we give a formal statement of these results.

Theorem 7.2.6 *The vector space $\mathbb{R}^n$ over $\mathbb{R}$ has dimension n. The vector space $\mathbb{C}^n$ has dimension n over $\mathbb{C}$, and dimension $2n$ over $\mathbb{R}$.*

For example, $(1, 0)$, $(i, 0)$, $(0, 1)$ and $(0, i)$ form a basis of the vector space $\mathbb{C}^2$ over $\mathbb{R}$ for

$$(x + iy, u + iv) = x(1, 0) + y(i, 0) + u(0, 1) + v(0, i),$$

where the scalars x, y, u and v are all real. We give two more examples.

Example 7.2.7 If a plane in $\mathbb{R}^3$ is a vector space, then it must contain the zero vector in $\mathbb{R}^3$. Let Π be the plane given by $x + y + z = 0$. As the condition $x + y + z = 0$ is preserved under the formation of linear combinations, and as $(0, 0, 0)$ is in Π, we see that Π is a vector space (formally, there is more to check here, but we omit the details). The vectors $v_1 = (-1, 1, 0)$ and $v_2 = (-1, 0, 1)$ are in Π, and they span Π because if $(x, y, z) \in \Pi$ then $x = -(y + z)$ and so

$$(x, y, z) = (-y - z, y, z) = y(-1, 1, 0) + z(-1, 0, 1) = yv_1 + zv_2.$$

On the other hand, if $0 = \lambda_1 v_1 + \lambda_2 v_2$ then $(-\lambda_1 - \lambda_2, \lambda_1, \lambda_2) = (0, 0, 0)$ and hence $\lambda_1 = \lambda_2 = 0$; thus v_1 and v_2 are linearly independent. We deduce that $\dim(\Pi) = 2$. $\square$

Example 7.2.8 Let V be the vector space of solutions of the equation

$$\frac{d^2 y}{dx^2} - 2\frac{dy}{dx} - 3y = 0;$$

this equation is said to be 'linear' *precisely because* any linear combination of solutions is again a solution. Now e^{kx} is a solution if and only if $k^2 - 2k - 3 = 0$

(that is, k is 3 or -1) so that $y = ae^{3x} + be^{-x}$ is a solution for any a and b. In particular,

$$y(x) = \left(\frac{A+B}{4}\right)e^{3x} + \left(\frac{3A-B}{4}\right)e^{-x}$$

is a solution with $y(0) = A$ and $y'(0) = B$. We shall take for granted the fact that any solution $y(x)$ of the given equation is completely determined by the values $y(0)$ and $y'(0)$ (for the equation gives $y''(0)$ and the higher derivatives of y are obtained by differentiating both sides of the equation). This means that the general solution of the equation is a linear combination $\lambda e^{3x} + \mu e^{-x}$. However, these functions are linearly independent for if $\lambda e^{3x} + \mu e^{-x} = 0$ for all x, then (putting this function and its derivative equal to zero when $x = 0$) we see that $\lambda = \mu = 0$. It follows that the space of solutions is a vector space of dimension two with basis e^{3x} and e^{-x}. These ideas apply to any linear differential equation with constant coefficients, and they show that the solution space of an n-th order *linear* differential equation has dimension n. This discussion places the so-called n 'arbitrary constants' in the solution on a firm foundation, for we now see that they are just the scalar coefficients in a general linear combination of the n basis elements. □

We end this section with the following very important result.

Theorem 7.2.9 *Let V be any finite dimensional vector space.*

(a) *Suppose that the vectors $v_1, \ldots, v_n$ span V. Then $\dim(V) \leq n$, and some subset of $v_1, \ldots, v_n$ is a basis of V.*
(b) *Suppose that $u_1, \ldots, u_m$ are linearly independent vectors in V. Then $m \leq \dim(V)$, and $u_1, \ldots, u_m$ is a subset of some basis of V.*

This shows that $\dim(V)$ is the smallest number of vectors in any spanning set, and that any spanning set can be contracted (by deleting some of its members) to a basis of V. Likewise, it shows that $\dim(V)$ is the largest number of linearly independent vectors in V, and that every linearly independent set of vectors can be extended to a basis of V. There are two corollaries of Theorem 7.2.9 that are worthy of explicit mention.

Corollary 7.2.10 *Suppose that V has dimension k. If $v_1, \ldots, v_k$ span V, or are linearly independent, then they are a basis of V.*

Corollary 7.2.11 *If a vector space has a finite spanning set, then it is finite dimensional.*

The proof of Theorem 7.2.9 We prove (a). The vectors $v_1, \ldots, v_n$ span V. If they are linearly independent they are a basis of V and $\dim(V) = n$. We may suppose, then, that they are not linearly independent so there are scalars $\lambda_1, \ldots, \lambda_n$, not all zero, such that $\lambda_1 v_1 + \cdots + \lambda_n v_n = 0$. By relabelling, we may assume that $\lambda_n \neq 0$, and then

$$v_n = (-\lambda_1/\lambda_n)v_1 + \cdots + (-\lambda_{n-1}/\lambda_n)v_{n-1}.$$

This implies that $v_1, \ldots, v_{n-1}$ span V, and this process can be continued until we obtain a subset of k vectors from $v_1, \ldots, v_k$ which span V and which are also linearly independent. In this case $\dim(V) = k \leq n$.

We now prove (b). As V is finite dimensional, it has a basis, say $w_1, \ldots, w_k$. Now suppose that $u_1, \ldots, u_m$ are linearly independent vectors in V. If w_1 is not a linear combination of $u_1, \ldots, u_m$ then the vectors $u_1, \ldots, u_m, w_1$ are linearly independent because if

$$\lambda_1 u_1 + \cdots + \lambda_m u_m + \mu w_1 = 0,$$

then $\mu = 0$ (else w_1 is a linear combination of the u_j contrary to our assumption), and then the linear independence of $u_1, \ldots, u_m$ shows that $\lambda_1 = \cdots = \lambda_m = 0$. This argument shows that either (a) w_1 is a linear combination of $u_1, \ldots, u_m$, or (b) $u_1, \ldots, u_m, w_1$ are linearly independent. In both cases, $u_1, \ldots, u_m$ can be extended to a linearly independent set of vectors such that some linear combination of these is w_1. The process continues, considering w_2, then w_3 and so on, until we eventually arrive at a set of linearly independent vectors, say $u'_1, \ldots, u'_r$, which includes $u_1, \ldots, u_m$, and various linear combinations of which give each of $w_1, \ldots, w_n$. As $w_1, \ldots, w_n$ span V so do $u'_1, \ldots, u'_r$, and as $u'_1, \ldots, u'_r$ are linearly independent we conclude that are a basis of V. This implies that $r = k$, and that $m \leq r = k = \dim(V)$. □

Exercise 7.2

1. Show that the vector space $\mathbb{C}^n$ over $\mathbb{R}$ has dimension $2n$.
2. Let $V = \{(t, t^2) \in \mathbb{R}^2 : t \in \mathbb{R}\}$ (a parabola), and define addition and scalar multiplication by $(t, t^2) + (s, s^2) = (t+s, (t+s)^2)$ and $\lambda(t, t^2) = (\lambda t, \lambda^2 t^2)$. Show that V is a real vector space of dimension one.
3. Show that the set of points (x_1, x_2, x_3) in $\mathbb{R}^3$ that satisfy the simultaneous equations $x_1 + x_2 = 0$ and $2x_1 - 3x_2 + 4x_3 = 0$ is a vector space. Find a basis for this space, and hence find its dimension.
4. Show that $\{(x_1, x_2, x_3, x_4) \in \mathbb{R}^4 : 2x_1 - 3x_2 + x_3 - x_4 = 0\}$ is a real vector space of dimension three. Find a basis for this space.
5. Show that the set of complex polynomials p of degree at most three that satisfy $p'(0) = 0$ is a vector space over $\mathbb{C}$. What is its dimension?

7.3 Subspaces

A subset U of a vector space V over $\mathbb{F}$ is a *subspace* of V if U is a vector space in its own right (with the same operations as in V, and over the same field $\mathbb{F}$). We also say that a subspace U of V is a *proper subspace* of V if $U \neq V$. If U is a subspace of V, then

(a) $0_V \in U$, and
(b) if $u_1, \ldots, u_n$ are in U, and $\lambda_1, \ldots \lambda_n$ are in $\mathbb{F}$, then $\lambda_1 u_1 + \cdots + \lambda_n u_n$ is in U.

The second of these conditions is clear, but the first might not be (for perhaps U has a zero vector of its own, distinct from the zero vector 0_V of V). However, the zero scalar multiple of any vector in U is 0_V, so this is certainly in U. As U is an additive group, its identity element is unique so it must be 0_V; thus 0_V is indeed the zero vector for U. In fact, (a) and (b) are also sufficient for U to be a subspace for, as is easily checked, all the other requirements for U to be a vector space hold automatically in U because they hold in V. Note that if $u \in U$ and if (b) holds, then $(-1)u \in U$ and so $-u \in U$ because $(-1)u = -u$ (thus the additive inverse of a vector in U coincides with its additive inverse in V). Because of their importance, we give a formal statement of these facts.

Theorem 7.3.1 *A subset U of a vector space V is a subspace of V if and only if* (a) $0_V \in U$, *and* (b) *any linear combination of vectors in U is also in U.*

We remark that (a) here can be replaced by the statement that U is non-empty, for then there is some u in U and hence, by (b), $u + (-u) \in U$. However, (b) alone is not sufficient, for (b) is satisfied when U is the empty set and this is not a vector space (for it has no zero vector).

Now suppose that U is a subspace of a finite dimensional vector space V and let $u_1, \ldots, u_m$ be a basis of U. Then $u_1, \ldots, u_m$ are linearly independent vectors in U, and hence also in V, and so, by Theorem 7.2.9,

$$\dim(U) = m \leq \dim(V).$$

If equality holds here, then $u_1, \ldots, u_m$ must be a basis of V for otherwise, we could extend $u_1, \ldots u_m$ to be a basis of V with more than m elements in the new basis. This argument, together with Theorem 7.2.9, proves the following result.

Theorem 7.3.2 *Let U be a subspace of a finite dimensional space V. Then $\dim(U) \leq \dim(V)$ with equality if and only if $U = V$. Moreover, any basis of U can be extended to a basis of V.*

It is clear from Theorem 7.3.1 that if U and W are subspaces of V then so is $U \cap W$, where

$$U \cap W = \{v \in V : v \in U \text{ and } v \in W\}.$$

Indeed, given vectors $v_1, \ldots, v_r$ in $U \cap W$; they are in U and hence so is any linear combination of them. Moreover, as they are in W, this same linear combination is in W and hence also in $U \cap W$. As this argument is valid for any collection of subspaces (even an infinite collection) we have proved the next result.

Theorem 7.3.3 *The intersection of any collection of subspaces of V is a subspace of V.*

A simple and familiar illustration of this can be found in $\mathbb{R}^3$. The non-trivial subspaces of $\mathbb{R}^3$ are the lines through the origin, and the planes through the origin. The intersection of two such planes is either a plane (when the two planes are identical) or a line (when they are distinct). The intersection of two distinct, parallel planes is empty, but this does not contradict Theorem 7.3.3 because one of these planes does not contain the zero vector so it is not a subspace of $\mathbb{R}^3$.

Given two subspaces U and W of V, their union

$$U \cup W = \{v \in V : v \in U \text{ or } v \in W\}$$

is not, in general, a subspace: for example, the union of two lines in $\mathbb{R}^3$ is never a plane (and hardly ever a line). Because of this, we consider the sum $U + W$ (instead of the union) of U and W which is defined as follows.

Definition 7.3.4 Given two subspaces U and W of V, the *sum*, or *join*, of U and V is

$$U + W = \{u + w : u \in U, w \in W\}.$$

In the same way, for subspaces U_i, $U_1 + \cdots + U_k$ is the set of vectors of the form $u_1 + \cdots + u_k$, where $u_j \in U_j$. □

It is clear from Theorem 7.3.1 that $U_1 + \cdots + U_k$ is a subspace of V. It is also clear that $U + W$ *is the smallest subspace of V that contains both U and W*. Indeed, if V_0 is any subspace of V that contains U and W, then V_0 contains all vectors u in U, all vectors w in W, and hence all sums of the form $u + w$. Thus V_0 contains $U + W$. The dimensions of the four subspaces U, W, $U \cap W$ and $U + W$ of V are related by a very simple formula.

7.3 Subspaces

Theorem 7.3.5 *Suppose that U and W are subspaces of a finite dimensional vector space V. Then*

$$\dim(U) + \dim(W) = \dim(U \cap W) + \dim(U + W). \qquad (7.3.1)$$

Proof The smallest of these four subspaces is $U \cap W$, so we begin by taking any basis $(v_1, \ldots, v_p)$ of this. As $v_1, \ldots, v_p$ are linearly independent we can extend $v_1, \ldots, v_p$ to a basis $(v_1, \ldots, v_p, u_1, \ldots, u_r)$ of U, and also to a basis $(v_1, \ldots, v_p, w_1, \ldots, w_s)$ of W. We claim that

$$v_1, \ldots, v_p, u_1, \ldots, u_r, w_1, \ldots, w_s \qquad (7.3.2)$$

is a basis of $U + W$. If this is so, then (7.3.1) follows immediately as we have $\dim(U) = p + r$, $\dim(W) = p + s$, $\dim(U \cap W) = p$, and $\dim(U + W) = p + r + s$.

Now any vector in $U + W$ is of the form $u + w$, where $u \in U$ and $w \in W$, and this shows that the vectors in (7.3.2) span $U + W$. To show that these vectors are linearly independent, we suppose that there are scalars α_j, β_j and γ_j such that

$$(\alpha_1 v_1 + \cdots + \alpha_p v_p) + (\beta_1 u_1 + \cdots + \beta_r u_r) + (\gamma_1 w_1 + \cdots + \gamma_s w_s) = 0. \qquad (7.3.3)$$

Now this implies that the vector $\gamma_1 w_1 + \cdots + \gamma_s w_s$ is in U, and as it is (by definition) in W, it is in $U \cap W$. This means that it is of the form $\delta_1 v_1 + \cdots + \delta_p v_p$ for some scalars δ_i, and from this we deduce that

$$(\alpha_1 v_1 + \cdots + \alpha_p v_p) + (\beta_1 u_1 + \cdots + \beta_r u_r) + (\delta_1 v_1 + \cdots + \delta_p v_p) = 0.$$

As this is a linear relationship between the linearly independent vectors $v_1, \ldots, v_p, u_1, \ldots, u_r$, we can now deduce that $\beta_1 = \cdots = \beta_r = 0$. If we now use this in conjunction with (7.3.3) and the fact that the vectors $v_1, \ldots, v_p, w_1, \ldots, w_s$ are linearly independent, we see that each α_i and each γ_k in (7.3.3) is zero. This shows that the vectors in (7.3.2) are indeed linearly independent, and it completes the proof. $\square$

We illustrate Theorem 7.3.5 with two examples.

Example 7.3.6 Let U and V be subspaces of $\mathbb{R}^4$, and suppose that $\dim(U) = 2$ and $\dim(V) = 3$. What are the different possibilities? Now by Theorem 7.3.5, $\dim(U \cap V) + \dim(U + V) = 5$, and, of course,

$$\dim(U \cap V) \leq \dim(U) = 2 \leq \dim(U + V) \leq \dim(\mathbb{R}^4) = 4.$$

It follows that either $\dim(U \cap V) = 1$ and $U + V = \mathbb{R}^4$, or $\dim(U \cap V) = 2$ and $\dim(U + V) = 3$. In the latter case $U = U \cap V$ and $V = U + V$, and each of these imply that $U \subset V$. Thus the two possibilities are (a) $U \cap V$ is a line in $\mathbb{R}^4$ and $U + V = \mathbb{R}^4$, or (b) $U \subset V$. □

Example 7.3.7 Let

$$U = \{(x_1, x_2, x_3, x_4, x_5) \in \mathbb{R}^5 : x_1 + x_2 - x_3 + x_4 - x_5 = 0\},$$
$$W = \{(x_1, x_2, x_3, x_4, x_5) \in \mathbb{R}^5 : 2x_1 + 3x_2 - x_3 = 0\}.$$

Now $W = \{(a, b, 2a + 3b, c, d) : a, b, c, d \in \mathbb{R}\}$, and it is easy to see from this that the four vectors $(1, 0, 2, 0, 0), (0, 1, 3, 0, 0), (0, 0, 0, 1, 0)$ and $(0, 0, 0, 0, 1)$ is a basis of W. This shows that $\dim(W) = 4$, and a similar argument shows that $\dim(U) = 4$. Moreover,

$$\begin{aligned} 8 &= \dim(U) + \dim(W) \\ &= \dim(U \cap W) + \dim(U + W) \\ &\leq \dim(U \cap W) + \dim(\mathbb{R}^5) \\ &= \dim(U \cap W) + 5. \end{aligned}$$

It follows that either

(a) $\dim(U \cap W) = 3$ and $\dim(U + W) = 5$, or
(b) $\dim(U \cap W) = \dim(U + W) = 4$.

As $U \cap W \subset U \subset U + W$, (b) would imply that $U \cap W = U = U + W$ and similarly, $U \cap W = W$. This implies that $U = W$ and as this is clearly not so, we see that (b) cannot hold. We deduce that (a) holds, and hence U and W intersect in a subspace of dimension three. In essence, we have shown that the common solutions of two (independent) equations in five unknowns is a vector space of dimension three; this idea will be discussed in more detail in Chapter 9. □

Two linearly independent vectors in $\mathbb{R}^3$ determine a plane through the origin, and this plane is characterized as being the smallest subspace of $\mathbb{R}^3$ that contains the two vectors. We shall now show that this idea can be used in a general vector space. Given any *set* X of vectors in a vector space V, we consider the class of all subspaces of V that contain X (of course, V itself is one such subspace). The intersection of all of these subspaces is again a subspace of V (Theorem 7.3.3), and clearly it is the smallest subspace of V that contains X. The important points here are that the intersection of these subspaces *does exist*, and *it is a subspace*, and these make the following definition legitimate.

Definition 7.3.8 Let X be any non-empty subset of a vector space V. Then there is a smallest subspace of V that contains X, and we say that this is the subspace of V *generated by* X.

There is another more constructive approach to the subspace generated by X. Let W be the set of all finite linear combinations of vectors in X. Then W is clearly a subspace of V and, by construction, $X \subset W$. As any subspace of V that contains X must obviously contain W, we see that W is indeed the subspace of V generated by X. We end this section by recording this result.

Theorem 7.3.9 *Let X be a non-empty subset of a vector space V. Then there is a smallest subspace of V that contains X, and this is the set of all finite linear combinations of vectors chosen from X.*

As an example, the polynomials $1, x^2, x^4, \ldots$ generate the subspace of all even polynomials of the vector space of all polynomials.

Exercise 7.3

1. Let L be a line and Π be a plane in $\mathbb{R}^3$, both containing the origin. Use Theorem 7.3.5 to show that $L \subset \Pi$ or $L + \Pi = \mathbb{R}^3$.
2. Show that the vectors $(1, 0, 2, 0, -1)$, $(0, 1, 3, 0, -2)$, $(0, 0, 0, 1, 1)$ form a basis of $U \cap W$ in Example 7.3.7, so that $\dim(U \cap W) = 3$.
3. Let U be the subspace of $\mathbb{R}^4$ spanned by the vectors $(1, -1, 1, 2)$, $(3, 1, 2, 1)$ and $(7, 9, 3, -6)$. What is $\dim(U)$? Find a subspace W of $\mathbb{R}^4$ such that $U \cap W = \{0\}$ and $\dim(U) + \dim(W) = \dim(\mathbb{R}^4)$.
4. Let $V = \mathbb{R}$, and consider V as a vector space over $\mathbb{R}$. Show that if U is a subspace of V then either $U = \{0\}$ or $U = V$.
5. Let V be a vector space over $\mathbb{R}$ of dimension n, and let U be a subspace of dimension m, where $m < n$. Show that if $m = n - 1$ then there are only two subspaces of V that contain U (namely U and V), whereas if $m < n - 1$ then there are infinitely many distinct subspaces of V that contain U.

7.4 The direct sum of two subspaces

If W is a plane through the origin in $\mathbb{R}^3$, and if L is a line through the origin but not lying in W, then every vector in $\mathbb{R}^3$ can be expressed *uniquely* in the form $u + w$, where $u \in L$ and $w \in W$. This idea extends easily to the general situation.

Definition 7.4.1 Suppose that U and W are subspaces of a vector space V. Then we say that the subspace $U + W$ is the *direct sum* of U and W if every vector in $U + W$ can be expressed *uniquely* in the form $u + w$, where $u \in U$

and $w \in W$. If $U + W$ is the direct sum, we write $U \oplus W$ instead of $U + W$. More generally, if $U_1, \ldots, U_k$ are subspaces of V, then the sum $U_1 + \cdots + U_k$ is a *direct sum*, and we write $U_1 \oplus \cdots \oplus U_k$, if every vector in $U_1 \cdots + U_k$ can be expressed uniquely in the form $u_1 + \cdots + u_k$, where $u_j \in U_j$. □

The following result shows how we can check whether or not $U + W$ is the direct sum $U \oplus W$ of U and W.

Theorem 7.4.2 *Suppose that U and W are subspaces of a vector space V. Then the following are equivalent:*

(1) $U + W$ is the direct sum $U \oplus W$ of U and W;
(2) $U \cap W = \{0\}$;
(3) $\dim(U) + \dim(W) = \dim(U + W)$.

Proof Suppose that (1) holds, and take any v in $U \cap W$. Then we can write $v = 0_V + v = v + 0_V$, where in each of these sums the first vector is in U and the second vector is in W. By the uniqueness asserted in Definition 7.4.1, we must have $v = 0_V$ so that (2) follows. Next, Theorem 7.3.5 shows that (2) and (3) are equivalent. Now assume that (2) holds and, for any vector v in $U + W$, let us suppose that $v = u_1 + w_1 = u_2 + w_2$, where each u_i is in U and each w_j is in W. Then $u_1 - u_2 = w_2 - w_1$ and so, as (2) holds, $u_1 - u_2 = 0 = w_2 - w_1$. We deduce that v in $U + W$ has a unique expression in the form $u + w$ so that (1) holds. □

Theorem 7.4.2 has the following corollary.

Corollary 7.4.3 *Suppose that U and W are subspaces of V, and that $U \cap V = \{0\}$, and $\dim(U) + \dim(W) = \dim(V)$. Then $V = U \oplus W$.*

Proof Theorem 7.4.2 shows that $U + W$ is a direct sum, and that $\dim(U + W) = \dim(V)$. Theorem 7.3.2 implies that $V = U \oplus W$. □

We give one example; other examples occur in the exercises.

Example 7.4.4 Let U be the subspace of $\mathbb{R}^4$ spanned by the vectors $u = (1, 1, 1, 1)$ and $u' = (1, 1, -1, -1)$. How can we find a subspace W of $\mathbb{R}^4$ such that $U \oplus W = \mathbb{R}^4$? It seems reasonable to guess that we can find two vectors that are 'orthogonal' to U and that such vectors might generate a choice of W (note that for a given U, the subspace W is not unique). Now (by analogy with $\mathbb{R}^3$) a vector (x_1, x_2, x_3, x_4) is orthogonal to U if the x_j satisfy the two equations

$$x_1 + x_2 + x_3 + x_4 = 0, \quad x_1 + x_2 - x_3 - x_4 = 0,$$

so we can take, say, $w = (1, -1, 0, 0)$ and $w' = (0, 0, 1, -1)$. Certainly w and w' generate a two-dimensional subspace W of $\mathbb{R}^4$ so it is only matter of showing

7.4 The direct sum of two subspaces

that $U \cap W = \{0\}$. To do this we take any vector z, say, in this intersection and write $z = \lambda u + \mu u' = \rho w + \sigma w'$. This gives

$$(\lambda, \lambda, \lambda, \lambda) + (\mu, \mu, -\mu, -\mu) = (\rho, -\rho, 0, 0) + (0, 0, \sigma, -\sigma),$$

so that $\lambda = \mu = \rho = \sigma = 0$. It follows that $U \oplus W = \mathbb{R}^4$. $\square$

We have seen that if a plane Π and a line L in $\mathbb{R}^3$ meet only at the origin, then $\mathbb{R}^3$ is the direct sum $L \oplus W$. This means that given W we can construct a line L such that $\mathbb{R}^3 = L \oplus W$ and, incidentally, it is clear that for a given W, the line L *is not unique*. Likewise, given L we can construct W such that $\mathbb{R}^3 = L \oplus W$. The same is true for any proper subspace of any vector space.

Theorem 7.4.5 *Let U be proper subspace of a vector space V. Then there is a subspace W such that $U \oplus W = V$.*

Proof Take any basis $u_1, \ldots, u_k$ of U and extend it to a basis, say $u_1, \ldots, u_k, w_1, \ldots, w_\ell$, of V. Now let W be the subspace spanned by $w_1, \ldots, w_\ell$. By definition, $V = U + W$, and by Theorem 7.4.2, $U + W = U \oplus W$. $\square$

Exercise 7.4

1. Find a subspace W_0 of $\mathbb{R}^4$ such that in Example 7.4.3, we have $U \oplus W_0 = \mathbb{R}^4$ yet $W \neq W_0$. Show that there are infinitely many distinct choices of a subspace W_1 such that $U \oplus W_1 = \mathbb{R}^4$.
2. Let V be the vector space of all complex polynomials, and let V_o and V_e be the subspaces of even polynomials and odd polynomials, respectively. Show that $V = V_e \oplus V_o$.
3. Let V be the vector space of real solutions of the differential equation $\ddot{x} + x = 0$. Let S be the subspace of solutions $x(t)$ such that $x(0) = 0$, and let C be the subspace of solutions $x(t)$ such that $\dot{x}(0) = 0$. Show that $V = S \oplus C$.
4. Suppose that U, V and W are subspaces of some given vector space. With the obvious definition of $U + V + W$, show that every element of $U + V + W$ can be expressed uniquely in the form $u + v + w$, where $u \in U, v \in V$ and $w \in W$, if and only if

$$\dim(U + V + W) = \dim(U) + \dim(V) + \dim(W).$$

Note that this is *not* equivalent to $U \cap V = V \cap W = W \cap U = \{0\}$, as can be seen by taking U, V and W to be three distinct but coplanar lines through the origin in $\mathbb{R}^3$. In this case, $\dim(U + V + W) = 2$.

7.5 Linear difference equations

A second order linear *difference equation*, or *recurrence relation*, is a relation of the form

$$x_{n+2} + a_n x_{n+1} + b_n x_n = 0, \quad n \geq 0, \qquad (7.5.1)$$

where the a_n and b_n are given real or complex numbers. A sequence $y_0, y_1, \ldots$ is a *a solution* of (7.5.1) if, for all n, $y_{n+2} + a_n y_{n+1} + b_n y_n = 0$. The reader may have met difference equations before in which the a_n and b_n are constants (for example, $x_{n+2} + x_{n+1} - 2x_n = 0$) but here *we allow these coefficients to depend on n*. Likewise, the reader may have only met real difference equations before, but there is no reason not to consider complex difference equations (in fact, because of the Fundamental Theorem of Algebra, there is every reason to do so). The key observation is that the set of solutions of (7.5.1) is a subspace of dimension two of the vector space of real, or complex, sequences.

Theorem 7.5.1 *The set S of solutions $x_0, x_1, \ldots$ of (7.5.1) is a vector space of dimension two.*

Proof First, notice that the sequence $0, 0, \ldots$ is a solution of (7.5.1). Suppose that y_n and z_n are solutions, and that λ is a scalar. Then it is obvious that both λy_n and $y_n + z_n$ are solutions. This shows that a linear combination of solutions is again a solution, and we leave the reader to verify that the rest of the requirements for a vector space are satisfied.

We must now show that $\dim(S) = 2$. Let $u_0 = 1$ and $u_1 = 0$, and then determine $u_2, u_3, \ldots$ inductively from (7.5.1) so that $1, 0, u_2, u_3, \ldots$ is a solution. Similarly, we can find $v_2, v_3, \ldots$ so that $0, 1, v_2, v_3, \ldots$ is a solution. It is evident that neither of these solutions is a scalar multiple of the other, so they are linearly independent. However, if $y_0, y_1, \ldots$ is any solution, then

$$(y_0, y_1, y_2, \ldots) = y_0(1, 0, u_2, \ldots) + y_1(0, 1, v_2, \ldots),$$

because any solution of (7.5.1) is plainly completely determined by its first two elements. This shows that the solutions u_n and v_n form a basis of the space of solutions, so this has dimension two. $\square$

The next result give a simple test to determine whether or not two solutions form a basis of the solution space.

Theorem 7.5.2 *A pair u_n and v_n of solutions of (7.5.1) forms a basis of S if and only if $u_0 v_1 - u_1 v_0 \neq 0$.*

Proof Because $\dim(S) = 2$, u_n and v_n *fail* to form a basis of S if and only if they are linearly dependent; that is, there are scalars λ and μ, not both zero,

7.5 Linear difference equations

such that $\lambda u_n = \mu v_n$ for all n. Clearly this implies that $\lambda u_n = \mu v_n$ for $n = 0, 1$. Conversely, if $\lambda u_n = \mu v_n$ for $n = 0, 1$ then, by induction, $\lambda u_n = \mu v_n$ for all n; for example,

$$\lambda u_2 = -[a_1(\lambda u_1) + b_1(\lambda u_0)]$$
$$= -[a_1(\mu v_1) + b_1(\mu v_0)]$$
$$= \mu v_2.$$

Finally, it is clear that there exist λ and μ, not both zero, with $\lambda u_n = \mu v_n$, $n = 1, 2$, if and only if $u_0 v_1 - u_1 v_0 = 0$. □

Let us now consider the constant coefficient case; that is, when (7.5.1) becomes

$$x_{n+2} + a x_{n+1} + b x_n = 0, \tag{7.5.2}$$

where a and b are constants. In this case $(1, t, t^2, \ldots)$, where $t \neq 0$, is a solution if and only if t satisfies the *auxiliary equation* $t^2 + at + b = 0$. For the moment we will work in the space of *complex solutions*; then the auxiliary equation has two roots, say r and s. If $r \neq s$ then we have two solutions $(1, r, r^2, \ldots)$ and $(1, s, s^2, \ldots)$ of (7.5.1) and Theorem 7.5.2 shows that these form a basis of the space of all solutions. Thus *if the auxiliary equation has distinct roots r and s, then the most general solution of (7.5.1) is of the form $\lambda r^n + \mu s^n$.*

Now suppose that the auxilary equation has a repeated root r; then, as before, $(1, r, r^2, \ldots)$ is a solution. We need another solution that is not a scalar multiple of this one, and we can obtain this in the following way. First, as r is a repeated root of the auxiliary equation, this equation must be of the form $t^2 - 2rt + r^2 = 0$. Thus the difference equation must be $x_{n+1} - 2r x_{n+1} + r^2 x_n = 0$. We write $x_n = r^n y_n$, and find that x_n is a solution if and only if $y_{n+2} - 2 y_{n+1} + y_n = 0$ or, equivalently, $y_{n+2} - y_{n+1} = y_{n+1} - y_n$. Clearly this is satisfied by $y_n = n$; thus we now have a second solution namely $(0, r, 2r^2, 3r^3, \ldots)$. Theorem 7.5.2 shows that these two solutions are linearly independent so we have shown that *if the auxiliary equation has a non-zero repeated root r, then the most general solution of (7.5.1) is of the form $r^n(\lambda + \mu n)$.*

Finally, suppose that we are seeking real solutions of a real difference equation. If the discussion above yields real solutions, then we are done. If not, then there is a complex, *non-real*, root w of the auxiliary equation, and as the coefficients are real, the other root is $\bar{w}$, where $w \neq \bar{w}$. This means that the two sequences w^n $\bar{w}^n$ are complex solutions, and hence so too are $(w^n + \bar{w}^n)/2$ and $(w^n - \bar{w}^n)/2i$. We have now obtained two real solutions $R^n \cos(n\theta)$ and $R^n \sin(n\theta)$, where $w = Re^{i\theta}$, and these form a basis for the real solutions of (7.5.2).

Exercise 7.5

1. The *Fibonacci sequence* $F_0, F_1, \ldots$ is given by $F_0 = 0$, $F_1 = 1$ and $F_{n+2} = F_{n+1} + F_n$. Show that

$$F_n = \frac{1}{\sqrt{5}} \left(\frac{1+\sqrt{5}}{2} \right)^n - \frac{1}{\sqrt{5}} \left(\frac{1-\sqrt{5}}{2} \right)^n.$$

   Show directly from this formula that F_n is a positive integer.

2. The *Pell sequence* P_n is defined by $P_1 = 1$, $P_2 = 2$, and $P_{n+2} = 2P_{n+1} + P_n$. Show that

$$P_n = \frac{1}{2\sqrt{2}} \left[(1+\sqrt{2})^n - (1-\sqrt{2})^n \right].$$

3. Find the solution u_n of $x_{n+2} - 3x_{n+1} + 2x_n = 0$ with $u_0 = u_1 = 1$. Find the solution v_n with $v_0 = 1$ and $v_1 = 2$.
4. Find the general solution of $x_{n+2} - 2x_{n+1} + 2x_n = 0$.
5. By assuming that $x_n = (-1)^n(\lambda n + \mu)$ find a solution of the difference equation $x_{n+2} + (n+2)x_{n+1} + (n+2)x_n = 0$.

7.6 The vector space of polynomials

The set of complex polynomials is a vector space over $\mathbb{C}$, and clearly this is an infinite dimensional vector space.

Theorem 7.6.1 *The space* $P(\mathbb{C}, d)$ *of complex polynomials of degree at most d is a complex vector space of dimension $d + 1$.*

Proof Every polynomial of degree at most d is a linear combination of the polynomials

$$p_0(z) = 1, \; p_1(z) = z, \; p_2(z) = z^2, \ldots, p_d(z) = z^d; \tag{7.6.1}$$

thus these polynomials span $P(\mathbb{C}, d)$. In order to show that these polynomials are linearly independent, we suppose that there are complex numbers λ_j such that $\lambda_0 p_0 + \cdots + \lambda_d p_d$ is the zero polynomial (that is, it is zero at every z). Now the Fundamental Theorem of Algebra guarantees that if one of $\lambda_1, \ldots, \lambda_d$ is non-zero, then $\sum_j \lambda_j z^j$ has only finitely many zeros. As every z is a zero of $\sum_j \lambda_j p_j$, we see that $\lambda_1 = \cdots = \lambda_d = 0$, and hence that $\lambda_0 = 0$ also. Thus $p_0, p_1, \ldots, p_d$ are linearly independent and so form a basis of $P(\mathbb{C}, d)$. □

There is nothing canonical about this choice of basis.

7.6 The vector space of polynomials

Theorem 7.6.2 *Suppose that $q_0, q_1, \ldots, q_d$ are polynomials, where q_m has degree m and q_0 is not the zero polynomial. Then $q_0, q_1, \ldots, q_d$ is a basis of $P(\mathbb{C}, d)$.*

Proof Take a polynomial f of degree at most d. Choose a constant λ_d such that $f - \lambda_d q_d$ has degree at most $d-1$; then a constant λ_{d-1} such that $(f - \lambda_d q_d) - \lambda_{d-1} q_{d-1}$ has degree at most $d - 2$, and so on. This shows that f is a linear combination of the q_j, so the q_j span $P(\mathbb{C}, d)$. Corollary 7.2.10 now shows that the q_j form a basis of $P(\mathbb{C}, d)$. □

A basis of a vector space is, in essence, a coordinate system in the space, and there are always infinitely many bases to choose from. Bases are a tool for solving problems, and the art lies in choosing a basis that is best suited to the problem under consideration. We chose the basis in (7.6.1) because this was best suited to an application of the Fundamental Theorem of Algebra; we shall now consider situations in which other bases are more appropriate.

Example 7.6.3: Lagrange's interpolation formula We know that $P(\mathbb{C}, d)$ has dimension $d + 1$. Choose $d + 1$ distinct points $z_0, \ldots z_d$ in $\mathbb{C}$ and, for $j = 0, 1, \ldots, d$, let

$$q_j(z) = \prod_{i=0, i \neq j}^{d} \frac{z - z_i}{z_j - z_i}.$$

Here, the Π-sign denotes a product, just as the Σ-sign denotes a sum; for example, if $d = 2$ then

$$q_1(z) = \frac{(z - z_0)(z - z_2)}{(z_1 - z_0)(z_1 - z_2)}.$$

Each q_j is in $P(\mathbb{C}, d)$, and the q_j have been chosen to have the property

$$q_j(z_k) = \begin{cases} 1 & \text{if } k = j; \\ 0 & \text{if } k \neq j. \end{cases}$$

Now consider any polynomial p in $P(\mathbb{C}, d)$; then

$$p(z) = p(z_0)q_0(z) + p(z_1)q_1(z) + \cdots + p(z_d)q_d(z) \qquad (7.6.2)$$

for each side takes the same value at the $d + 1$ distinct points z_j, and so, by the Fundamental Theorem of Algebra, their difference is the zero polynomial. This argument shows that $q_0, \ldots, q_d$ span $P(\mathbb{C}, d)$, and it follows from Corollary 7.2.10 that they form a basis of $P(\mathbb{C}, d)$. As an example of this formula, we observe that the (unique) quadratic polynomial p that satisfies $p(0) = a$, $p(1) = b$

and $p(2) = c$ is
$$a\frac{(z-1)(z-2)}{(0-1)(0-2)} + b\frac{(z-0)(z-2)}{(1-0)(1-2)} + c\frac{(z-0)(z-1)}{(2-0)(2-1)}.$$

The expression for p given in (7.6.2) is known as *Lagrange's interpolation formula*. □

Example 7.6.4: Legendre polynomials The vectors **i**, **j** and **k** in $\mathbb{R}^3$ are mutually orthogonal unit vectors that form a basis of $\mathbb{R}^3$, and they have the property that the coefficients x_j of a vector **x** can be obtained by taking scalar products, namely $x_1 = \mathbf{x}\cdot\mathbf{i}$, and so on. These important properties generalize to many other systems, and we shall illustrate one of these generalizations here. Given two real polynomials p and q, we define

$$(p, q) = \int_{-1}^{1} p(x)q(x)\,dx.$$

We call this the *scalar product* of p and q, and it shares many of the important properties of the scalar product of vectors. For example, it is linear in p, and in q, and $(p, p) > 0$ unless p is the zero polynomial. By analogy with vectors we define the 'length' $\|p\|$ of p by

$$\|p\|^2 = (p, p) = \int_{-1}^{1} [p(x)]^2\,dx,$$

and we say that p and q are *orthogonal*, and write $p \perp q$, if $(p, q) = 0$.

The question we now address is whether or not we can find a basis of polynomials in the space of real polynomials of degree at most d which have unit length and which are orthogonal to each other? Suppose for the moment that we can, and let these be $p_0, \ldots p_d$. Then for any polynomial q of degree at most d there exist real numbers a_j such that $q(x) = \sum_{j=0}^{d} a_j p_j(x)$. If we now multiply both sides of this identity by $p_k(x)$ and integrate over the interval $[-1, 1]$, we find that

$$(q, p_k) = \sum_{j=0}^{d} a_j(p_j, p_k) = a_k(p_k, p_k) = a_k$$

because $(p_j, p_k) = 0$ when $j \neq k$. This mimics the situation in $\mathbb{R}^3$.

The *existence* of such a basis p_k can be proved by an elementary argument in calculus, and the resulting polynomials are scalar multiples of the *Legendre polynomials* p_n which are defined by

$$p_n(x) = \frac{1}{2^n\, n!}\frac{d^n}{dx^n}\left[(x^2 - 1)^n\right].$$

The first three of these polynomials are

$$p_0(x) = 1, \quad p_1(x) = x, \quad p_2(x) = \tfrac{3}{2}x^2 - \tfrac{1}{2}, \qquad (7.6.3)$$

as the reader can easily verify. In fact, the p_n are mutually orthogonal polynomials with $\|p_n\| = \sqrt{2/(2n+1)}$. $\square$

Example 7.6.5: common zeros of two polynomials Consider two complex polynomials f and g of degrees n and m, respectively, and form the polynomials

$$f(z), zf(z), \ldots, z^{m-1}f(z), g(z), zg(z), \ldots z^{n-1}g(z). \qquad (7.6.4)$$

As these $m+n$ polynomials lie in $P(\mathbb{C}, m+n-1)$ it is natural to ask when they form a basis of this space. They form a basis if and only if they span $P(\mathbb{C}, m+n-1)$, and the next result gives a perhaps surprising condition for this to be so.

Theorem 7.6.6 *The polynomials in (7.6.4) form a basis of the space $P(\mathbb{C}, m+n-1)$ if and only if f and g have no common zero.*

Proof Let V be the subspace spanned by $f(z), zf(z), \ldots z^{m-1}f(z)$, and let W be the subspace spanned by $g(z), zg(z), \ldots z^{n-1}g(z)$. We claim that $\dim(V) = m$ and $\dim(W) = n$. As the polynomials $z^r f(z)$, $r = 0, \ldots, m-1$, span V we only need show that they are linearly independent. Now any linear combination of them is of the form $q(z)f(z)$, where q is a polynomial of degree at most $m-1$, and if this is to be the zero polynomial then, as f is non-constant, q must be the zero polynomial. It follows that the given polynomials are linearly independent and hence that $\dim(V) = n$. Similarly, $\dim(W) = n$. By Theorem 7.3.5,

$$\dim(V+W) + \dim(V \cap W) = \dim(V) + \dim(W) = m+n,$$

so that the polynomials in (7.6.4) form a basis of $P(\mathbb{C}, m+n-1)$ if and only if $V \cap W = \{0\}$.

It remains to show that f and g have no common zero if and only if $V \cap W = \{0\}$ or, equivalently, that f and g have a common zero if and only if $V \cap W \neq \{0\}$. Suppose first that f and g have a common zero, say z_1. We write $f(z) = (z-z_1)f_1(z)$ and $g(z) = (z-z_1)g_1(z)$, and then $f_1(z)g(z) = f(z)g_1(z) = h(z)$, say. Clearly, $h \in V \cap W$ so $V \cap W \neq \{0\}$. Conversely, suppose that $V \cap W \neq \{0\}$. Then there are polynomials $f_2(z)$ and $g_2(z)$ with $\deg(f_2) < n$, $\deg(g_2) < m$ and $g_2(z)f(z) = f_2(z)g(z)$. Now f has n zeros and f_2 has at most $n-1$ zeros; thus f and g have a common zero. $\square$

For more details on this topic, see Section 9.6.

Exercise 7.6

1. Use Lagrange's interpolation formula to find the most general cubic polynomial that vanishes at 0, 1 and −1, and verify this directly.
2. Verify that the polynomials in (7.6.3) are the first three Legendre polynomials, and show that for these polynomials, $p_i \perp p_j$ if $i \neq j$. Show that $p_3(x) = \frac{5}{2}x^3 - \frac{3}{2}x$.
3. Choose distinct points $z_1, \ldots, z_k$ in $\mathbb{C}$. Show that the space V_d of polynomials of degree at most d that are zero at $z_1, \ldots, z_k$ is a vector space. What is the dimension of V_d? [You should consider the cases $k \leq d$, $k = d$ and $k > d$ separately.]
4. Find a basis, and the dimension, of the space of polynomials spanned by the polynomials in (7.6.4) when $f(z) = z^2(z-1)$ and $g(z) = z^2(z+1)$.

7.7 Linear transformations

The crucial property enjoyed by vector spaces is that a linear combination of vectors is again a vector. The defining property of a linear map is that it preserves linear combinations in the natural way (that is, the image of a linear combination is the same linear combination of the images).

Definition 7.7.1 A map $\alpha : V \to W$ between vector spaces V and W (over the same field $\mathbb{F}$ of scalars) is *linear* if, for all scalars $\lambda_1, \ldots, \lambda_n$, and all vectors $v_1, \ldots, v_n$,

$$\alpha(\lambda_1 v_1 + \cdots + \lambda_n v_n) = \lambda_1 \alpha(v_1) + \cdots + \lambda_n \alpha(v_n).$$

If α is linear we say that it is a *linear transformation*, or a *linear map*, and this will be so if, for all scalars λ, and all vectors u and v, $\alpha(\lambda x) = \lambda \alpha(x)$ and $\alpha(x+y) = \alpha(x) + \alpha(y)$. □

Each linear map $\alpha : V \to W$ maps 0_V to 0_W (the zero vectors in V and W), for if $0_\mathbb{F}$ is the scalar zero, then $\alpha(0_V) = \alpha(0_\mathbb{F} 0_V) = 0_\mathbb{F} \alpha(0_V) = 0_W$. Similarly, $\alpha(-v) = -\alpha(v)$, so that α maps the inverse of v to the inverse of $\alpha(v)$.

The scalar and vector products in $\mathbb{R}^3$ give examples of linear maps. For each a in $\mathbb{R}^3$, $\sigma_a(x) = x \cdot a$ is a linear map and (Theorem 4.2.5) every linear map of $\mathbb{R}^3$ to $\mathbb{R}$ is of this form. Likewise, $\beta_a(x) = a \times x$ is a linear map but, as $\beta_a(a) = 0$, no linear injective map can be of this form. Differentiation and integration provide examples of linear maps in analysis. Assuming the existence of derivatives, we have

$$\frac{d}{dx}(f+g) = \frac{df}{dx} + \frac{dg}{dx}, \quad \frac{d}{dx}(\lambda f) = \lambda \frac{df}{dx},$$

and these are the properties required of a linear map. Without going into details, the same is true of the higher derivatives $f \mapsto d^k f/dx^k$, and of all linear combinations of these derivatives. Thus we have the concept of a *linear operator*

$$f \mapsto a_0 f + a_1 \frac{df}{dx} + \cdots + a_k \frac{d^k f}{dx^k}$$

and hence of a linear differential equation. The familiar formulae

$$\int_a^b (f+g) = \int_a^b f + \int_a^b g, \quad \int_a^b (\lambda f) = \lambda \int_a^b f$$

show that the definite integral is a linear map between two vector spaces of functions.

The next shows that a linear map $\alpha : V \to W$ can be defined by, and is uniquely determined by, its action on a basis of V.

Theorem 7.7.2 *Suppose that V and W are finite-dimensional vector spaces over $\mathbb{F}$, and let $v_1, \ldots, v_n$ be a basis of V. Choose any vectors $w_1, \ldots, w_n$ in W. Then there exists a unique linear map $\alpha : V \to W$ such that $\alpha(v_j) = w_j$ for each j.*

Proof Each x in V can be written as $x = \sum_j x_j v_j$, where the x_j are uniquely determined by x; thus we can define a map $\alpha : V \to W$ by $\alpha(x) = \sum_j x_j w_j$. Clearly, for each j, $\alpha(v_j) = w_j$. Further, α is linear for

$$\alpha(\lambda x) = \alpha\left(\sum_j \lambda x_j v_j\right) = \sum_j (\lambda x_j) w_j = \lambda \alpha(x),$$

$$\alpha(x+y) = \alpha\left(\sum_j (x_j + y_j) v_j\right) = \sum_j (x_j + y_j) w_j = \alpha(x) + \alpha(y).$$

Now suppose that β is a linear map with $\beta(v_j) = w_j$. Then for any x, $\beta(x) = \sum_j x_j \beta(v_j) = \sum_j x_j \alpha(v_j) = \alpha(x)$, so that $\alpha = \beta$. □

Finally, we remark that a function $f : V \to W$ between vector spaces V and W is *additive* if $f(\sum_j v_j) = \sum_j f(v_j)$ for all vectors v_j. The map $\alpha(z) = \bar{z}$ from $\mathbb{C}$ to itself is additive, but as

$$\alpha(iz) = \overline{iz} = -i\bar{z} \neq i\bar{z} = i\alpha(z),$$

(unless $z = 0$), α is not linear. This shows that the condition $f(\lambda v) = \lambda f(v)$ is an essential part of the definition of linear maps. It is a remarkable fact that there exist additive functions $f : \mathbb{R} \to \mathbb{R}$ that are *not* continuous anywhere and, as any linear map from $\mathbb{R}$ to itself is continuous, these maps are also additive but not linear. A linear map $g : \mathbb{R} \to \mathbb{R}$ is continuous, for if $x_n \to x$, then $g(x_n) = x_n g(1) \to x g(1) = g(x)$. Finally, note that for any additive map we have, with $v = mu$,

$$(n/m) f(v) = (n/m) f(mu) = n f(u) = f(nu) = f\left(\tfrac{n}{m} v\right),$$

so that $f(\lambda v) = \lambda f(v)$ for every *rational* λ. Thus every additive map is linear with respect to the field $\mathbb{Q}$ of rational numbers, though not necessarily linear with respect to $\mathbb{R}$.

Exercise 7.7

1. Show that the most general linear map of $\mathbb{R}^n$ to $\mathbb{R}$ is of the form $x \mapsto x \cdot a$, where $x \cdot a = x_1 a_1 + \cdots + x_n a_n$.
2. Suppose that $\alpha : \mathbb{R}^m \to \mathbb{R}^n$ is a linear map, and for each x in $\mathbb{R}^m$, let $\alpha(x) = (\alpha_1(x), \ldots, \alpha_n(x))$. Show that each $\alpha_j : \mathbb{R}^m \to \mathbb{R}$ is a linear map. Use Exercise 1 to show that the most general linear map of $\mathbb{R}^2$ to itself is given by $(x, y) \mapsto (ax + by, cx + dy)$ for some real numbers a, b, c and d.
3. Let Π be the plane in $\mathbb{R}^3$ that contains $\mathbf{0}$, $\mathbf{j}$ and $\mathbf{k}$. Show that the orthogonal projection α of $\mathbb{R}^3$ onto Π is a linear map.
4. Let S be the vector space of real sequences and, for $x = (x_1, x_2, \ldots)$ define $\alpha(x) = (0, x_1, x_2, \ldots)$ and $\beta(x) = (x_2, x_3, \ldots)$. Show that α and β are linear maps. Show also that $\beta\alpha = I$ but that $\alpha\beta \neq I$, where I is the identity map on S. This shows that in general, $\beta\alpha = I$ does it not imply that $\beta = \alpha^{-1}$. Show that α is not injective, and β is not surjective.
5. Let U be a subspace of V, and suppose that $\alpha : U \to W$ is linear. Show that α *extends* to a linear map of V to W; that is, there exists a linear map $\hat{\alpha} : V \to W$ such that $\hat{\alpha} = \alpha$ on U.
6. Let V and W be vector spaces over $\mathbb{F}$.
   (a) Suppose that $\alpha : V \to W$ is an injective linear map. Show that if $v_1, \ldots, v_n$ are linearly independent vectors in V, then $\alpha(v_1), \ldots, \alpha(v_n)$ are linearly independent vectors in W. Deduce that if such an α exists, then $\dim(V) \leq \dim(W)$.
   (b) Suppose that $\beta : V \to W$ is a surjective linear map. Show that if $v_1, \ldots, v_n$ span V then $\beta(v_1), \ldots, \beta(v_n)$ span W. Deduce that if such a β exists, then $\dim(V) \geq \dim(W)$.
   (c) Suppose that $\gamma : V \to W$ is a bijective linear map. Show that if $v_1, \ldots, v_n$ is a basis of V then $\gamma(v_1), \ldots, \gamma(v_n)$ is a basis of W. Deduce that if such a γ exists, then $\dim(V) = \dim(W)$.

7.8 The kernel of a linear transformation

A linear map $\alpha : V \to W$ automatically provides us with a particular subspace of V, and a particular subspace of W.

Theorem 7.8.1 *Suppose that* $\alpha : V \to W$ *is a linear map. Then*

(a) $\{v \in V : \alpha(v) = 0_W\}$ *is a subspace of* V, *and*
(b) $\{\alpha(v) : v \in V\}$ *is a subspace of* W.

Proof Let $K = \{v \in V : \alpha(v) = 0_W\}$, and $U = \{\alpha(v) : v \in V\}$. By Theorem 7.3.1, a subset X of a vector space is a subspace if $0 \in X$ and if X contains any linear combination of vectors in X. As $\alpha(0_V) = 0_W$, we see that $0_V \in K$ and $0_W \in U$. If $v_1, \ldots, v_r$ are in K, then so is $\sum_j \lambda_j v_j$ because $\alpha(\sum_j \lambda_j v_j) = \sum_j \lambda_j \alpha(v_j) = 0$. Similarly, if $w_1, \ldots, w_r$ are in U with, say, $\alpha(v_j) = w_j$, then so is $\sum_j \lambda_j w_j$ because $\sum_j \lambda_j w_j = \sum_j \lambda_j \alpha(v_j) = \alpha(\sum_j \lambda_j v_j)$. □

There are standard names for these subspaces, and their dimensions are related.

Definition 7.8.2 Let $\alpha : V \to W$ be a linear map. The *kernel* $\ker(\alpha)$ of α is the subspace $\{v \in V : \alpha(v) = 0_W\}$ of V, and its dimension is the *nullity* of α. The *range* $\alpha(V)$ of α is the subspace $\{\alpha(v) : v \in V\}$ of W, and its dimension is the *rank* of α.

Theorem 7.8.3 *Let* $\alpha : V \to W$ *be a linear map. Then*

$$\dim(V) = \dim \ker(\alpha) + \dim \alpha(V).$$

In particular, $\dim \alpha(V) \leq \dim(V)$.

The proof of Theorem 7.8.3 Choose a basis $v_1, \ldots, v_k$ of $\ker(\alpha)$ and use Theorem 7.2.9 to extend this to a basis $v_1, \ldots, v_k, v_{k+1}, \ldots, v_n$ of V. We want to show that $\dim \alpha(V) = n - k$ and so it suffices to show that the vectors $\alpha(v_{k+1}), \ldots, \alpha(v_n)$ are a basis of $\alpha(V)$. Now any vector in $\alpha(V)$ is of the form $\alpha(\lambda_1 v_1 + \cdots + \lambda_n v_n)$, and as

$$\alpha(\lambda_1 v_1 + \cdots + \lambda_n v_n) = \lambda_1 \alpha(v_1) + \cdots + \lambda_n \alpha(v_n)$$
$$= \lambda_{k+1} \alpha(v_{k+1}) + \cdots + \lambda_n v_n,$$

we see that $\alpha(v_{k+1}), \ldots, \alpha(v_n)$ span $\alpha(V)$. To show that these vectors are linearly independent, suppose that for some scalars $\mu_{k+1}, \ldots, \mu_n$,

$$\mu_{k+1} \alpha(v_{k+1}) + \cdots + \mu_n \alpha(v_n) = 0.$$

Then $\alpha(\mu_{k+1}v_{k+1} + \cdots + \mu_n v_n) = 0$, so that $\mu_{k+1}v_{k+1} + \cdots + \mu_n v_n$ is in ker(α), and hence there are scalars $\mu_1, \ldots, \mu_k$ such that

$$\mu_{k+1}v_{k+1} + \cdots + \mu_n v_n = \mu_1 v_1 + \cdots + \mu_k v_k.$$

Because $v_1, \ldots, v_n$ is a basis of V, we see that $\mu_1 = \cdots = \mu_n = 0$ and this shows that $\alpha(v_{k+1}), \ldots, \alpha(v_n)$ are linearly independent. □

The idea of the kernel of a linear map leads to an important technique which arises frequently throughout mathematics, and which is an essential feature of *linear* maps. Let $\alpha : V \to W$ be a linear map and suppose that, for a given w in W, we have two solutions v_1 and v_2 of the equation $\alpha(v) = w$. Then $\alpha(v_1 - v_2) = \alpha(v_1) - \alpha(v_2) = w - w = 0$ and so $v_1 - v_2 \in$ ker(α). It follows that if v_0 is one solution of $\alpha(v) = w$, then the general solution is of the form $v_0 + v$, where $v \in$ ker(α). Using the natural notation $v_0 +$ ker(α) = $\{v_0 + v : v \in$ ker(α)$\}$, we have proved the following result.

Theorem 7.8.4 *Suppose that $\alpha : V \to W$ is linear, that $w \in W$, and that $\alpha(v_0) = w$. Then $\alpha(v) = w$ if and only if $v \in v_0 +$ ker(α).*

This shows that the general solution of $\alpha(v) = w$ is any one solution plus the general solution of $\alpha(v) = 0$. This general principle is usually met first in the context of linear differential equations (where the general solution is described as the sum of a 'particular integral' and the 'complementary function'), and an example will suffice to illustrate this.

Example 7.8.5 Let us discuss the solutions of the equation

$$\alpha(y) = 6, \quad \alpha(y) = \frac{d^2 y}{dx^2} - 3\frac{dy}{dx} + 2y.$$

Now α is a linear map (from the space of twice differentiable real functions on $\mathbb{R}$ to the space of real functions on $\mathbb{R}$), and the general solution is $y_p + y_c$, where y_p is any solution of $\alpha(y) = 6$, and y_c is the general solution of $\alpha(y) = 0$. In this example we can take $y_p = 3$ (that is, the constant function with value 3), and $y_c = \lambda e^x + \mu e^{2x}$ for any constants λ and μ. In the context of Theorem 7.8.3, let V to be the three-dimensional vector space of functions spanned by the functions $1, e^x, e^{2x}$ (that is, the class of functions $A + Be^x + Ce^{2x}$, for any real A, B and C). Then α is a linear map from V to itself, and ker(α) is the two-dimensional subspace spanned by the functions e^x and e^{2x}. The space $\alpha(V)$ is the one-dimensional subspace of constant functions in V. □

Example 7.8.6 We illustrate Theorems 7.8.1 and 7.8.3 by reviewing the scalar product and the vector product in $\mathbb{R}^3$. Let $\sigma_a(x) = x \cdot a$, where a is a fixed non-zero vector. Then ker(σ_a) has dimension 2 (it is the plane Π given by $x \cdot a = 0$), while $\sigma_a(\mathbb{R}^3) = \mathbb{R}$, which has dimension one. The general solution of $x \cdot a = d$ is the sum of a particular solution, for example $x_0 = (d/||a||^2)a$, and the kernel Π. Thus the solutions of $x \cdot a = d$ are the 'translated plane' $x_0 + \Pi$.

Now let $\beta_a(x) = a \times x$, where $a \neq 0$. As $\beta_a(x) = 0$ if and only if $x = \lambda a$, say, we see that $\ker(\beta)$ has dimension one, and hence that $\beta(\mathbb{R}^3)$ has dimension two. Further, as $\beta_a(x) \perp a$, we see that $\beta_a(\mathbb{R}^3) = W$, where W is the plane given by $x \cdot a = 0$. This shows that *if $b \perp a$, then there is some solution of $x \times a = b$* (and the set of all solutions is a line); *otherwise, there is no solution*. These results were obtained by direct means in Chapter 3. We recall from Chapter 3 that if $b \in W$, then one solution of $a \times x = b$ is $-(a \times b)/||a||^2$. Thus, by Theorem 7.8.3, the general solution is this solution plus the general element of $\ker(\beta_a)$, namely a scalar multiple of a. □

Exercise 7.8

1. Suppose that $\alpha : V \to W$ is linear. Show that for any subspace U of V, $\alpha(U)$ is a subspace of W, and $\dim \alpha(U) \leq \dim(U)$.
2. Let $a = (1, 1, 1)$. The vector product $x \mapsto a \times x$ is the map that takes (x_1, x_2, x_3) to $(-x_2 + x_3, x_1 - x_3, -x_1 + x_2)$. By making explicit calculations, find $\ker(\alpha)$, and a basis for $\alpha(\mathbb{R}^3)$.
3. Suppose that $\alpha : V \to W$ is a linear map and that W_0 is a subspace of W. Let $U = \{v \in V : \alpha(v) \in W_0\}$. Show that U is a subspace of V. [This generalizes Theorem 7.8.1(a).]
4. Let $\alpha : \mathbb{R}^4 \to \mathbb{R}^4$ be defined by

$$\begin{pmatrix} x_1 \\ x_2 \\ x_3 \\ x_4 \end{pmatrix} \mapsto \begin{pmatrix} x_1 + x_2 + x_3 \\ x_2 + x_3 + x_4 \\ x_3 + x_4 + x_1 \\ x_4 + x_1 + x_2 \end{pmatrix}.$$

Find a basis for $\ker(\alpha)$, and a basis for $\alpha(\mathbb{R}^4)$, and verify Theorem 7.8.3 in this case.
5. Construct a linear map $\alpha : \mathbb{R}^4 \to \mathbb{R}^4$ whose kernel is spanned by $(1, 0, 0, 1)$ and $(0, 1, 1, 0)$.
6. Given a vector space V and a subspace U of V, show that there is a linear map $\alpha : V \to V$ such that $U = \ker(\alpha)$.
7. Let V be of dimension n, and suppose that $\alpha : V \to V$ is linear. Show that if $\dim \alpha^m(V) = \dim \alpha^{m+1}(V)$, then

$$\dim \alpha^m(V) = \dim \alpha^{m+1}(V) = \dim \alpha^{m+2}(V) = \cdots,$$

and hence that $\alpha^m(V) = \alpha^{m+1}(V) = \alpha^{m+2}(V) = \cdots$. Deduce that there exists an integer k with $k \leq n = \dim(V)$, such that
$$\dim(V) > \dim \alpha(V) > \cdots > \dim \alpha^k(V),$$
and $\alpha^k(V) = \alpha^{k+1}(V) = \alpha^{k+1}(V) = \cdots$. Deduce that if $\dim \alpha^k(V) = 0$ for some k, then $\dim \alpha^n(V) = 0$.

Suppose now that $\dim \alpha^{n-1}(V) = 1$ and $\dim \alpha^n(V) = 0$. Show that
$$\dim \alpha^k(V) = \begin{cases} n-k & \text{if } k \leq n; \\ 0 & \text{if } k \geq n. \end{cases}$$

Illustrate these results in the specific case when V is the space of polynomials of degree less than n, and α is differentiation.

7.9 Isomorphisms

The composition of two linear maps is linear, and the inverse (when it exists) of a linear map is linear. These are basic and important facts.

Theorem 7.9.1 *Let U, V and W be vector spaces. If $\alpha : U \to V$ and $\beta : V \to W$ are linear, then so is $\beta\alpha : U \to W$.*

Theorem 7.9.2 *Suppose that $\alpha : V \to W$ is a linear bijection of a vector space V onto a vector space W. Then $\alpha^{-1} : W \to V$ is linear.*

The proof of Theorem 7.9.1 This is clear for
$$\beta\alpha(\lambda_1 v_1 + \lambda_2 v_2) = \beta\Big(\alpha(\lambda_1 v_1 + \lambda_2 v_2)\Big)$$
$$= \beta\Big(\lambda_1 \alpha(v_1) + \lambda_2 \alpha(v_2)\Big)$$
$$= \lambda_1 \beta\alpha(v_1) + \lambda_2 \beta\alpha(v_2).$$
□

The proof of Theorem 7.9.2 As α is a bijection, $\alpha^{-1} : W \to V$ exists as a map, but we do need to show that it is linear. Take any two vectors w_j in W, and let $v_j = \alpha^{-1}(w_j)$. Then, for all scalars μ_j,
$$\alpha^{-1}(\mu_1 w_1 + \mu_2 w_2) = \alpha^{-1}\big(\mu_1 \alpha(v_1) + \mu_2 \alpha(v_2)\big)$$
$$= \alpha^{-1}\big(\alpha(\mu_1 v_1 + \mu_2 v_2)\big)$$
$$= \mu_1 v_1 + \mu_2 v_2$$
$$= \mu_1 \alpha^{-1}(w_1) + \mu_2 \alpha^{-1}(w_2),$$
so that α^{-1} is linear. □

7.9 Isomorphisms

Roughly speaking, two vector spaces are said to be isomorphic if they are the same in every sense except for the notation that describes them. Formally, this is described as follows.

Definition 7.9.3 Two vector spaces V and W are said to be *isomorphic* if there exists a bijective linear map α of V onto W. Any such α is an *isomorphism* of V onto W.

Theorems 7.9.1 and 7.9.2 show that the composition of isomorphisms is again an isomorphism, and that the inverse of an isomorphism, is also an isomorphism. It is important to understand that all intrinsic structural properties of vector spaces transfer without change under an isomorphism; that is, if a general result (involving the concepts we have been discussing) about vector spaces is true for V, then it (or an appropriate re-statement of it) will also be true for any vector space W isomorphic to V. For example, suppose that $\alpha : V \to W$ is an isomorphism and take a basis $v_1, \ldots, v_k$ of V. As every vector in W is the image of a linear combination of the v_j, and as α preserves linear combinations, it follows that every vector in W is some linear combination of the $\alpha(v_j)$; thus $\alpha(v_1), \ldots, \alpha(v_k)$ span W. However, if $\lambda_1 \alpha(v_1) + \cdots + \lambda_k \alpha(v_k) = 0$, then $\alpha(\lambda_1 v_1 + \cdots + \lambda_k v_k) = 0$ and hence, as α is injective, $\lambda_1 v_1 + \cdots + \lambda_k v_k = 0$. But as the v_j are linearly independent, this gives $\lambda_1 = \cdots = \lambda_k = 0$ and so we see that $\alpha(v_1), \ldots, \alpha(v_k)$ is a basis of W. It follows that under an isomorphism, *a basis of V maps to a basis of W* and so $\dim(V) = \dim(W)$. The converse of this is also true.

Theorem 7.9.4 *Two finite dimensional vector spaces over the same field are isomorphic if and only if they have the same dimension.*

Proof Suppose that U and V are vector spaces over $\mathbb{F}$ of dimension n. Choose a basis $u_1, \ldots, u_n$ of U and a basis $v_1, \ldots, v_n$ of V, and use Theorem 7.7.2 to define a linear map $\alpha : U \to V$ by

$$\alpha(\lambda_1 u_1 + \cdots + \lambda_n u_n) = \lambda_1 v_1 + \cdots + \lambda_n v_n.$$

It is obvious that α is injective, and surjective, so it is an isomorphism of U onto V. $\square$

Theorem 7.9.4 shows that *any real vector space of dimension n is isomorphic to $\mathbb{R}^n$*; thus, in effect, there is *only one* real vector space of dimension n. In particular (and this should be obvious), the space $\mathbb{R}^{n,t}$ of column vectors is isomorphic to $\mathbb{R}^n$. We end with a result which identifies isomorphisms (see Exercise 7.7.4).

Theorem 7.9.5 *Let $\alpha : U \to V$ be a linear map between vector spaces U and V of the same finite dimension. Then the following are equivalent:*

(1) α *is injective;*
(2) α *is surjective;*
(3) α *is bijective;*
(4) $\ker(\alpha) = \{0\}$.
(5) $\alpha^{-1} : V \to U$ *exists.*
(6) α *is an isomorphism.*

Proof As $\alpha(u) = \alpha(v)$ if and only if $u - v \in \ker(\alpha)$, we see that (1) is equivalent to (4). Next, as $\dim(U) = \dim(V)$, Theorem 7.8.3 shows that $\alpha(U) = V$ if and only if $\ker(\alpha) = \{0\}$. Thus (2) is equivalent to (4), and hence also to (1). If (1) holds, then so does (2) and these imply (3). Conversely, (3) implies both (1) and (2). Finally, (3) is obviously equivalent to (5), and, by definition, (6) is equivalent to (3). □

Exercise 7.9

1. Let $\alpha : V \to W$ be a linear map. Show that α is injective if and only if $\dim \alpha(V) = \dim V$, and that α is surjective if and only if $\dim \alpha(V) = \dim(W)$.
2. Let S be the vector space of solutions of the differential equation $\ddot{x} + 4x = 0$. Show that $\dim(S) = 2$, and that $\alpha : S \to S$ defined by $\alpha(x) = \dot{x}$ is an isomorphism of S onto itself.
3. Construct an isomorphism from $\mathbb{R}^{2n}$ over $\mathbb{R}$ onto $\mathbb{C}^n$ over $\mathbb{R}$.
4. Let $v_1, \ldots, v_n$ be a basis of a vector space V, and let ρ be a permutation of $\{1, \ldots, n\}$. Let $\alpha : V \to V$ be the unique linear map that satisfies $\alpha(\sum_j \lambda_j v_j) = \sum_j \lambda_j v_{\rho(j)}$. Show that α is an isomorphism of V onto itself.

7.10 The space of linear maps

For given vector spaces V and W over the same field $\mathbb{F}$, let $\mathcal{L}(V, W)$ be the set of all linear maps $\alpha : V \to W$. If α and β are in $\mathcal{L}(V, W)$, and if λ is a scalar, there are natural definitions of the linear maps $\alpha + \beta$ and $\lambda \alpha$, namely

$$(\alpha + \beta)(v) = \alpha(v) + \beta(v), \quad (\lambda \alpha)(v) = \alpha(\lambda v),$$

and from these it is easy to see that $\mathcal{L}(V, W)$ is a vector space over $\mathbb{F}$. We leave the reader to provide the details (the zero vector in $\mathcal{L}(V, W)$ is the 'zero map' which takes every v in V to 0_W).

7.10 The space of linear maps

Theorem 7.10.1 *The space $\mathcal{L}(V, W)$ of linear maps from a vector space V of dimension n to a vector space W of dimension m is a vector space of dimension mn.*

Proof Choose a basis $v_1, \ldots, v_n$ of V, and a basis $w_1, \ldots, w_m$ of W. Then, by Theorem 7.7.2, we can define nm linear maps $\epsilon_{ij} : V \to W$ by requiring that ϵ_{ij} maps v_i to w_j and all other v_k to 0 (in W); for example, ϵ_{23} acts as follows:

$$v_1 \to 0, \quad v_2 \to w_3, \quad v_3 \to 0, \quad \ldots, \quad v_n \to 0.$$

These maps are linearly independent, for if $\sum_{i,j} \lambda_{ij} \epsilon_{ij} = 0$ (the zero map), then

$$0 = \sum_{i,j} \lambda_{ij} \epsilon_{ij}(v_1) = \sum_j \lambda_{1j} w_j,$$

so that $\lambda_{11} = \cdots = \lambda_{1m} = 0$ (because the w_i are linearly independent). By applying the same function to $v_2, \ldots, v_n$ we see that $\lambda_{ij} = 0$ for all i and j. To see that the ϵ_{ij} span $\mathcal{L}(V, W)$ we recall from Theorem 7.7.2 that a general linear map $\alpha : V \to W$ is uniquely determined by its action on the basis $v_1, \ldots, v_n$. For each j, let $\alpha(v_j) = \sum_k \mu_{jk} w_k$. Then $\alpha = \sum_{rs} \mu_{rs} \epsilon_{rs}$ because, for each j,

$$\sum_{r,s} \mu_{rs} \epsilon_{rs}(v_j) = \sum_s \mu_{js} \epsilon_{js}(v_j) = \sum_s \mu_{js} w_s = \alpha(v_j).$$

$\square$

We end with an application of Theorem 7.10.1 to linear maps *of V into itself*. Suppose that $\alpha : V \to V$ is linear, where V has dimension n. We can apply α repeatedly, and we denote the k-th iterate of α (α applied k times) by α^k. As the linear maps $I, \alpha, \alpha^2, \ldots, \alpha^{n^2}$ are all in $\mathcal{L}(V, V)$ it follows from Theorem 7.10.1 that they must be linearly dependent. This shows that there are constants $a_0, \ldots, a_{n^2}$, not all zero, such that

$$a_0 I + a_1 \alpha + \cdots + a_{n^2} \alpha^{n^2} = 0$$

(the zero map). It we now define the polynomial p by

$$p(z) = a_0 + a_1 z + \cdots + a_{n^2} z^{n^2} = 0,$$

then $p(\alpha)(v) = 0$ for every v in V. Thus we have proved the following result.

Theorem 7.10.2 *For any linear map $\alpha : V \to V$ there is a polynomial p of degree at most n^2 such that $p(\alpha)(v) = 0$ for all v in V.*

Later we shall see that we can take p to be of degree *at most n*.

Exercise 7.10

1. The space $\mathcal{L}(V, \mathbb{R})$ is the space of all real-valued linear maps on V. Let $v_1, \ldots, v_n$ be a basis of V, and define the maps ϵ_j by $\epsilon_i(v_j) = 1$ if $i = j$ and 0 otherwise. Work through the proof of Theorem 7.10.1 in this case to show that $\epsilon_1, \ldots, \epsilon_n$ is a basis of of $\mathcal{L}(V, \mathbb{R})$. What is ϵ_j when $V = \mathbb{R}^3$ and the basis is **i, j, k**?
2. Let Π be a plane through the origin in $\mathbb{R}^3$, and let β be the orthogonal projection of $\mathbb{R}^3$ onto Π. Find a polynomial p such that $p(\beta) = 0$.
3. Let $\alpha(x) = x \times a$ (the vector product in $\mathbb{R}^3$). Find a polynomial p such that $p(\alpha) = 0$.

8
Linear equations

8.1 Hyperplanes

A plane in $\mathbb{R}^3$ that passes through the origin can be characterized as the solutions of a single linear homogeneous equation in three variables, say $a_1x_1 + a_2x_2 + a_3x_3 = 0$, as a subspace whose dimension is one less than that of the underlying space $\mathbb{R}^3$, and as the set of vectors that are orthogonal to a given vector (the normal to the plane). We now generalize these ideas to $\mathbb{R}^n$. We recall that $e_1, \ldots, e_n$ is the standard basis of $\mathbb{R}^n$, and if $x = \sum x_j e_j$ and $y = \sum y_j e_j$, then the *scalar product* $x \cdot y$ is given by

$$x \cdot y = \sum_{j=1}^{n} x_j y_j.$$

We say that x and y are *orthogonal*, and write $x \perp y$, when $x \cdot y = 0$.

Definition 8.1.1 Let V be a vector space of dimension n. A *line* in V is a subspace of dimension one. A *hyperplane* in V is a subspace of dimension $n - 1$.

Theorem 8.1.2 *For any subset W of $\mathbb{R}^n$ the following are equivalent:*

(1) W *is a hyperplane in* $\mathbb{R}^n$;
(2) *There is a non-zero a in $\mathbb{R}^n$ such that* $W = \{x \in \mathbb{R}^n : x \cdot a = 0\}$;
(3) *There are scalars $a_1, \ldots, a_n$, not all zero, such that*

$$W = \{(x_1, \ldots, x_n) \in \mathbb{R}^n : x_1 a_1 + \cdots + x_n a_n = 0\}.$$

Proof Obviously, (2) is equivalent to (3). Now let W be as in (2). Then $\alpha(x) = x \cdot a$ is a surjective linear map of $\mathbb{R}^n$ onto $\mathbb{R}$ with kernel W, and (1) follows as

$$n = \dim(\mathbb{R}^n) = \dim \alpha(\mathbb{R}^n) + \dim \ker(\alpha) = 1 + \dim(W).$$

Now suppose that W is a hyperplane in V. Take a basis $w_1, \cdots, w_{n-1}$ of W and extend this to a basis $w_1, \cdots, w_{n-1}, w_n$ of $\mathbb{R}^n$. Thus, for each j, there are scalars a_{ij} such that

$$e_j = a_{j1}w_1 + \cdots + a_{jn}w_n, \tag{8.1.1}$$

so that, for any x in $\mathbb{R}^n$,

$$x = \sum_{j=1}^n x_j e_j = \sum_{i=1}^n \left(\sum_{j=1}^n x_j a_{ji} \right) w_i.$$

Now $x \in W$ if and only if x is a linear combination of $w_1, \ldots, w_{n-1}$, and this is so if and only if $x_1 a_{1n} + \cdots + x_n a_{nn} = 0$. As not all of the numbers $a_{1n}, \ldots, a_{nn}$ can be zero (for, if they were then, from (8.1.1), $w_1, \ldots, w_{n-1}$ would span $\mathbb{R}^n$), we see that (1) implies (3). □

Exercise 8.1

1. Find a basis for the hyperplane $2x_1 + 5x_2 = 0$ in $\mathbb{R}^2$.
2. Find a basis for the hyperplane $2x_1 - x_2 + 5x_3 = 0$ in $\mathbb{R}^3$.
3. Find a basis for the hyperplane $x_1 - x_2 + 3x_3 - 2x_4 = 0$ in $\mathbb{R}^4$.
4. Show that $e_1 - e_n, e_2 - e_n, \ldots, e_{n-1} - e_n$ is a basis of the hyperplane $x_1 + \cdots x_n = 0$ in $\mathbb{R}^n$.

8.2 Homogeneous linear equations

We shall now discuss the space of solutions of a set of m simultaneous homogeneous linear equations in the n real variables $x_1, \ldots, x_n$, say

$$\begin{aligned} a_{11}x_1 + \cdots a_{1n}x_n &= 0, \\ &\vdots \\ a_{m1}x_1 + \cdots a_{mn}x_n &= 0. \end{aligned} \tag{8.2.1}$$

It is sometimes convenient to write these in *matrix form*

$$\begin{pmatrix} a_{11} & \cdots & a_{1n} \\ \vdots & & \vdots \\ a_{m1} & \cdots & a_{mn} \end{pmatrix} \begin{pmatrix} x_1 \\ \vdots \\ x_n \end{pmatrix} = \begin{pmatrix} 0 \\ \vdots \\ 0 \end{pmatrix}, \tag{8.2.2}$$

where the array (a_{ij}) is known as a *matrix*. For the moment, this is just a convenient notation; later we shall see that the left-hand side here is the product of two matrices.

8.2 Homogeneous linear equations

As the solution set of each single equation is a subspace of $\mathbb{R}^n$, the set of solutions of the system (8.2.1) is the intersection of subspaces, and so is itself a subspace of $\mathbb{R}^n$. To complete our discussion we need to find the dimension of this subspace, and show how to find it (that is, how to solve the given equations). As each of the m equations places a single constraint on the n numbers $x_1, \ldots, x_n$, we might expect the space of solutions to have dimension $n - m$; however, a little more care is needed as the equations may not be independent of each other (indeed, there may even be identical equations in the list).

Any solution $(x_1, \ldots, x_n)$ is a point in $\mathbb{R}^n$ and in order to exploit the geometry of $\mathbb{R}^n$ we use (8.2.2) to define the *row vectors* $r_1, \ldots, r_m$ in $\mathbb{R}^n$ by

$$r_j = (a_{j1}, \ldots, a_{jn}) = a_{j1}e_1 + \cdots + a_{jn}e_n, \quad j = 1, \ldots, m. \quad (8.2.3)$$

The equations (8.2.1) and (8.2.2) can now be written as

$$x \cdot r_1 = 0, \quad \ldots, \quad x \cdot r_m = 0;$$

thus we are seeking those vectors x that are orthogonal to each of the vectors $r_1, \ldots, r_m$. If, for example, r_m is a linear combination of $r_1, \ldots, r_{m-1}$, then any x that is orthogonal to $r_1, \ldots, r_{m-1}$ is automatically orthogonal to r_m, and so we may discard the equation $x \cdot r_m = 0$. Thus, by discarding redundant equations as necessary, *we may assume that the vectors* $r_1, \ldots, r_m$ *are linearly independent vectors in* $\mathbb{R}^n$. We can now state our main result.

Theorem 8.2.1 *Suppose that* $r_1, \ldots, r_m$ *are linearly independent vectors in* $\mathbb{R}^n$. *Then the set*

$$S = \{x \in \mathbb{R}^n : x \cdot r_1 = 0, \ldots, x \cdot r_m = 0\}$$

of solutions of (8.2.1) *is a subspace of* $\mathbb{R}^n$ *of dimension* $n - m$.

Proof We know that S is a subspace of $\mathbb{R}^n$, and we show first that $\dim(S) \leq n - m$. Let W be the subspace spanned by the linear independent vectors $r_1, \ldots, r_m$; we claim that $S \cap W = \{0\}$. Indeed, if $x \in S \cap W$ then x is in S and so is orthogonal to each r_j, and hence also to any linear combination of the r_j. However, as $x \in W$, x is itself a linear combination of the r_j; thus x is orthogonal to itself and so $x = 0$. We deduce that

$$\begin{aligned}
\dim(S) + m &= \dim(S) + \dim(W) \\
&= \dim(S + W) + \dim(S \cap W) \\
&= \dim(S + W) \\
&\leq \dim(\mathbb{R}^n) = n.
\end{aligned}$$

To show that $\dim(S) \geq n - m$, define a linear map $\alpha : \mathbb{R}^n \to \mathbb{R}^m$ by

$$\alpha(x) = (x \cdot r_1, \ldots, x \cdot r_m). \qquad (8.2.4)$$

Clearly, $S = \ker(\alpha)$. Thus

$$n = \dim \ker(\alpha) + \dim \alpha(\mathbb{R}^n) \leq \dim(S) + \dim(\mathbb{R}^m), \qquad (8.2.5)$$

so that $\dim(S) \geq n - m$. Thus $\dim(S) = n - m$. □

If we now use the conclusion of Theorem 8.2.1, namely that S has dimension $n - m$, in (8.2.5), we see that equality holds throughout (8.2.5), so that $\alpha(\mathbb{R}^n) = \mathbb{R}^m$. Thus we have the following corollary.

Corollary 8.2.2 *Suppose that $r_1, \ldots, r_m$ are linearly independent vectors in $\mathbb{R}^n$. Then $\alpha : \mathbb{R}^n \to \mathbb{R}^m$ given by (8.2.4) is surjective.*

We end this section with an example which illustrates Theorem 8.2.1 by the elimination of variables. In general, if we have m equations in n variables, we can eliminate, say x_n, from the equations to leave $m - 1$ equations in $n - 1$ variables. This process can be repeated until we are left with one equation in $n - m + 1$ variables. We then treat $n - m$ of these variables as 'parameters' and, retracing our steps, we obtain the remaining variables in terms of these $n - m$ parameters.

Example 8.2.3 We solve the system of equations

$$x_1 + 2x_2 = 0,$$
$$3x_1 + 2x_2 - 4x_3 + x_4 = 0,$$
$$x_1 + x_2 + x_3 - x_4 = 0.$$

Geometrically, we have to find the intersection of three hyperplanes in $\mathbb{R}^4$. According to Theorem 8.2.1, if the equations (or, equivalently, the normals to the hyperplanes) are linearly independent then the set of solutions will be a one-dimensional subspace (that is, a line) in $\mathbb{R}^4$. The normals to the hyperplanes are the vectors $(1, 2, 0, 0)$, $(3, 2, -4, 1)$ and $(1, 1, 1, -1)$, and (although it is not necessary to do so) we leave the reader to check that these are linearly independent. In this example, we can add the last two equations to obtain $4x_1 + 3x_2 - 3x_3 = 0$; equivalently, we can replace the equations $x \cdot r_2 = x \cdot r_3 = 0$ by the equations $x \cdot (r_2 + r_3) = x \cdot r_3 = 0$. Either way, we now need to solve the system of equations

$$x_1 + 2x_2 = 0,$$
$$4x_1 + 3x_2 - 3x_3 = 0,$$
$$x_1 + x_2 + x_3 - x_4 = 0.$$

Writing $x_1 = 6t$, we find x_2, x_3, and x_4 in this order, and we see that the general solution is $(6t, -3t, 5t, 8t)$ for any real t. Thus S is the line in $\mathbb{R}^4$ that is in the direction $(6, -3, 5, 8)$. □

Exercise 8.2

1. Solve the system of equations
$$x_1 + x_2 - x_3 - x_4 = 0,$$
$$4x_1 - x_2 + 4x_3 - x_4 = 0,$$
$$3x_1 - 7x_2 + x_3 + 3x_4 = 0.$$

2. Solve the system of equations
$$x_1 - x_2 + 3x_3 + x_4 = 0,$$
$$2x_1 + 2x_3 - 5x_4 = 0,$$
$$x_1 + 3x_2 - 5x_3 - 13x_4 = 0.$$

3. Solve the system of equations
$$x_1 + x_2 - 2x_3 + 3x_4 = 0,$$
$$2x_1 + 3x_2 - x_3 + 4x_4 = 0.$$

4. Solve the equation $x_1 + x_2 + x_3 + x_4 = 0$.

8.3 Row rank and column rank

Let A be the matrix in (8.2.2). Each row and each column of A provides us with a vector in $\mathbb{R}^n$ and $\mathbb{R}^m$, respectively, and we are going to explore the subspaces spanned by these vectors.

Definition 8.3.1 Let A be the matrix in (8.2.2). The j-th *row vector* r_j of A is the vector formed from the j-th row of A, namely

$$r_j = (a_{j1}, \ldots, a_{jn}), \quad j = 1, \ldots, m. \tag{8.3.1}$$

The j-th *column vector* c_j of A is the vector formed from the j-th column of A, namely

$$c_j = \begin{pmatrix} a_{1j} \\ \vdots \\ a_{mj} \end{pmatrix} = (a_{1j}, \ldots, a_{mj})^t, \quad j = 1, \ldots, n. \tag{8.3.2}$$

Definition 8.3.2 The m row vectors $r_1, \ldots, r_m$ span a subspace R of $\mathbb{R}^n$, and the *row rank* of A is $\dim(R)$. The n column vectors $c_1, \ldots, c_n$ span a subspace C of $\mathbb{R}^{m,t}$, and the *column rank* of A is $\dim(C)$. We denote these ranks by $\rho_{\text{row}}(A)$ and $\rho_{\text{col}}(A)$ respectively.

The basic result is that the row rank of any matrix is the same as its column rank.

Theorem 8.3.3 *For any matrix A, the row rank of A equals the column rank of A.*

Proof Let α be as in (8.2.4), where we are no longer assuming that the vectors $r_1, \ldots, r_m$ are linearly independent. Let $\rho_{\text{row}}(A) = t$. Then there is a set $r'_1, \ldots, r'_t$ of linearly independent vectors chosen from $r_1, \ldots, r_m$ such that every r_j is a linear combination of the r'_i. This shows that

$$\ker(\alpha) = \{x \in \mathbb{R}^n : x \cdot r'_1 = 0, \ldots, x \cdot r'_t = 0\},$$

and so, by Theorem 8.2.1, $\dim \ker(\alpha) = n - t$. It follows that

$$\rho_{\text{row}}(A) = t = \dim(\mathbb{R}^n) - \dim \ker(\alpha) = \dim \alpha(\mathbb{R}^n).$$

Next, if $x = (x_1, \ldots, x_n)$ then $x \cdot r_j = x_1 a_{j1} + x_2 a_{j2} + \cdots + x_n a_{jn}$, so that $e_i \cdot r_j = a_{ji}$. It follows that

$$\alpha(e_i) = (e_i \cdot r_1, \ldots, e_i \cdot r_m) = (a_{1i}, a_{2i}, \ldots, a_{mi}) = c_i^t, \quad i = 1, \ldots, n;$$

thus $\alpha(\mathbb{R}^n)$ is spanned by the row vectors $c_1^t, \ldots, c_n^t$ in $\mathbb{R}^m$. Obviously, this space has the same dimension of the space spanned by the column vectors $c_1, \ldots, c_n$ in $\mathbb{R}^{m,t}$ (for this is the same result written in columns rather that rows); thus $\rho_{\text{col}}(A) = \dim \alpha(\mathbb{R}^n) = \rho_{\text{row}}(A)$. □

Exercise 8.3

1. Verify Theorem 8.3.3 for each of the following matrices:

$$\begin{pmatrix} 3 & 2 & 1 \\ 1 & 2 & 3 \end{pmatrix}, \quad \begin{pmatrix} 1 & 2 & 3 \\ 4 & 5 & 6 \\ 5 & 4 & 3 \end{pmatrix}.$$

8.4 Inhomogeneous linear equations

We now discuss the space of solutions of a set of m simultaneous inhomogeneous linear equations in the real variables $x_1, \ldots, x_n$; for example.

$$\begin{aligned} a_{11}x_1 + \cdots + a_{1n}x_n &= b_1, \\ \vdots \quad\quad\quad \vdots \quad\quad &\quad \vdots \\ a_{m1}x_1 + \cdots + a_{mn}x_n &= b_m. \end{aligned} \quad (8.4.1)$$

There are two issues of concern here, namely the *existence of at least one solution*, and *the characterization of the general solution*. We shall see how these issues can be reduced to finding (i) the subspace spanned by a given set of vectors, and (ii) the kernel of a linear map.

We recall the n column vectors $c_1, \ldots c_n$ in $\mathbb{R}^{m,t}$ defined in (8.3.2), and we define the vector b (also in $\mathbb{R}^{m,t}$) by $b = (b_1, \ldots, b_m)^t$. Then the system (8.4.1) is equivalent to the single vector equation

$$x_1 c_1 + \cdots + x_n c_n = b,$$

in $\mathbb{R}^{m,t}$ in which the c_j and b are column vectors and the x_i are scalars. This observation constitutes the proof of the following theorem which settles the question of the existence of a solution.

Theorem 8.4.1 *Let* $b = (b_1, \ldots, b_m)^t$. *Then the system* (8.4.1) *has a solution if and only if b lies in the subspace of $\mathbb{R}^{m,t}$ spanned by the column vectors $c_1, \ldots, c_n$ of the matrix a_{ij}.*

Theorem 8.4.1 gives a condition that can, in principle, be checked in any given example, for consider the matrices

$$A = \begin{pmatrix} a_{11} & a_{12} & \cdots & a_{1n} \\ \vdots & & & \vdots \\ a_{m1} & a_{m2} & \cdots & a_{mn} \end{pmatrix}, \quad B = \begin{pmatrix} a_{11} & a_{12} & \cdots & a_{1n} & b_1 \\ \vdots & & & & \vdots \\ a_{m1} & a_{m2} & \cdots & a_{mn} & b_m \end{pmatrix}.$$

The matrix B is often called the *augmented matrix*. The condition that b lies in the space spanned by the column vectors c_j is precisely the condition that A and B have the same column ranks and, by Theorem 8.3.1, this is so if and only if A and B have the same row ranks.

In order to characterize the set of all solutions (when a solution exists) we again use the linear map $\alpha : \mathbb{R}^n \to \mathbb{R}^m$ defined in (8.2.4). In this case the solution set of (8.4.1) is precisely the set $S = \{x \in \mathbb{R}^n : \alpha(x) = b^t\}$. Let us now suppose that at least one solution, say x^*, of (8.4.1) exists. Then $\alpha(x) = b^t$ if and only if $\alpha(x) - \alpha(x^*) = 0$ or, equivalently, if and only if $\alpha(x - x^*) = 0$, or $x - x^* \in \ker(\alpha)$. Thus we have proved the following theorem (which is, in fact, a special case of Theorem 7.8.4).

Theorem 8.4.2 *Suppose that x^* is a solution of (8.4.1). Then x is a solution of (8.4.1) if and only if $x = x^* + y$, where y is a solution of the homogeneous system (8.2.1).*

We remark that Theorem 8.4.2 shows that the set of solutions of an inhomogeneous system of linear equations is (if it is not empty) the *translation of a subspace of* $\mathbb{R}^n$. We give an example to illustrate this.

Example 8.4.3 First, consider the system of equations

$$\begin{aligned} y_1 - 2y_2 + 5y_3 &= 0 \\ 2y_1 - 4y_2 + 8y_3 &= 0 \\ -3y_1 + 6y_2 + 7y_3 &= 0. \end{aligned} \quad (8.4.2)$$

The first two equations imply that $y_3 = 0$, and then the system reduces to the single equation $y_1 - 2y_2 = 0$. Thus the space of solutions of this system is the line L given by $\{(2t, t, 0) : t \in \mathbb{R}\}$.

Now consider the system

$$\begin{aligned} x_1 - 2x_2 + 5x_3 &= 1 \\ 2x_1 - 4x_2 + 8x_3 &= 2 \\ -3x_1 + 6x_2 + 7x_3 &= -3. \end{aligned} \quad (8.4.3)$$

This has the same matrix as the system (8.4.2) and, in the notation used above, b is the first column vector, so that b is certainly in the space spanned by the column vectors. As $(1, 0, 0)$ is a solution, then general solution of (8.4.3) is the translated line $(1, 0, 0) + L$.

Finally, consider the system

$$\begin{aligned} x_1 - 2x_2 + 5x_3 &= 1 \\ 2x_1 - 4x_2 + 8x_3 &= 2 \\ -3x_1 + 6x_2 + 7x_3 &= 3, \end{aligned} \quad (8.4.4)$$

(also with the same A). For brevity, let us write column vectors as row vectors. If a solution exists (and we shall show it does not) then b must be a linear combination of c_1, c_2, c_3. As $c_2 = -2c_1$, this means that there are scalars λ

and μ such that $(1, 2, 3) = \lambda(1, 2, -3) + \mu(5, 8, 7)$. As no such λ and μ exist, (8.4.4) has no solutions. □

Exercise 8.4

1. Solve the equations
$$\begin{pmatrix} 1 & 2 & 1 \\ 1 & 3 & 0 \\ 2 & 2 & 1 \end{pmatrix} \begin{pmatrix} x_1 \\ x_2 \\ x_3 \end{pmatrix} = \begin{pmatrix} 9 \\ 10 \\ 10 \end{pmatrix}.$$

2. Find all values of t for which the system of equations
$$2x_1 + x_2 + 4x_3 + 3x_4 = 1$$
$$x_1 + 3x_2 + 2x_3 - x_4 = 3t$$
$$x_1 + x_2 + 2x_3 + x_4 = t^2$$
has a solution, and in each case give the general solution of the system.

8.5 Determinants and linear equations

So far, we have concentrated on the geometric theory behind the solutions of linear equations. In the case when $m = n$, A is a square matrix and there is a simple algebraic criterion for the existence of a unique solution of (8.4.1) in terms of an $n \times n$ determinant (which we have not yet defined). In fact, there is a formula, known as *Cramer's rule*, which expresses the solution as ratios of determinants. The interested reader can find this in many texts; however it is of less interest than it used to be as there are now many computer packages available for solving systems of linear equations. We briefly sketch the approach through determinants.

Theorem 8.5.1 *Let A be the matrix (a_{ij}) in (8.2.2) and suppose that $m = n$. Then the system of equations (8.4.1) has a unique solution if and only if the row (and column) rank of A is n.*

Proof Let $\alpha(x) = (x \cdot r_1, \ldots, x \cdot r_n)$. Then x is a solution of (8.4.1) if and only if $\alpha(x) = (b_1, \ldots, b_n)$. If this has a unique solution, then the kernel of α can only contain the zero vector, so that $\alpha(\mathbb{R}^n) = \mathbb{R}^n$. Conversely, if this is so, then (8.4.1) has a solution, and it must be unique as then $\ker(\alpha) = \{0\}$. This argument shows that (8.4.1) has a unique solution if and only if $\alpha(\mathbb{R}^n) = \mathbb{R}^n$. However, this is so if and only if the row rank, and the column rank, of A is n (this is part of the proof of Theorem 8.3.3). □

Let us assume for the moment that we have defined an $n \times n$ determinant, namely

$$\det(A) = \begin{vmatrix} a_{11} & \cdots & a_{1n} \\ \vdots & \ddots & \vdots \\ a_{n1} & \cdots & a_{nn} \end{vmatrix},$$

where this is a real (or complex) number. We shall see that the definition of a determinant (which we give in the next section) leads to the conclusion that a determinant is non-zero if and only if its rows (or columns) are linearly independent; equivalently, the row (and column) rank of the matrix is n. Assuming this, we have the following result.

Theorem 8.5.2 *Let A be the matrix (a_{ij}) in (8.2.2) and suppose that $m = n$. Then the system of equations (8.4.1) has a unique solution if and only if $\det(A) \neq 0$.*

The point about this result is that it gives an *algorithm* for checking whether or not the system of equations (8.4.1) has a unique solution (namely whether $\det(A)$ is zero or not) and this algorithm can be implemented on a computer.

Exercise 8.5

1. Solve the equations

$$3x + 2y + z = 5$$
$$-x - y + 4z = 1$$
$$2x + 3y - 2z = 6.$$

   Find $\det(A)$ for this system of equations.
2. Consider the system of equations

$$3x + 2y + z = a$$
$$-x - y + 4z = b$$
$$2x + 3y - 2z = c.$$

   What is $\det(A)$? Find x, y and z in terms of a, b and c.

8.6 Determinants

This section is devoted to the definition and properties of an $n \times n$ determinant, and our definition is a direct generalization of the 3×3 case.

8.6 Determinants

Definition 8.6.1 Suppose that A is an $n \times n$ matrix, say (a_{ij}). Then
$$\det(A) = \sum_{\sigma \in S_n} \epsilon(\sigma) a_{1\sigma(1)} \cdots a_{n\sigma(n)}, \qquad (8.6.1)$$
where the sum is over all permutations σ of $\{1, 2, \ldots, n\}$, and where $\epsilon(\sigma)$ is the sign of the permutation σ. We write
$$\det(A) = \begin{vmatrix} a_{11} & \cdots & a_{1n} \\ \vdots & & \vdots \\ a_{m1} & \cdots & a_{mn} \end{vmatrix}.$$

We illustrate this with two examples. □

Example 8.6.2 Suppose that A is an $n \times n$ upper-triangular matrix; that is, $a_{ij} = 0$ whenever $i > j$ (equivalently, all of the entries 'below' the diagonal are zero); then $\det(A) = a_{11} \cdots a_{nn}$. For example, if
$$A = \begin{pmatrix} a & b \\ 0 & d \end{pmatrix}, \quad B = \begin{pmatrix} b_{11} & b_{12} & b_{13} \\ 0 & b_{22} & b_{23} \\ 0 & 0 & b_{33} \end{pmatrix},$$
then $\det(A) = ad$ and $\det(B) = b_{11}b_{22}b_{33}$. To prove the general result, we consider (8.6.1) for an upper-triangular matrix A. Take any permutation σ, and note that $a_{n\sigma(n)} = 0$ when $\sigma(n) < n$; that is, unless $\sigma(n) = n$. Thus we may restrict the sum in (8.6.1) to those σ that fix n. Next, $a_{n-1,\sigma(n-1)} = 0$ unless $\sigma(n-1)$ is $n-1$ or n and, as it cannot be n, σ must also fix $n-1$. This argument can be continued to show that only the identity permutation contributes to the sum in (8.6.1) and the result follows. □

Example 8.6.3 For another example, suppose that
$$A = \begin{pmatrix} 1 & 2 & 0 & 0 \\ 3 & 4 & 0 & 0 \\ 0 & 0 & 5 & 0 \\ 0 & 0 & 0 & 6 \end{pmatrix}.$$

Now for any permutation σ, $a_{3\sigma(3)} = 0$ unless $\sigma(3) = 3$, and similarly for $a_{4\sigma(4)}$, so the sum in (8.6.1) reduces to the sum over the two permutations ι (the identity) and the transposition $\tau = (1\ 2)$. Thus
$$\det(A) = \epsilon(\iota)a_{11}a_{22}a_{33}a_{44} + \epsilon(\tau)a_{12}a_{21}a_{33}a_{44}$$
$$= (a_{11}a_{22} - a_{12}a_{21})a_{33}a_{44}$$
$$= (1 \times 4 - 3 \times 2) \times 5 \times 6$$
$$= -60.$$

□

Without further discussion, we shall now list and prove the main results on determinants; these are valid for both real and complex entries.

Definition 8.6.4 The *transpose* A^t of a matrix A is obtained by 'turning the matrix over'. Thus if $A = (a_{ij})$ and $B = (b_{ij}) = A^t$, then $b_{ij} = a_{ji}$; explicitly,

$$A = \begin{pmatrix} a_{11} & \cdots & a_{1j} & \cdots & a_{1n} \\ \vdots & & \vdots & & \vdots \\ a_{m1} & \cdots & a_{mj} & \cdots & a_{mn} \end{pmatrix}, \quad A^t = \begin{pmatrix} a_{11} & \cdots & a_{m1} \\ \vdots & & \vdots \\ a_{1j} & \cdots & a_{mj} \\ \vdots & & \vdots \\ a_{1n} & \cdots & a_{mn} \end{pmatrix}.$$

Note that A and A^t have different sizes unless $m = n$. □

Theorem 8.6.5 *For any square matrix A, $\det(A^t) = \det(A)$.*

Proof It is clear that for any permutations ρ and σ,

$$a_{1\sigma(1)} \cdots a_{n\sigma(n)} = a_{\rho(1)\sigma\rho(1)} \cdots a_{\rho(n)\sigma\rho(n)},$$

because the product on the right is simply a re-ordering of the product on the left. If we now put $\rho = \sigma^{-1}$, we obtain

$$a_{1\sigma(1)} \cdots a_{n\sigma(n)} = a_{\sigma^{-1}(1)1} \cdots a_{\sigma^{-1}(n)n}.$$

and hence, as $\epsilon(\sigma^{-1}) = \epsilon(\sigma)$,

$$\det(A) = \sum_{\sigma^{-1} \in S_n} \epsilon(\sigma^{-1}) a_{\sigma^{-1}(1)1} \cdots a_{\sigma^{-1}(n)n}.$$

Now summing over σ is the same as summing over σ^{-1}, so that

$$\det(A) = \sum_{\sigma \in S_n} \epsilon(\sigma) a_{\sigma(1)1} \cdots a_{\sigma(n)n},$$

If $B = A^t$, then $b_{ij} = a_{ji}$, and this shows that $\det(A) = \det(B)$. □

Theorem 8.6.6 *If the matrix B is obtained from A by interchanging two rows, or two columns, then $\det(B) = -\det(A)$. In particular, if two rows, or two columns, are identical, then $\det(A) = 0$.*

Proof It is clear from Theorem 8.6.5 that it suffices to prove the result for rows. Suppose, then, that B is obtained from A by interchanging the r-th and s-th rows. We let $A = (a_{ij})$ and $B = (b_{ij})$, so that for all j, $b_{ij} = a_{ij}$ if $i \neq r, s$, and $b_{rj} = a_{sj}$, $b_{sj} = a_{rj}$. Now let τ be the transposition $(r\ s)$; then, for each permutation σ,

$$b_{1\sigma(1)} \cdots b_{n\sigma(n)} = a_{1\sigma\tau(1)} \cdots a_{n\sigma\tau(n)}.$$

It follows that

$$\det(B) = \sum_{\sigma \in S_n} \epsilon(\sigma) b_{1\sigma(1)} \cdots b_{n\sigma(n)}$$
$$= \sum_{\sigma \in S_n} \epsilon(\sigma) a_{1\sigma\tau(1)} \cdots a_{n\sigma\tau(n)}$$
$$= -\sum_{\sigma \in S_n} \epsilon(\sigma\tau) a_{1\sigma\tau(1)} \cdots a_{n\sigma\tau(n)}.$$

As σ ranges over S_n, so does $\sigma\tau$ (for a fixed τ), so $\det(B) = -\det(A)$. The last statement follows because if we interchange two identical rows we both leave the determinant unaltered and change its sign. □

Theorem 8.6.7 *The function* $\det(A)$ *of A is a linear function of each row of A, and of each column of A,*

Proof Theorem 8.6.5 shows that we need only prove the result for rows, and Theorem 8.6.6 then shows that we need only prove the result for the first row. Now $\det(A)$ is a finite sum $\sum_\sigma \epsilon(\sigma) a_{1\sigma(1)} A_\sigma$, say, where the terms A_σ are independent of the entries $a_{11}, \ldots, a_{1,n}$. As each term in this sum is a linear function of the first row, so is the sum. □

Theorem 8.6.8 *If the rows of A are linearly dependent, then* $\det(A) = 0$. *A similar statement holds for the columns of A.*

Proof We suppose that the rows of A are linearly dependent. Then some row, say r_j, is a linear combination of the other rows, and so when we replace r_j by this linear combination and then use the linearity of the determinant function as a function of the rows (Theorem 8.6.7), we see that $\det(A)$ is a linear combination of determinants, each of which has two identical rows. The result now follows from Theorem 8.6.6. □

We remark that this shows that if $\det(A) \neq 0$, then the rows of A are linearly independent so that, in the context of Theorems 8.5.1 and 8.5.2, $\alpha(\mathbb{R}^n) = \mathbb{R}^n$. Thus we have proved the following result.

Theorem 8.6.9 *If* $\det(A) \neq 0$, *then the system of equations* (8.4.1) *has a unique solution.*

The converse (and less useful) result will follow from results to be proved in Chapter 9 (see Theorem 9.4.4).

Exercise 8.6

1. Show that
$$\begin{vmatrix} a & b & p & q \\ c & d & r & s \\ 0 & 0 & e & f \\ 0 & 0 & g & h \end{vmatrix} = (ad - bc)(eh - gf).$$

2. Prove that
$$\begin{vmatrix} 1 & 1 & 1 \\ a^2 & b^2 & c^2 \\ a^3 & b^3 & c^4 \end{vmatrix} = (b - c)(c - a)(a - b)(bc + ca + ab).$$

3. Suppose that
$$a_1 x + b_1 y + c_1 z = 0,$$
$$a_2 x + b_2 y + c_2 z = 0,$$
$$a_3 x + b_3 y + c_3 z = 0,$$

    and let A be associated the 3×3 matrix. Show that $xD = yD = zD = 0$, where $D = \det(A)$.

4. Show that for every matrix A, and every scalar λ, $(A^t)^t = A$ and $(\lambda A)^t = \lambda(A^t)$.

9
Matrices

9.1 The vector space of matrices

An $m \times n$ *matrix* A is a rectangular array of numbers, with m rows and n columns, that is written in the form

$$A = \begin{pmatrix} a_{11} & a_{12} & \cdots & a_{1n} \\ \vdots & \vdots & & \vdots \\ a_{m1} & a_{m2} & \cdots & a_{mn} \end{pmatrix}. \qquad (9.1.1)$$

We frequently use the notation $A = (a_{ij})$ when the values of m and n are understood from the context (or not important). The a_{ij} are the *entries*, or *coefficients*, of A; the matrix A is a *real matrix* if each a_{ij} is real, and a *complex matrix* if each a_{ij} is complex. The matrix A is *square* if $m = n$, and then the *diagonal elements* of A are $a_{11}, \ldots, a_{nn}$. The $m \times n$ *zero matrix* is the matrix with all entries zero; we should perhaps denote this by some symbol such as 0_{mn}, but (like everyone else) we shall omit the suffix and use 0 instead. The *rows*, or *row vectors*, of A are the vectors $(a_{j1}, a_{j2}, \ldots, a_{jn})$, and the *columns*, or *column vectors*, are defined similarly (and usually written vertically). We need a notation for the set of matrices of a given size.

Definition 9.1.1 The set of $m \times n$ matrices with entries in $\mathbb{F}$ is denoted by $M^{m \times n}(\mathbb{F})$.

There is a natural definition of addition, and scalar multiplication, of matrices in $M^{m \times n}(\mathbb{F})$, namely

$$\begin{pmatrix} a_{11} & \cdots & a_{1n} \\ \vdots & \ddots & \vdots \\ a_{m1} & \cdots & a_{mn} \end{pmatrix} + \begin{pmatrix} b_{11} & \cdots & b_{1n} \\ \vdots & \ddots & \vdots \\ b_{m1} & \cdots & b_{mn} \end{pmatrix} = \begin{pmatrix} c_{11} & \cdots & c_{1n} \\ \vdots & \ddots & \vdots \\ c_{m1} & \cdots & c_{mn} \end{pmatrix},$$

where $c_{ij} = a_{ij} + b_{ij}$, and

$$\lambda \begin{pmatrix} a_{11} & \cdots & a_{1n} \\ \vdots & \ddots & \vdots \\ a_{m1} & \cdots & a_{mn} \end{pmatrix} = \begin{pmatrix} \lambda a_{11} & \cdots & \lambda a_{1n} \\ \vdots & \ddots & \vdots \\ \lambda a_{m1} & \cdots & \lambda a_{mn} \end{pmatrix}.$$

It is clear from these definitions that a linear combination of matrices A_j in $M^{m \times n}(\mathbb{F})$ is also in $M^{m \times n}(\mathbb{F})$, so (after checking some trivial facts) we see that this space is a vector space. The zero 'vector' in this space is the zero matrix, and the inverse of A is $-A$.

Theorem 9.1.2 *The vector space $M^{m \times n}(\mathbb{F})$ has dimension mn.*

Proof For integers r and s satisfying $1 \leq r \leq m$ and $1 \leq s \leq n$, let E_{rs} be the $m \times n$ matrix (e_{ij}) which has $e_{rs} = 1$ and all other entries zero; for example, if $m = n = 2$ then

$$E_{11} = \begin{pmatrix} 1 & 0 \\ 0 & 0 \end{pmatrix}, \quad E_{12} = \begin{pmatrix} 0 & 1 \\ 0 & 0 \end{pmatrix},$$
$$E_{21} = \begin{pmatrix} 0 & 0 \\ 1 & 0 \end{pmatrix}, \quad E_{22} = \begin{pmatrix} 0 & 0 \\ 0 & 1 \end{pmatrix}. \tag{9.1.2}$$

There are precisely mn matrices E_{rs} in $M^{m \times n}(\mathbb{F})$. It is clear that

$$(a_{ij}) = \sum_{i,j} a_{ij} E_{ij}, \tag{9.1.3}$$

so the E_{ij} span $M^{m \times n}(\mathbb{F})$. It is also easy to see that the matrices E_{rs} are linearly independent for

$$\sum_{r=1}^{m} \sum_{s=1}^{n} \lambda_{rs} E_{rs} = (\lambda_{ij}).$$

If this is the zero matrix, then $\lambda_{ij} = 0$ for all i and j. $\square$

We shall ignore the distinction in punctuation between $(x_1, \ldots, x_n)$ (for vectors) and $(x_1 \cdots x_n)$ (for matrices); then $(x_1, \ldots, x_n)$ in $\mathbb{R}^n$ is a $1 \times n$ matrix and $M^{1 \times n}(\mathbb{R}) = \mathbb{R}^n$. Further, the definitions of addition and scalar multiplication of matrices coincides with that used previously for $\mathbb{R}^n$. Thus a special case of Theorem 9.1.2 is that $\dim(\mathbb{R}^n) = n$ (and in this case E_{1j} coincides with e_j). We recall Definition 8.6.4 of the transpose of a matrix, and we use this to define what is meant by a symmetric, and a skew-symmetric, square matrix.

Definition 9.1.3 *A square matrix A is symmetric if $A^t = A$, and skew-symmetric if $A^t = -A$.*

9.1 The vector space of matrices

A real-valued function $g(x)$ is *even* if $g(-x) = g(x)$, and is *odd* if $g(-x) = -g(x)$), and every function $f(x)$ can be written uniquely as the sum of an even function and an odd function, namely

$$f(x) = \tfrac{1}{2}[f(x) + f(-x)] + \tfrac{1}{2}[f(x) - f(-x)].$$

An analogous statement holds for matrices: *every square matrix can be written uniquely as the sum of a symmetric matrix and a skew-symmetric matrix*. Indeed, if A is a square matrix, then

$$A = \tfrac{1}{2}(A + A^t) + \tfrac{1}{2}(A - A^t),$$

and $\tfrac{1}{2}(A + A^t)$ is symmetric and $\tfrac{1}{2}(A - A^t)$ is skew-symmetric. This expression is unique for if $A_1 + B_1 = A_2 + B_2$, where the A_i are symmetric and the B_j are skew-symmetric, then $A_1 - A_2 = B_2 - B_1 = X$, say, where $X = X^t$ and $X = -X^t$. It follows that X is the zero matrix, and hence that $A_1 = A_2$ and $B_1 = B_2$. Finally, notice that if (x_{ij}) is skew-symmetric then every diagonal element x_{ii} is zero because $x_{ii} = -x_{ii}$. It should be obvious to the reader that the space of symmetric $n \times n$ matrices is a vector space, as is the space of skew-symmetric matrices.

Theorem 9.1.4 *The vector space $M^{m \times n}(\mathbb{F})$ is the direct sum of the subspace of symmetric matrices, of dimension $n(n+1)/2$, and the subspace of skew-symmetric matrices of dimension $n(n-1)/2$.*

Proof We leave the reader to check that the set of symmetric matrices, and the set of skew-symmetric matrices, do indeed form subspaces of $M^{m \times n}(\mathbb{F})$. We have already seen that every matrix can be expressed uniquely as the sum of a symmetric matrix and a skew-symmetric matrix, and this shows that $M^{m \times n}(\mathbb{F})$ is the direct sum of these two subspaces. Next, the matrices $E_{ij} - E_{ji}$ (see the proof of Theorem 9.1.2), where $1 \leq i < j \leq n$, are skew-symmetric, and as the diagonal elements of a skew-symmetric matrix are zero, it is clear that these matrices span the subspace of skew-symmetric matrices. As they are also linearly independent, we see that the subspace of skew-symmetric matrices has dimension $(n^2 - n)/2$. By Theorems 7.4.1 and 9.1.2, the subspace of symmetric matrices has dimension $n^2 - (n^2 - n)/2$. $\square$

We give two more examples.

Example 9.1.5 Consider the space $Z^{m \times n}$ of real $m \times n$ matrices X with the property that the sum of the elements over each row, and over each column, is zero. It should be clear that $Z^{m \times n}$ is a real vector space, and we claim that $\dim(Z^{m \times n}) = (m-1)(n-1)$. We start the proof in the case when $m = n = 3$, but only to give the general idea.

We start with an 'empty' 3×3 matrix and then make an arbitrary choice of the entries, say a, b, c and d, that are not in the last row or column; this gives us a matrix

$$\begin{pmatrix} a & b & * \\ c & d & * \\ * & * & * \end{pmatrix},$$

where $*$ represents an as yet undetermined entry in the matrix. We now impose the condition that the first two columns must sum to zero, and after this we impose the condition that all rows sum to zero; thus the 'matrix' becomes

$$\begin{pmatrix} a & b & -a-b \\ c & d & -c-d \\ -a-c & -b-d & a+b+c+d \end{pmatrix}.$$

Notice that the last column automatically sums to zero (because the sum over all elements is zero, as is seen by summing over rows, and the first two columns sum to zero). Exactly the same argument can be used for any 'empty' $m \times n$ matrix.

The choice of elements not in the last row or last column is actually a choice of an arbitrary matrix in $M^{(m-1) \times (n-1)}(\mathbb{F})$, so this construction actually creates a surjective map Φ from $M^{(m-1) \times (n-1)}(\mathbb{F})$ onto $Z^{m \times n}$. It should be clear that this map is linear, and that the only element in its kernel is the zero matrix. Thus

$$\dim(Z^{m \times n}) = \dim \ker(\Phi) + \dim M^{(m-1) \times (n-1)}(\mathbb{F}) = (m-1)(n-1)$$

as required. $\square$

Example 9.1.6 This example contains a discussion of magic squares (this is a 'popular' item, but it is not important). For any $n \times n$ matrix X, the *trace* tr(X) of X is the sum $x_{11} + \cdots + x_{nn}$ of the diagonal elements, and the *anti-trace* tr*(X) of X is the sum over the 'other' diagonal, namely $x_{1n} + \cdots + x_{n1}$. A real $n \times n$ matrix A is a *magic square* if the sum over each row, the sum over each column, and the sum over each of the two diagonals (that is, tr(A) and tr*(X)), all give the same value, say $\mu(A)$. We note that $\mu(A) = n^{-1} \sum_{i,j} a_{ij}$. It is easy to see that the space $S^{n \times n}$ of $n \times n$ magic squares is a real vector space so, naturally, we ask what is its dimension? It is easy to see that $\dim(S^{n \times n}) = 1$ when n is 1 or 2, and we shall now show that for $n \geq 3$, $\dim(S^{n \times n}) = n(n-2)$.

Let $S_0^{n \times n}$ be the subspace of matrices A for which $\mu(A) = 0$. This subspace is the kernel of the linear map $A \mapsto \mu(A)$ from $S^{n \times n}$ to $\mathbb{R}$, and as this map is surjective (consider the matrix A with all entries x/n) we see that $\dim(S^{n \times n}) = \dim(S_0^{n \times n}) + 1$.

Next, the space $Z^{n \times n}$ of $n \times n$ matrices all of whose rows and columns sum to zero has dimension $(n-1)^2$ (see Example 9.1.6). Now define $\Phi : Z^{n \times n} \to \mathbb{R}^2$ by $\Phi(X) = (\text{tr}(X), \text{tr}^*(X))$. Then Φ is a linear map, and $\ker(\Phi) = S_0^{n \times n}$. It is not difficult to show that Φ is surjective (we shall prove this shortly), and with this we see that

$$(n-1)^2 = \dim(Z^{n \times n}) = \dim S_0^{n \times n} + 2,$$

so that $\dim(S^{n \times n}) = (n-1)^2 - 1 = n(n-2)$. It remains to show that Φ is surjective, and it is sufficient to construct matrices P and Q in $Z^{n \times n}$ such that $\Phi(P) = (a, 0)$ and $\Phi(Q) = (0, b)$ for all (or just some non-zero) a and b. If $n = 3$, we let

$$P = (a/3) \begin{pmatrix} 1 & 1 & -2 \\ -2 & 1 & 1 \\ 1 & -2 & 1 \end{pmatrix}, \quad Q = (b/3) \begin{pmatrix} -2 & 1 & 1 \\ 1 & 1 & -2 \\ 1 & -2 & 1 \end{pmatrix},$$

and then $\Phi(P) = (a, 0)$ and $\Phi(Q) = (0, b)$.

If $n \geq 4$ we can take $p_{11} = p_{22} = a/2$, $p_{12} = p_{21} = -a/2$ and all other $p_{ij} = 0$; then $t(P) = a$ and $t^*(P) = 0$, so that $\Phi(P) = (a, 0)$. Similarly, we choose $q_{1,n-1} = q_{2n} = -b/2$ and $q_{1n} = q_{2,n-1} = b/2$, so that $\Phi(Q) = (0, b)$. $\square$

Exercise 9.1

1. A matrix (a_{ij}) is a *diagonal matrix* if $a_{ij} = 0$ whenever $i \neq j$. Show that the space D of real $n \times n$ diagonal matrices is a vector space of dimension n.
2. A matrix (a_{ij}) is an *upper-triangular matrix* if $a_{ij} = 0$ whenever $i > j$. Show that the space U of real $n \times n$ upper-triangular matrices is a vector space. What is its dimension?
3. Define what it means to say that a matrix (a_{ij}) is a *lower-triangular matrix* (see Exercise 2). Let L be the vector space of real lower-triangular matrices, and let D and U be as in Exercises 1 and 2. Show, *without* calculating any of the dimensions, that $\dim(U) + \dim(L) = \dim(D) + n^2$. Now verify this by calculating each of the dimensions.
4. Show that the space of $n \times n$ matrices with trace zero is a vector space of dimension $n^2 - 1$.
5. Show (in Example 9.1.7) that $\dim(S^{1 \times 1}) = \dim(S^{2 \times 2}) = 1$.
6. Show that if X is a 3×3 magic square, then $x_{22} = \Sigma(X)/3$. Deduce that if $\Sigma(X) = 0$ then X is of the form

$$X = \begin{pmatrix} a & -a-b & b \\ b-a & 0 & a-b \\ -b & a+b & -a \end{pmatrix}.$$

Let A, B, C be the matrices

$$\begin{pmatrix} 1 & -1 & 0 \\ -1 & 0 & 1 \\ 0 & 1 & -1 \end{pmatrix}, \quad \begin{pmatrix} 0 & -1 & 1 \\ 1 & 0 & -1 \\ -1 & 1 & 0 \end{pmatrix}, \quad \begin{pmatrix} 1 & 1 & 1 \\ 1 & 1 & 1 \\ 1 & 1 & 1 \end{pmatrix},$$

respectively. Show that
(a) $\{A, B, C\}$ is a basis of $S^{3 \times 3}$;
(b) $\{A, B\}$ is a basis of $S_0^{3 \times 3}$;
(c) $\{A, C\}$ is a basis of the space of symmetric 3×3 magic squares;
(d) $\{B\}$ is a basis of the space of skew-symmetric 3×3 magic squares.

9.2 A matrix as a linear transformation

A matrix is a rectangular array of numbers, say A given in (9.1.1), and we shall now describe how A acts naturally as a linear map. We recall (Example 7.1.2) that $\mathbb{R}^{n,t}$ is the space of real column vectors of dimension n, and similarly for $\mathbb{C}^{n,t}$.

Definition 9.2.1 The *action* of A as a linear map from $\mathbb{F}^{n,t}$ to $\mathbb{F}^{m,t}$ is defined to be the map

$$A : \begin{pmatrix} x_1 \\ \vdots \\ x_n \end{pmatrix} \mapsto \begin{pmatrix} a_{11}x_1 + \cdots + a_{1n}x_n \\ \vdots \\ a_{m1}x_1 + \cdots + a_{mn}x_n \end{pmatrix}, \quad (9.2.1)$$

which we write as $x \mapsto A(x)$. It is evident that for scalars λ and μ, $A(\lambda x + \mu y) = \lambda A(x) + \mu A(y)$, so that $A : \mathbb{F}^{n,t} \to \mathbb{F}^{m,t}$ is *linear*. □

Definition 9.2.1 prompts us to define the *product* of an $m \times n$ matrix with an $n \times 1$ by matrix (or column vector) by

$$\begin{pmatrix} a_{11} & \cdots & a_{1n} \\ \vdots & & \vdots \\ a_{m1} & \cdots & a_{mn} \end{pmatrix} \begin{pmatrix} x_1 \\ \vdots \\ x_n \end{pmatrix} = \begin{pmatrix} a_{11}x_1 + \cdots + a_{1n}x_n \\ \vdots \\ a_{m1}x_1 + \cdots + a_{mn}x_n \end{pmatrix}. \quad (9.2.2)$$

Shortly, we shall define the product of $m \times n$ matrix and an $n \times r$ matrix (in this order) to be an $m \times r$ matrix in a way that is consistent with this definition. One way to remember the formula (9.2.1) is to let $r_1, \ldots, r_m$ be the row vectors of A; if x is a column vector, then

$$A(x) = \begin{pmatrix} r_1 \cdot x^t \\ \vdots \\ r_m \cdot x^t \end{pmatrix},$$

where the entries here are the usual scalar products of two row vectors. Alternatively, we can describe the action of A in terms of standard bases. Let $e_1 = (1, 0, \ldots, 0), \ldots, e_n = (0, \ldots, 0, 1)$ be the standard basis of $\mathbb{F}^n$, and (here we must use different symbols because we may have $m \neq n$) let $\epsilon_1 = (1, 0, \ldots, 0), \ldots, \epsilon_m = (0, \ldots, 0, 1)$ be the standard basis of $\mathbb{F}^m$. Then, directly from (9.2.1),

$$A(e_i{}^t) = \begin{pmatrix} a_{1i} \\ \vdots \\ a_{mi} \end{pmatrix} = \sum_{j=1}^{m} a_{ji} \epsilon_j{}^t.$$

This shows that *the first column of A is the vector of coefficients of $A(e_1{}^t)$, the second column is the vector of coefficients of $A(e_2{}^t)$*, and so on, all in terms of the basis $\epsilon_1{}^t, \ldots, \epsilon_m{}^t$. More generally, if

$$x = (x_1, \ldots, x_n)^t = \sum_{k=1}^{n} x_k e_k{}^t,$$

then

$$A(x) = \sum_{k=1}^{n} x_k A(e_k{}^t) = \sum_{j=1}^{m} \left(\sum_{k=1}^{n} a_{jk} x_k \right) \epsilon_j{}^t$$

which agrees with (9.2.1). We now give some examples a matrix action.

Example 9.2.2 The matrix $(a_1 \cdots a_n)$ gives the linear map of $\mathbb{R}^{n,t}$ to $\mathbb{R}$ that takes the column vector x to the scalar product $a \cdot x^t$. □

Example 9.2.3 The matrix action

$$x = \begin{pmatrix} x_1 \\ x_2 \\ x_3 \end{pmatrix} \mapsto \begin{pmatrix} 0 & -a_3 & a_2 \\ a_3 & 0 & -a_1 \\ -a_2 & a_1 & 0 \end{pmatrix} \begin{pmatrix} x_1 \\ x_2 \\ x_3 \end{pmatrix} = \begin{pmatrix} a_2 x_3 - a_3 x_2 \\ a_3 x_1 - a_1 x_3 \\ a_1 x_2 - a_2 x_1 \end{pmatrix}$$

is the vector product map $x^t \mapsto a \times x^t$ of $\mathbb{R}^3$ into itself. □

Example 9.2.4 *The orthogonal projection onto a plane* Consider the map of $\mathbb{R}^3$ onto the plane Π with equation $x_1 + x_2 + x_3 = 0$ obtained by orthogonal projection (that is, $x \mapsto y$, where y is the point of Π that is nearest to x). It is clear that $y = x + t(1, 1, 1)$, where t is determined by the condition that $y \in \Pi$, and this gives

$$y = (x_1, x_2, x_3) - \left(\frac{x_1 + x_2 + x_3}{3} \right) (1, 1, 1).$$

Thus $y^t = A(x^t)$, where
$$A = \frac{1}{3}\begin{pmatrix} 2 & -1 & -1 \\ -1 & 2 & -1 \\ -1 & -1 & 2 \end{pmatrix}.$$
□

Let us now consider the product of two matrices. An $m \times n$ matrix A and an $r \times m$ matrix B provide two linear maps that are illustrated in the mapping diagram
$$\mathbb{F}^{n,t} \xrightarrow{A} \mathbb{F}^{m,t} \xrightarrow{B} \mathbb{F}^{r,t}.$$

It follows that the composition $x \mapsto B(A(x))$ is a linear map from $\mathbb{F}^{n,t}$ to $\mathbb{F}^{r,t}$, and we shall now show that this is also given by a matrix action. Suppose that $A = (a_{ij})$ and $B = (b_{ij})$, and that $x = (x_1, \cdots, x_n)^t$. Then

$$B(A(x)) = \begin{pmatrix} b_{11} & \cdots & b_{1m} \\ \vdots & & \vdots \\ b_{r1} & \cdots & b_{rm} \end{pmatrix} \begin{pmatrix} a_{11}x_1 + \cdots + a_{1n}x_n \\ \vdots \\ a_{m1}x_1 + \cdots + a_{mn}x_n \end{pmatrix}$$

$$= \begin{pmatrix} c_{11}x_1 + \cdots + c_{1n}x_n \\ \vdots \\ c_{r1}x_1 + \cdots + c_{rn}x_n \end{pmatrix},$$
(9.2.3)

where the c_{ij} are given by

$$c_{ij} = b_{i1}a_{1j} + \cdots b_{im}a_{mj}.$$
(9.2.4)

As an example of this, we see that the product
$$\begin{pmatrix} b_{11} & b_{12} & b_{13} \\ b_{21} & b_{22} & b_{23} \end{pmatrix} \begin{pmatrix} a_{11} & a_{12} \\ a_{21} & a_{22} \\ a_{31} & a_{32} \end{pmatrix}$$
is
$$= \begin{pmatrix} b_{11}a_{11} + b_{12}a_{21} + b_{13}a_{31} & b_{11}a_{12} + b_{12}a_{22} + b_{13}a_{32} \\ b_{21}a_{11} + b_{22}a_{21} + b_{23}a_{31} & b_{21}a_{12} + b_{22}a_{22} + b_{23}a_{32} \end{pmatrix}.$$

Given that the composite map is given by the matrix (9.2.3), it is natural to define the product BA of the matrices B and A (in this order) to be the matrix (c_{ij}) of this combined action.

Definition 9.2.5 The *matrix product* BA of an $r \times m$ matrix B and an $m \times n$ matrix A is the $r \times n$ matrix of the combined action $x \mapsto B(A(x))$. Explicitly, if $A = (a_{ij})$ and $B = (b_{ij})$, then $BA = (c_{ij})$, where the c_{ij} are given in (9.4.2). Informally, c_{ij} is the scalar product of the *i*-th row of B with the *j*-th column of A. □

Sometimes (9.2.3) is taken as the definition of the matrix product without any reference to the matrix action; however, there can be little (if any) justification for this definition except by consideration of the composite map. Note that the product BA of an $p \times q$ matrix B and a $r \times s$ matrix A is *only defined when* $q = r$ (as otherwise, the composite map is not defined). Accordingly, when we write a matrix product we shall always implicitly assume that the matrices are of the correct 'size' for the product to exist.

A tedious argument (which we omit) shows that *the matrix product is associative*: that is, $A(BC) = (AB)C$. For each n, let I_n be the $n \times n$ identity matrix; that is

$$I_n = \begin{pmatrix} 1 & 0 & \cdots & 0 \\ 0 & 1 & \cdots & 0 \\ \vdots & \vdots & \ddots & \vdots \\ 0 & 0 & \cdots & 1 \end{pmatrix}, \qquad (9.2.5)$$

where the entries in the diagonal are 1, and all other entries are zero. If X is an $m \times n$ matrix, then $I_m X = X$ and $X I_n = X$ but, of course, the products $X I_m$ and $I_n X$ are not defined (unless $m = n$). The matrix product is not commutative; for example, as the reader should check, if

$$A = \begin{pmatrix} 1 & 2 \\ 2 & 4 \end{pmatrix}, \qquad B = \begin{pmatrix} -2 & -6 \\ 1 & 3 \end{pmatrix},$$

then

$$AB = \begin{pmatrix} 0 & 0 \\ 0 & 0 \end{pmatrix}, \qquad BA = \begin{pmatrix} -14 & -28 \\ 7 & 14 \end{pmatrix}.$$

This example also shows that *the product of two matrices can be the zero matrix without either being the zero matrix*, and it follows from this that *there are non-zero matrices that do not have a multiplicative inverse*. Finally, we consider the transpose of a product.

Theorem 9.2.6 *The transpose $(AB)^t$ of the product AB is $B^t A^t$.*

Proof Let $A = (a_{ij})$, $A^t = (\alpha_{ij})$, $B = (b_{ij})$, $B^t = (\beta_{ij})$, $AB = (c_{ij})$ and $B^t A^t = (\gamma_{ij})$. Note that $\alpha_{ij} = a_{ji}$, and $\beta_{ij} = b_{ji}$, and we want to show that $\gamma_{ij} = c_{ji}$. Now $\gamma_{ij} = \sum_r \beta_{ir} \alpha_{rj} = \sum_r b_{ri} a_{jr} = c_{ji}$, so that $B^t A^t = (AB)^t$. □

Exercise 9.2

1. Show that if A is a square matrix, then AA^t is symmetric. Choose any 2×2 matrix and verify this directly.

2. Show that two symmetric $n \times n$ matrices A and B commute ($AB = BA$) if and only if their product AB is symmetric.
3. Show that if a 2×2 complex matrix X commutes with every 2×2 complex matrix, then $X = \lambda I$ for some complex λ, where I is the 2×2 identity matrix.
4. Let $A = \begin{pmatrix} 1 & 2 \\ 3 & 6 \end{pmatrix}$, and for each 2×2 real matrix X let $\alpha(X) = AX$; thus α is a map of the space $M^{2 \times 2}(\mathbb{R})$ of real 2×2 matrices into itself. Show that α is a linear map. Find a basis of the kernel, and of the range, of α. [These bases should be made up of 2×2 matrices, and the sum of the dimensions of these two subspaces should be four.]
5. Let
$$A = \begin{pmatrix} 1 & 2 \\ 3 & 4 \end{pmatrix}, \quad B = \begin{pmatrix} 0 & 2 \\ 3 & 3 \end{pmatrix}.$$
Show that B commutes with A ($AB = BA$). Show that the set of 2×2 real matrices X that commute with A is a subspace M_0 of the space of real 2×2 matrices. Show also that $\dim(M_0) = 2$, and that I_2 and B form a basis of M_0. As A and A^2 commute with A, this implies that A and A^2 are linear combinations of I and B. Find these linear combinations.
6. Let A be a real $n \times n$ matrix. Show that the map $\alpha : X \mapsto AX - XA$ is a linear map of $M^{n \times n}(\mathbb{R})$ to itself. Show that the set of matrices X that commute with A is a subspace $M(A)$ of $M^{n \times n}(\mathbb{R})$. You should do this (a) by a direct argument, and (b) by considering the kernel of α. Is A in $M(A)$? Now use a dimension argument to show that the map α is not surjective, thus there is some matrix B such that $B \neq AX - XA$ for any X.

9.3 The matrix of a linear transformation

In this section we introduce the *matrix representation of a linear transformation*. Matrices play exactly the same role for linear transformations as co-ordinates play for vectors, so it may be helpful to begin with a few comments on vectors and their co-ordinates. First, it is essential to understand that a vector space V is just a set of objects (with a structure), and that a vector x in V does not have any co-ordinates *until* an ordered basis of V has been specified. In particular, the vector space $\mathbb{R}^2$ is the set of ordered pairs of real numbers, and points in $\mathbb{R}^2$, for instance $x = (3, 2)$, do not have any co-ordinates until we choose an ordered basis of $\mathbb{R}^2$.

A basis is a frame of reference (rather like a set of co-ordinate axes), and once we are given an ordered basis $\mathcal{B} = \{v_1, \ldots, v_n\}$ of V, we can take any x in V and then write $x = \sum_j x_j v_j$, where the scalars x_j are the *co-ordinates*

9.3 The matrix of a linear transformation

of x relative to $\mathcal{B}$. These co-ordinates x_j are uniquely determined by the x and $\mathcal{B}$, so we should perhaps write this as $x_\mathcal{B} = (x_1, \ldots, x_n)$. To return to our example, if we take the 'natural' basis $\mathcal{E} = \{e_1, e_2\}$ of $\mathbb{R}^2$, where $e_1 = (1, 0)$ and $e_2 = (0, 1)$, and let $x = (3, 2)$, then $x = 3e_1 + 2e_2$ so that $x_\mathcal{E} = (3, 2)$. On the other hand, if we decide to use the basis $\mathcal{B} = \{u_1, u_2\}$, where $u_1 = (1, 0)$ and $u_2 = (1, 1)$, then $x = u_1 + 2u_2$ so that $x_\mathcal{B} = (1, 2)$. It should be clear that if we keep the point x fixed (as an 'abstract vector'), and vary the choice of basis, the the co-ordinates of x will also change; equally, if we insist on keeping the co-ordinates fixed, the vector will change as the basis changes.

We turn now to linear transformations, where the situation is similar. Given two vector spaces V and W, we wish to assign 'co-ordinates' to each linear map $\alpha : V \to W$, and we can do this once we have chosen an ordered basis of V and an ordered basis of W. Here, the 'co-ordinates' of α form a *matrix*, and if we keep α fixed and vary the bases in V and W then the matrix for α will change. In the same way, if we start with a (fixed) matrix, and then vary the bases, we will obtain different linear maps. This discussion makes it clear that it will be helpful to have a convenient notation for a basis of a vector space. Given a vector space V, we shall use $\mathcal{B}_V$ (with a suffix V) to denote a typical basis of V. When we want to consider a linear map $\alpha : V \to W$, where V and W have prescribed bases $\mathcal{B}_V$ and $\mathcal{B}_W$, respectively, we shall write

$$\alpha : \mathcal{B}_V \to \mathcal{B}_W. \tag{9.3.1}$$

We shall now show that the linear map (9.3.1) can be specified in terms of co-ordinates, and that the 'co-ordinates' for α form a matrix.

Let $\mathcal{B}_V = \{v_1, \ldots, v_n\}$ and $\mathcal{B}_W = \{w_1, \ldots, w_m\}$. Then there are scalars a_{ij} (uniquely determined by α, $\mathcal{B}_V$ and $\mathcal{B}_W$) such that

$$\alpha(v_j) = \sum_{i=1}^{m} a_{ij} w_i, \quad j = 1, \ldots, n, \tag{9.3.2}$$

Next, let $y = \alpha(x)$, where $x = \sum_j x_j v_j$, and $y = \sum_i y_i w_i$. Then, as α is linear,

$$\sum_{k=1}^{m} y_k w_k = \alpha \left(\sum_{j=1}^{n} x_j v_j \right)$$

$$= \sum_{j=1}^{n} x_j \alpha(v_j)$$

$$= \sum_{j=1}^{n} x_j \left(\sum_{i=1}^{m} a_{ij} w_i \right)$$

$$= \sum_{i=1}^{m} \left(\sum_{j=1}^{n} a_{ij} x_j \right) w_i.$$

As the w_i are linearly independent, this means that

$$y_i = \sum_{j=1}^n a_{ij} x_j = a_{i1} x_1 + \cdots + a_{in} x_n,$$

and this is precisely the effect of applying the matrix

$$A = \begin{pmatrix} a_{11} & \cdots & a_{1n} \\ \vdots & \ddots & \vdots \\ a_{m1} & \cdots & a_{mn} \end{pmatrix}.$$

to the vector $(x_1, \ldots, x_n)^t$ as given in (9.2.2). We can summarize this discussion as follows: *a linear map $\alpha : \mathcal{B}_V \to \mathcal{B}_W$ determines a matrix A as above, and that if $y = \alpha(x)$, then the action of this matrix is to map the co-ordinates of x relative to $\mathcal{B}_V$ to the co-ordinates of $\alpha(x)$ relative to $\mathcal{B}_W$.*

Definition 9.3.1 Given $\alpha : \mathcal{B}_V \to \mathcal{B}_W$, the matrix A constructed above is the *matrix representation of α relative to $\mathcal{B}_V$ and $\mathcal{B}_W$*. □

If we keep α fixed but change the bases, then the matrix representation of α will change, so that *a linear map has many different matrix representations* (one for each pair of bases). For any basis $\mathcal{B}_V$ of V, the matrix representation of the identity map $I : \mathcal{B}_V \to \mathcal{B}_V$ is the identity matrix; this follows from (9.3.2). However the matrix representation of I relative to a pair of *different* bases of V will *not* be the identity matrix.

Finally, suppose we are given a real $m \times n$ matrix A: this acts as a linear map from $\mathbb{R}^{n,t}$ to $\mathbb{R}^{m,t}$ and so has a matrix representation relative to every pair of bases of these spaces. Moreover, as for any other linear map, this matrix representation will change if we change the bases, so the matrix representation for A (as a linear map) will not always be the matrix A. On the other hand, the matrix representation for A relative the standard bases $(1, 0, \ldots, 0), \ldots$ of these spaces will be A itself. This is analogous to the distinction between the point $(3, 2)$ in $\mathbb{R}^2$, and its co-ordinate row $(3, 2)_{\mathcal{E}}$, and the distinction between the role of a matrix as a linear map, and as a matrix representation of a linear map, must be kept clearly in mind. It is time to clarify these ideas with some specific examples.

Example 9.3.2 The vector space V of solutions of the differential equation $\ddot{x} + 4x = 0$ has basis $\mathcal{B} = \{c, s\}$, where c and s are the functions given by $c(t) = \cos 2t$ and $s(t) = \sin 2t$. Now consider the linear map $\alpha : V \to V$ given by $\alpha(x)(t) = x(t) + \dot{x}(t) + \ddot{x}(t)$. As $\alpha(c) = -3c - 2s$ and $\alpha(s) = 2c - 3s$, we see that the matrix representation of $\alpha : \mathcal{B} \to \mathcal{B}$ is

$$\begin{pmatrix} -3 & 2 \\ -2 & -3 \end{pmatrix}.$$

9.3 The matrix of a linear transformation

The solution $x = Ac + Bs$ is mapped by α to $(-3A + 2B)c + (-2A - 3B)s$ and, as predicted, the matrix action is given by

$$\begin{pmatrix} A \\ B \end{pmatrix} \mapsto \begin{pmatrix} -3A + 2B \\ -2A - 3B \end{pmatrix} = \begin{pmatrix} -3 & 2 \\ -2 & -3 \end{pmatrix} \begin{pmatrix} A \\ B \end{pmatrix}.$$ □

Example 9.3.3 We recall from Example 9.2.4 that the orthogonal projection, say α, of $\mathbb{R}^3$ onto the plane Π given by $x_1 + x_2 + x_3 = 0$ has matrix representation

$$A = \frac{1}{3} \begin{pmatrix} 2 & -1 & -1 \\ -1 & 2 & -1 \\ -1 & -1 & 2 \end{pmatrix}$$

with respect to the standard basis $\{(1, 0, 0), (0, 1, 0), (0, 0, 1)\}$ in both domain and the range. Let us now consider the matrix representation of α in terms of a more sensible choice of basis. As α fixes every vector on Π, we choose a basis u and v of Π and then extend this to a basis of $\mathbb{R}^3$ by adding the normal to Π, namely $w = (1, 1, 1)$. Then $\alpha(u) = u$, $\alpha(v) = v$ and $\alpha(w) = 0$, so that $\alpha(\lambda u + \mu v + vw) = \lambda u + \mu v$, and the action from co-ordinates to co-ordinates is given by

$$\begin{pmatrix} \lambda \\ \mu \\ 0 \end{pmatrix} \mapsto \begin{pmatrix} \lambda \\ \mu \\ 0 \end{pmatrix} = \begin{pmatrix} 1 & 0 & 0 \\ 0 & 1 & 0 \\ 0 & 0 & 0 \end{pmatrix} \begin{pmatrix} \lambda \\ \mu \\ v \end{pmatrix}.$$

It follows that the matrix representation of α with respect to $\mathcal{B}$ is

$$\begin{pmatrix} 1 & 0 & 0 \\ 0 & 1 & 0 \\ 0 & 0 & 0 \end{pmatrix}.$$

Of course, the simplicity of this new representation derives from a choice of basis that is intimately related to the geometric action of α. □

Example 9.3.4 Let u be a unit vector in $\mathbb{R}^3$, and let α be the map of $\mathbb{R}^3$ into itself given by $\alpha(x) = u \times x$ (the vector product). Next, let v be any unit vector orthogonal to u, and let $w = u \times v$. Then $\mathcal{B} = \{u, v, w\}$ is a basis of $\mathbb{R}^3$, and $\alpha(u) = 0$, $\alpha(v) = w$, $\alpha(w) = -v$. Thus $\alpha(\lambda u + \mu v + vw) = -vv + \mu w$, and the action from co-ordinates to co-ordinates is given by

$$\begin{pmatrix} \lambda \\ \mu \\ v \end{pmatrix} \mapsto \begin{pmatrix} 0 \\ -v \\ \mu \end{pmatrix} = \begin{pmatrix} 0 & 0 & 0 \\ 0 & 0 & -1 \\ 0 & 1 & 0 \end{pmatrix} \begin{pmatrix} \lambda \\ \mu \\ v \end{pmatrix}.$$

This is simpler than the previous representation, and again this is because of a better choice of basis. □

The following result is an immediate consequence of the ideas of a matrix representation.

Theorem 9.3.5 *Suppose that $\alpha : \mathcal{B}_V \to \mathcal{B}_W$ and $\beta : \mathcal{B}_V \to \mathcal{B}_W$ are represented by matrices A and B, respectively. Then $\alpha + \beta$, and $\lambda \alpha$, are represented by $A + B$ and λA, respectively.*

We shall now consider the matrix representation of a composition of two linear maps, say

$$(U, \mathcal{B}_U) \xrightarrow{\beta} (V, \mathcal{B}_V) \xrightarrow{\alpha} (W, \mathcal{B}_W),$$

Not surprisingly, the matrix product has been defined in such a way that the matrix representation of a composition of two maps is the product of the two matrix representations.

Theorem 9.3.6 *Suppose that the linear maps $\alpha : \mathcal{B}_V \to \mathcal{B}_W$ and $\beta : \mathcal{B}_U \to \mathcal{B}_V$ have matrix representations A and B, respectively. Then $\alpha \beta : \mathcal{B}_U \to \mathcal{B}_W$ has matrix representation AB.*

Proof Let

$$\mathcal{B}_U = \{u_1, \ldots, u_\ell\}, \quad \mathcal{B}_V = \{v_1, \ldots, v_m\}, \quad \mathcal{B}_W = \{w_1, \ldots, w_n\}.$$

Also, for $j = i, \ldots, \ell$, and $k = 1, \ldots, m$, let

$$\beta(u_j) = \sum_{i=1}^{m} b_{ij} v_i, \quad \alpha(v_k) = \sum_{q=1}^{n} a_{qk} w_q, \quad \alpha \beta(u_j) = \sum_{i=1}^{n} c_{ij} w_i.$$

Then

$$\sum_{i=1}^{n} c_{ij} w_i = \alpha(\beta(u_j))$$

$$= \alpha \left(\sum_{k=1}^{m} b_{kj} v_k \right)$$

$$= \sum_{k=1}^{m} b_{kj} \alpha(v_k)$$

$$= \sum_{k=1}^{m} b_{kj} \left(\sum_{q=1}^{n} a_{qk} w_q \right)$$

$$= \sum_{q=1}^{n} \left(\sum_{k=1}^{m} a_{qk} b_{kj} \right) w_q,$$

so that $c_{ij} = a_{i1} b_{1j} + \cdots + a_{im} b_{mj}$. $\square$

Exercise 9.3

1. Show that

$$\begin{pmatrix} a & * & * \\ 0 & b & * \\ 0 & 0 & c \end{pmatrix} \begin{pmatrix} a' & * & * \\ 0 & b' & * \\ 0 & 0 & c' \end{pmatrix} = \begin{pmatrix} aa' & * & * \\ 0 & bb' & * \\ 0 & 0 & cc' \end{pmatrix},$$

where $*$ refers to some unspecified entry in the matrix.

2. Let

$$A = \begin{pmatrix} 1 & 2 & 1 \\ 0 & 0 & 2 \\ 0 & 0 & 0 \end{pmatrix}, \quad B = \begin{pmatrix} 1 & 3 & 2 \\ 0 & 0 & 2 \\ 0 & 0 & 0 \end{pmatrix}, \quad C = \begin{pmatrix} 0 & 1 & 1 \\ 0 & 0 & 2 \\ 0 & 0 & 0 \end{pmatrix}.$$

Let $\mathcal{B}$ be the basis $\{1, z, z^2\}$ of the space P_2 of complex polynomials of degree at most two. Show that the map $\alpha : \mathcal{B} \to \mathcal{B}$ defined by $\alpha(p)(z) = p'(z) + p(1)$ is a linear map with matrix representation A. Now let $\mathcal{B}'$ be the basis $\{1, 1+z, 1+z+z^2\}$. Show that the matrix of $\alpha : \mathcal{B}' \to \mathcal{B}'$ is B. Show also that the linear map $\beta(p)(z) = p'(z) + p''(z)$ has matrix C relative to the basis $\mathcal{B}'$.

3. The vector space $M^{2 \times 2}(\mathbb{R})$ has basis $\mathcal{B} = \{E_{11}, E_{12}, E_{21}, E_{22}\}$ given in (9.1.2). Let $\tau(A)$ be the transpose of the matrix A. Show that $\tau : \mathcal{B} \to \mathcal{B}$ is a linear map with matrix representation

$$\begin{pmatrix} 1 & 0 & 0 & 0 \\ 0 & 0 & 1 & 0 \\ 0 & 1 & 0 & 0 \\ 0 & 0 & 0 & 1 \end{pmatrix}.$$

4. Let $\mathcal{E} = \{(1, 0), (0, 1)\}$ of $\mathbb{C}^2$, and let

$$\alpha(z_1, z_2) = (3z_2, iz_1), \quad \beta(z_1, z_2) = (z_1 + 2iz_2, (1+i)z_1).$$

Show that α and β are linear maps of $\mathbb{C}^2$ to itself. Find the matrix representations A of α, B of β, and C of $\alpha\beta$ (in each case relative to $\mathcal{E}$), and verify (by matrix multiplication) that $C = AB$.

9.4 Inverse maps and matrices

In this section we discuss the relationship between inverse maps and inverse matrices. Recall that, for each n, there is the $n \times n$ identity matrix I_n; it is usual to omit the suffix n and use I regardless of the value of n.

Definition 9.4.1 An $n \times n$ matrix A is *invertible*, or *non-singular*, if there is an $n \times n$ matrix B such that $AB = I = BA$. If so, we say that B is the *inverse* of A, and we write $B = A^{-1}$. □

Such a matrix B need not exist, but if it does then it is unique, for suppose that we also have $AC = I = CA$. Then $B = BI = B(AC) = (BA)C = IC = C$.

The following theorem is the major result of this section; however, its proof will be delayed until the end of the section.

Theorem 9.4.2 *Let V and W be vector spaces of the same finite dimension, and suppose that the linear map $\alpha : \mathcal{B}_V \to \mathcal{B}_W$ has matrix representation A. Then the following are equivalent:*

(1) $\alpha^{-1} : W \to V$ *exists*;
(2) A^{-1} *exists*;
(3) $\det(A) \neq 0$.

The first step is to link inverse matrices and inverse transformations. Suppose that $\alpha : V \to W$ is a bijective linear map (an isomorphism) between finite-dimensional vector spaces. Then the inverse map α^{-1} exists and is linear. Moreover, $\dim(V) = \dim(W)$, so that any matrix representation A of α is a square matrix. We shall now show (as we would hope, and expect) that A^{-1} exists and is the matrix representation of α^{-1} with respect to the same bases.

Theorem 9.4.3 *Suppose that $\alpha : V \to W$ is an isomorphism, and let A be the matrix representation of $\alpha : \mathcal{B}_V \to \mathcal{B}_W$. Then A^{-1} exists and is the matrix representation of $\alpha^{-1} : \mathcal{B}_W \to \mathcal{B}_V$.*

Proof Let B be the matrix representation of $\alpha^{-1} : \mathcal{B}_W \to \mathcal{B}_V$. According to Theorem 9.3.6, the linear map $\alpha^{-1}\alpha : (V, \mathcal{B}_V) \to (V, \mathcal{B}_V)$ has matrix representation BA. However, $\alpha^{-1}\alpha$ is the identity map, and this is represented by the identity matrix, irrespective of the choice of $\mathcal{B}_V$. Thus $BA = I$, and similarly, $AB = I$. □

Definition 9.4.1 of an inverse matrix B of A requires that $AB = BA = I$. If we only know that $AB = I$, it is not obvious that $BA = I$.

Theorem 9.4.4 *Suppose that A and B are $n \times n$ matrices. If $AB = I$ then $BA = I$.*

Proof Let $\mathcal{E}$ be the standard basis of $\mathbb{F}^{n,t}$ (that is, $e_1^t, \ldots, e_n^t$), and define the linear map α by matrix multiplication; that is, $\alpha(x) = Ax$. The matrix representation of α with respect to $\mathcal{E}$ is A. Similarly, the matrix representation of the linear map β defined by $\beta(x) = Bx$ is B. As the matrix representation

9.4 Inverse maps and matrices

of $\alpha\beta$ is AB, and as $AB = I$, we see that $\alpha\beta(x) = x$ for all x. This implies that α^{-1} exists and is β. As $\beta(\alpha(x)) = \alpha^{-1}(\alpha(x)) = x$ for all x, and as this linear map has matrix representation BA, we see that $BA = I$. □

The proof of Theorem 9.4.2 depends on the following result.

Theorem 9.4.5 *For any $n \times n$ matrices A and B,*

$$\det(AB) = \det(A)\det(B).$$

In particular, $\det(AB) = \det(BA)$.

Before we prove this, we record the following corollary whose proof is just an application of Theorem 9.4.5 to the equation $AA^{-1} = I$.

Corollary 9.4.6 *If A^{-1} exists, then $\det(A) \neq 0$, and $\det(A^{-1}) = 1/\det(A)$.*

The proof of Theorem 9.4.5 Let $A = (a_{ij})$, $B = (b_{ij})$ and $AB = (c_{ij})$ so that

$$c_{ij} = a_{i1}b_{1j} + \cdots + a_{in}b_{nj}. \tag{9.4.1}$$

Now, by definition,

$$\det(AB) = \begin{vmatrix} c_{11} & \cdots & c_{1n} \\ \vdots & & \vdots \\ c_{n1} & \cdots & c_{nn} \end{vmatrix} = \sum_{\rho \in S_n} \epsilon(\rho) c_{1\rho(1)} \cdots c_{n\rho(n)}. \tag{9.4.2}$$

However, by (9.4.1), the term $c_{1\rho(1)} \cdots c_{n\rho(n)}$ is

$$\left(a_{11}b_{1\rho(1)} + \cdots + a_{1n}b_{n\rho(1)}\right) \cdots \left(a_{n1}b_{1\rho(n)} + \cdots + a_{nn}b_{n\rho(n)}\right),$$

so, after expanding these products, we can rewrite (9.4.2) in the form

$$\det(AB) = \sum_{t_1,\ldots,t_n=1}^{n} a_{1t_1} \cdots a_{nt_n} \beta(B, t_1, \ldots, t_n), \tag{9.4.3}$$

where the terms $\beta(B, t_1, \ldots, t_n)$ depend only on the stated parameters, and do not depend on the a_{ij}. We now have to evaluate the terms $\beta(B, t_1, \ldots, t_n)$, and we do this by regarding (9.4.3) as an identity in all of the variables a_{pq} and b_{rs}, and making a specific choice of the a_{pq}.

Choose any integers $s_1, \ldots, s_n$ from $\{1, \ldots, n\}$, and then put $a_{1s_1} = \cdots = a_{ns_n} = 1$, and all other a_{ij} equal to zero. Let c'_{ij} be the value of c_{ij} corresponding to this choice of a_{ij}; thus, from (9.4.1) $c'_{ij} = b_{s_i j}$. As it is clear that

$$a_{1t_1} \cdots a_{nt_n} = \begin{cases} 1 & \text{if } t_1 = s_1, \ldots, t_n = s_n, \\ 0 & \text{otherwise,} \end{cases}$$

we see immediately from (9.4.2) and (9.4.3) that

$$\det\big((c'_{ij})\big) = \beta(B, s_1, \ldots, s_n).$$

Thus

$$\beta(B, s_1, \ldots, s_n) = \begin{vmatrix} b_{s_1 1} & \cdots & b_{s_1 n} \\ \vdots & & \vdots \\ b_{s_n 1} & \cdots & b_{s_n n} \end{vmatrix}.$$

Now by Theorem 8.6.6, this determinant is zero if $s_p = s_q$ for some $p \neq q$; in other words, $\beta(B, s_1, \ldots, s_n) = 0$ unless $s_1, \ldots, s_n$ is a permutation of $1, \ldots, n$. If we use this in conjunction with (9.4.3), we now see that

$$\det(AB) = \sum_{\rho \in S_n} a_{1\rho(1)} \cdots a_{n\rho(n)} \begin{vmatrix} b_{\rho(1)1} & \cdots & b_{\rho(1)n} \\ \vdots & & \vdots \\ b_{\rho(n)1} & \cdots & b_{\rho(n)n} \end{vmatrix}. \quad (9.4.4)$$

Finally, if ρ is a product of q transpositions, so that $\epsilon(\rho) = q$, then, by applying the operation of interchanging two rows of this determinant q times in an appropriate manner, we find that

$$\begin{vmatrix} b_{\rho(1)1} & \cdots & b_{\rho(1)n} \\ \vdots & & \vdots \\ b_{\rho(n)1} & \cdots & b_{\rho(n)n} \end{vmatrix} = \epsilon(\rho) \begin{vmatrix} b_{11} & \cdots & b_{1n} \\ \vdots & & \vdots \\ b_{n1} & \cdots & b_{nn} \end{vmatrix} = \epsilon(\rho)\det(B).$$

Thus

$$\det(AB) = \sum_{\rho \in S_n} a_{1\rho(1)} \cdots a_{n\rho(n)} \begin{vmatrix} b_{\rho(1)1} & \cdots & b_{\rho(1)n} \\ \vdots & & \vdots \\ b_{\rho(n)1} & \cdots & b_{\rho(n)n} \end{vmatrix}$$

$$= \sum_{\rho \in S_n} a_{1\rho(1)} \cdots a_{n\rho(n)} \epsilon(\rho)\det(B)$$

$$= \det(A)\det B. \qquad \square$$

It only remains to give the proof of Theorem 9.4.2.

The proof of Theorem 9.4.2 Theorem 9.4.3 shows that (1) implies (2), and Corollary 9.4.6 shows that (2) implies (3). Suppose now that (3) holds; then, from Theorem 7.9.5, it is sufficient to show that α is surjective. Now the matrix representation A of $\alpha : \mathcal{B}_V \to \mathcal{B}_W$ is such that

$$\alpha\Big(\sum_j \lambda_j v_j\Big) = \sum_i \mu_i w_i$$

9.5 Change of bases

if and only if

$$\begin{pmatrix} a_{11} & \cdots & a_{1n} \\ \vdots & \ddots & \vdots \\ a_{n1} & \cdots & a_{nn} \end{pmatrix} \begin{pmatrix} \lambda_1 \\ \vdots \\ \lambda_n \end{pmatrix} = \begin{pmatrix} \mu_1 \\ \vdots \\ \mu_n \end{pmatrix}, \tag{9.4.5}$$

and it follows from this that we have to show that given any μ_i there are λ_j such that (9.4.5) holds.

Now consider A as a linear map of $(\lambda_1, \ldots, \lambda_n)^t$ onto $(\mu_1, \ldots, \mu_n)^t$ given by (9.4.5). As the j-th column of A is $A(e_j^t)$, and as these columns are linearly independent, we see that the map given by (9.4.5) maps $\mathbb{R}^{n,t}$ onto itself. Thus it is the case that given any μ_i there are λ_j such that (9.4.5) holds, and we deduce that α^{-1} exists. □

Exercise 9.4

1. Let A be a 2×2 matrix with integer entries. Show that A^{-1} exists and has integer entries if and only if $\det(A) = \pm 1$.
2. Show that the set of $n \times n$ matrices A with entries in $\mathbb{F}$, and with $\det(A) \neq 0$, is a group with respect to matrix multiplication.
3. Let

$$A = \begin{pmatrix} a & b \\ c & d \end{pmatrix}.$$

   (i) Suppose that $ad - bc \neq 0$. Find A^{-1}.
   (ii) Suppose that $ad - bc = 0$. Find all column vectors x such that

$$\begin{pmatrix} a & b \\ c & d \end{pmatrix} \begin{pmatrix} x_1 \\ x_2 \end{pmatrix} = \begin{pmatrix} 0 \\ 0 \end{pmatrix}.$$

4. In each of the following examples, in which $\mathbb{R}^2$ is given the basis $e_1 = (1, 0)$ and $e_2 = (0, 1)$, determine whether each of the conditions in Theorem 9.4.2 holds:
   (i) $\alpha(xe_1 + ye_2) = (2x + y)e_1 + (-x + 3y)e_2$;
   (ii) $\alpha(xe_1 + ye_2) = (x - y)e_1 + (-2x + 2y)e_2$;
   (iii) $\alpha(xe_1 + ye_2) = (6x + 2y)e_1 + (12x + 4y)e_2$.

9.5 Change of bases

Suppose that α is a linear map of a vector space V into itself, and let $\mathcal{B}$ and $\mathcal{B}'$ be bases of V. Then $\alpha : \mathcal{B} \to \mathcal{B}$ and $\alpha : \mathcal{B}' \to \mathcal{B}'$ are represented by matrices, say A and A', respectively, and we want to find the relationship between A and A'. To find this relationship, we consider the following composition of maps, where I is the identity map of V onto itself:

$$\mathcal{B}' \xrightarrow{I} \mathcal{B} \xrightarrow{\alpha} \mathcal{B} \xrightarrow{I} \mathcal{B}'$$

Let P be the matrix representation of the map $I : \mathcal{B}' \to \mathcal{B}$. Then P^{-1} is the matrix representation of its inverse, namely $I : \mathcal{B} \to \mathcal{B}'$, and from Theorem 9.3.6, the composite map $I\alpha I : \mathcal{B}' \to \mathcal{B}'$ has matrix representation $P^{-1}AP$. As this composite map is just α, we have proved the following result.

Theorem 9.5.1 *Let $\mathcal{B}$ and $\mathcal{B}'$ be bases of a vector space V. If the linear map $\alpha : \mathcal{B} \to \mathcal{B}$ is represented by a matrix A, and the linear map $\alpha : \mathcal{B}' \to \mathcal{B}'$ is represented by A', then $A' = P^{-1}AP$, where P is the matrix representation of the identity map $I : \mathcal{B}' \to \mathcal{B}$.*

We illustrate this in the next example.

Example 9.5.2 Let $V = \mathbb{R}^2$, $\mathcal{B} = \{e_1, e_2\}$ and $\mathcal{B}' = \{v_1, v_2\}$, where $v_1 = (1, 1)$ and $v_2 = (1, 0)$. Now let α be the linear map defined by

$$\alpha(xe_1 + ye_2) = (x + y)e_1 + (x + 2y)e_2.$$

As $\alpha(e_1) = e_1 + e_2$ and $\alpha(e_2) = e_1 + 2e_2$, the matrix representation of $\alpha : \mathcal{B} \to \mathcal{B}$ is

$$A = \begin{pmatrix} 1 & 1 \\ 1 & 2 \end{pmatrix},$$

and as $\alpha(v_1) = 3v_1 - v_2$ and $\alpha(v_2) = v_1$ the matrix representation of $\alpha : \mathcal{B}' \to \mathcal{B}'$ is

$$A' = \begin{pmatrix} 3 & 1 \\ -1 & 0 \end{pmatrix}.$$

Finally, we must consider the matrix representation P of identity map $I : \mathcal{B}' \to \mathcal{B}$. Here, we have $I(v_1) = v_1 = e_1 + e_2$ and $I(v_2) = v_2 = e_1$, so that

$$P = \begin{pmatrix} 1 & 1 \\ 1 & 0 \end{pmatrix}.$$

Theorem 9.5.1 tells us that $A' = P^{-1}AP$ and we leave the reader to confirm that this is indeed so (as P^{-1} exists, it suffices to show that $PA' = AP$). □

Theorem 9.5.1 illuminates an important matter that we have already mentioned. Suppose that we want to study a linear map $\alpha : V \to V$. It may be convenient to represent α by a matrix A, but before we can do this we have to choose a basis of V. Some bases will lead to a simple form for A, while others

9.5 Change of bases

will lead to a more complicated form. Obviously we choose whichever basis of V suits our purpose best, but in any case, the degree of choice available here is given by Theorem 9.5.1.

There is another important issue here. Any result about the linear map α that is proved by resorting to a matrix representation of α *may possibly depend on the choice of the basis* (and the resulting matrix). In general, we would then want to show that the result does not, in fact, depend on the basis. In fact, it is far better to avoid this step altogether and to prove results about linear maps directly without resorting to their matrix representations. As an illustration of these ideas, let us consider the determinant of a linear map $\alpha : V \to V$. Let A and A' be two different matrix representations of α so that, by Theorem 9.5.1, there is a matrix P such that $A' = P^{-1}AP$. As $\det(XY) = \det(X)\det(Y)$ for any matrices X and Y, we see that

$$\begin{aligned}\det(A') &= \det(P^{-1}AP) \\ &= \det(P^{-1})\det(A)\det(P) \\ &= \det(P^{-1})\det(P)\det(A) \\ &= \det(I)\det(A) \\ &= \det(A).\end{aligned}$$

This gives the next result.

Theorem 9.5.3 *Suppose that $\alpha : V \to V$ is a linear map. Then there is a scalar $\det(\alpha)$ such that if A is any matrix representation of α, then $\det(A) = \det(\alpha)$.*

The scalar $\det(\alpha)$ is called the *determinant of* α, and the significance of Theorem 9.5.3 is that although $\det(\alpha)$ can be computed from any matrix representation A of α, it is *independent of the choice of A*.

Exercise 9.5

1. Let $\mathcal{B} = \{v_1, v_2, v_3\}$ be a basis of a vector space V, and let $\mathcal{B}' = \{v_1, v_1 + v_2, v_1 + v_2, v_3\}$. Show that $\mathcal{B}'$ is a basis for V. Now define a linear map $\alpha : V \to V$ by $\alpha(v_1) = v_2 + v_3$, $\alpha(v_2) = v_3 + v_1$ and $\alpha(v_3) = v_1 + v_2$. Find A, A' and P as in Theorem 9.5.1 and verify that $A' = P^{-1}AP$. What is $\det(\alpha)$ in this case?
2. Let a be any non-zero vector in $\mathbb{R}^3$ and let $\alpha : \mathbb{R}^3 \to \mathbb{R}^3$ be the linear map defined by $\alpha(x) = a \times x$ (the vector product). Without doing any calculations, explain why $\det(\alpha) = 0$.
3. Let a, b and c be linearly independent vectors in $\mathbb{R}^3$, and let $\alpha : \mathbb{R}^3 \to \mathbb{R}^3$ be defined by $\alpha(x) = (x \cdot a, x \cdot b, x \cdot c)$, where $x \cdot y$ is the scalar product in $\mathbb{R}^3$. What is $\det(\alpha)$?

4. In Example 9.3.3 we obtained two matrix representations, of the orthogonal projection α of $\mathbb{R}^3$ onto a plane Π. Verify all of the details of Theorem 9.5.1 in this example.

9.6 The resultant of two polynomials

This section is concerned with complex polynomials, and it extends the ideas discussed in Example 7.6.5. We showed there that if, say,

$$f(z) = a_0 + a_1 z + \cdots + a_n z^n,$$
$$g(z) = b_0 + b_1 z + \cdots + b_m z^m,$$

where $a_n b_m \neq 0$ and $n, m \geq 1$, then the $m+n-1$ polynomials

$$f(z), zf(z), \ldots, z^{m-1}f(z), g(z), zg(z), \ldots z^{n-1}g(z), \quad (9.6.1)$$

form a basis of the space of polynomials of degree at most $m+n-1$ if and only if f and g have no common zero. Now for some matrix M,

$$M \begin{pmatrix} 1 \\ z \\ \vdots \\ z^{m-1} \\ z^m \\ z^{m+1} \\ \vdots \\ \vdots \\ z^{m+n-1} \end{pmatrix} = \begin{pmatrix} f(z) \\ zf(z) \\ \vdots \\ z^{m-1}f(z) \\ g(z) \\ zg(z) \\ \vdots \\ z^{n-1}g(z) \end{pmatrix}, \quad (9.6.2)$$

and, by inspection, we see that

$$M = \begin{pmatrix} a_0 & a_1 & \cdot & \cdot & \cdot & a_n & 0 & 0 & \cdot & 0 \\ 0 & a_0 & a_1 & \cdot & \cdot & \cdot & a_n & 0 & \cdot & 0 \\ \vdots & \vdots & & & & & & & & \vdots \\ 0 & 0 & \cdots & 0 & a_0 & a_1 & \cdot & \cdot & \cdot & a_n \\ b_0 & b_1 & \cdot & \cdot & \cdot & b_m & 0 & 0 & \cdot & 0 \\ 0 & b_0 & b_1 & \cdot & \cdot & \cdot & b_m & 0 & \cdot & 0 \\ \vdots & \vdots & & & & & & & & \vdots \\ 0 & 0 & \cdots & 0 & b_0 & b_1 & \cdot & \cdot & \cdot & b_m \end{pmatrix}.$$

It is helpful to remember that diagonal of M comprises $\deg(g)$ terms a_0 followed by $\deg(f)$ terms b_m.

9.6 The resultant of two polynomials

Definition 9.6.1 The *resultant* $R(f, g)$ of the polynomials f and g is $\det M$. The *discriminant* $\Delta(f)$ of f is the resultant of $f(z)$ and its derivative $f'(z)$. □

Theorem 9.6.2 *The following are equivalent*:

(1) $\det(M) = 0$;
(2) *the polynomials in* (9.6.1) *are linearly dependent*;
(3) *the polynomials f and g have a common zero*.

Let us explore the consequences of this result before we give a proof. First, it shows that any two polynomials f and g have a common zero if and only if $R(f, g) = 0$, and the determinant $R(f, g)$ can be evaluated in any specific case even though we may be unable to find the zeros of f or g. Next, we know that f and f' have a common zero if and only if f has a repeated zero; thus we conclude that an arbitrary polynomial f *has distinct zeros if and only if* $\Delta(f) \neq 0$. For example, if $f(z) = z^2 + az + b$ then $f'(x) = 2z + a$ so that f has distinct zeros if and only if

$$\begin{vmatrix} b & a & 1 \\ a & 2 & 0 \\ 0 & a & 2 \end{vmatrix} \neq 0.$$

As this determinant is $4b - a^2$, this agrees with the familiar result obtained by an elementary argument.

Example 9.6.3 Given a cubic polynomial f, we can always find some z_0 such that the polynomial $f(z + z_0)$ has no term in z^2 (and $f(z + z_0)$ has distinct zeros if and only if f does), so there is no harm in assuming that $f(z) = z^3 + az + b$. Now a necessary and sufficient condition for $z^3 + az + b$ to have distinct roots is that

$$\Delta(f) = \begin{vmatrix} b & a & 0 & 1 & 0 \\ 0 & b & a & 0 & 1 \\ a & 0 & 3 & 0 & 0 \\ 0 & a & 0 & 3 & 0 \\ 0 & 0 & a & 0 & 3 \end{vmatrix} \neq 0,$$

and this is the condition $4a^3 + 27b^2 = 0$ (see Exercise 9.6.2). □

The proof of Theorem 9.6.2 First, Theorem 7.6.6 shows that (2) and (3) are equivalent. We shall show that (1) and (2) are equivalent by proving that $\det(M) \neq 0$ if and only if the polynomials in (9.6.1) are linearly independent. Throughout, P will be the vector space of complex polynomials of degree at most $m + n - 1$.

Suppose that $\det(M) \neq 0$; then, by Theorem 9.4.2, M^{-1} exists. If we multiply each side of (9.6.2) on the left by M^{-1} we see that each of the polynomials $1, z, z^2, \ldots, z^{m+n-1}$ is a linear combination of the polynomials in (9.6.1). This implies that the polynomials in (9.6.1) span P, and as there are exactly $m+n$ of them, they form a basis of P and so are linearly independent.

Now suppose that the polynomials in (9.6.1) are linearly independent. As the number of polynomials here is $\dim(P)$, they span P and so there is a matrix N such that

$$\begin{pmatrix} 1 \\ z \\ \vdots \\ z^{m-1} \\ z^m \\ z^{m+1} \\ \vdots \\ z^{m+n-1} \end{pmatrix} = N \begin{pmatrix} f(z) \\ zf(z) \\ \vdots \\ z^{m-1}f(z) \\ g(z) \\ zg(z) \\ \vdots \\ z^{n-1}g(z) \end{pmatrix}. \qquad (9.6.3)$$

With (9.6.2) this shows that

$$\begin{pmatrix} 1 \\ z \\ \vdots \\ z^{m-1} \\ z^m \\ z^{m+1} \\ \vdots \\ z^{m+n-1} \end{pmatrix} = NM \begin{pmatrix} 1 \\ z \\ \vdots \\ z^{m-1} \\ z^m \\ z^{m+1} \\ \vdots \\ z^{m+n-1} \end{pmatrix}.$$

As the polynomials z^j form a basis of P, this implies that $NM = I$, and hence that $\det(M) \neq 0$. $\square$

We end with two applications of Theorem 9.6.2.

Theorem 9.6.4 *Let f and g be nonconstant polynomials, and let h be the greatest common divisor of f and g. Then there are polynomials a and b, such that $a(z)f(z) + b(z)g(z) = h(z)$.*

Proof Suppose first that f and g have no common zero. Then the only polynomials that divide f and g are the constant polynomials, so we may take h (which is only determined up to a scalar multiple) to be the constant polynomial with value 1. By Theorem 9.6.2, the polynomials in (9.6.1) span P, and so some linear combination, say $a(z)f(z) + b(z)g(z)$, of them is the polynomial $h(z)$. For the general case, let $f_1 = f/h$ and $g_1 = g/h$, and apply the above result to f_1 and g_1. □

Theorem 9.6.5 *For each positive integer n there exists a non-constant polynomial $\Phi_n(w_1, w_2, \ldots, w_{n+1})$ in the $n+1$ complex variables w_j such that the polynomial $a_0 + a_1 z + \cdots + a_n z^n$ of degree n has a repeated zero if and only if $\Phi_n(a_0, \ldots, a_n) = 0$.*

Proof We take $\Phi(a_0, \ldots, a_n) = R(f, f')$. □

Exercise 9.6

1. Let $f(z) = a_0 + a_1 z + \cdots + a_n z^n$. Find z_0 such that the polynomial $f(z + z_0)$ has no term in z^{n-1}.
2. Suppose that $f(z) = z^3 + az + b$ has a repeated root z_1, so that $f(z_1) = f'(z_1) = 0$. Find z_1 from the equation $f'(z_1) = 0$ and then use $f(z_1) = 0$ to give a simple proof of the result in Example 9.6.3.

9.7 The number of surjections

In Chapter 1 (Theorem 1.5.9) we proved that if $m \geq n$, then there are

$$S(m, n) = \sum_{k=1}^{n} (-1)^{n-k} \binom{n}{k} k^m \tag{9.7.1}$$

surjections from a set with m elements to a set with n elements. We did this by showing that, in matrix notation (which was not available at the time),

$$\begin{pmatrix} \binom{1}{1} & 0 & 0 & \cdots & 0 \\ \binom{2}{1} & \binom{2}{2} & 0 & \cdots & 0 \\ \binom{3}{1} & \binom{3}{2} & \binom{3}{3} & \cdots & 0 \\ \vdots & \vdots & \vdots & \ddots & \vdots \\ \binom{n}{1} & \binom{n}{2} & \binom{n}{3} & \cdots & \binom{n}{n} \end{pmatrix} \begin{pmatrix} S(m, 1) \\ S(m, 2) \\ S(m, 3) \\ \vdots \\ S(m, n) \end{pmatrix} = \begin{pmatrix} 1^m \\ 2^m \\ 3^m \\ \vdots \\ n^m \end{pmatrix}, \tag{9.7.2}$$

In order to find $S(m, n)$ we have to find the inverse to the matrix on the left of (9.7.2), and this was, in effect, we did in Lemma 1.5.10. As an example, we have

$$\begin{pmatrix} 1 & 0 & 0 & 0 \\ 2 & 1 & 0 & 0 \\ 3 & 3 & 1 & 0 \\ 4 & 6 & 4 & 1 \end{pmatrix}^{-1} = \begin{pmatrix} 1 & 0 & 0 & 0 \\ -2 & 1 & 0 & 0 \\ 3 & -3 & 1 & 0 \\ -4 & 6 & -4 & 1 \end{pmatrix},$$

and this shows the general pattern of the inverse matrix.

Lemma 9.7.1 *Let $A = (a_{ij})$ be the square matrix in (9.7.2). Then $A^{-1} = (b_{ij})$, where $b_{ij} = (-1)^{i+j} a_{ij}$.*

With this result available, we multiply both sides of (9.7.2) on the left by A^{-1} and this yields (9.7.1). The proof of Lemma 9.7.1 is simply an exercise in matrix multiplication.

The proof of Lemma 9.7.1 Let $B = (b_{ij})$, where b_{ij} is defined in Lemma 9.7.1, and let $AB = (c_{ij})$. Then

$$c_{pq} = \sum_{k=1}^{n} a_{pk} b_{kq}.$$

If $p < q$ then, for each k, either $k > p$ (and $a_{pk} = 0$) or $k < q$ (so that $b_{kq} = 0$). Thus $c_{pq} = 0$ unless $p \geq q$. If $p \geq q$ then, with $r = p - q$ and $k = q + t$,

$$c_{pq} = \sum_{k=q}^{p} \binom{p}{k}\binom{k}{q}(-1)^{k+q}$$

$$= \binom{p}{q} \sum_{t=0}^{r} \binom{r}{t}(-1)^{t}$$

$$= \binom{p}{q}[1 + (-1)]^{r}$$

$$= \begin{cases} 1 & \text{if } p = q \text{ (and } r = 0\text{);} \\ 0 & \text{if } p > q \text{ (and } r > 0\text{).} \end{cases} \qquad \square$$

Exercise 9.7

1. Show that

$$\sum_{k=1}^{n} (-1)^{n-k} \binom{n}{k} k^n = n!.$$

10
Eigenvectors

10.1 Eigenvalues and eigenvectors

One of the key ideas that is used to analyse the behaviour of a linear map $\alpha : V \to V$ is that of an invariant subspace.

Definition 10.1.1 Suppose that $\alpha : V \to V$ is linear. Then a subspace U of V is said to be *invariant under* α, or an *α-invariant subspace*, if $\alpha(U) \subset U$. □

The simplest form of a linear map $\alpha : V \to V$ is multiplication by a fixed scalar; that is, for some scalar μ, $\alpha(v) = \mu v$. These maps are too simple to be of any interest, but it is extremely profitable to study subspaces on which a given linear map α has this particular form.

Definition 10.1.2 Suppose that $\alpha : V \to V$ is a linear map. Then, for any scalar λ, E_λ is the set of vectors v such that $\alpha(v) = \lambda v$. Clearly $0 \in E_\lambda$ for every λ. We say that λ is an *eigenvalue* of α if and only if E_λ contains a non-zero vector; any non-zero vector in E_λ is an *eigenvector* of α associated with λ. □

It is clear that E_λ is a subspace of V, for it is the kernel of the linear map $\alpha - \lambda I$, where I is the identity map (alternatively, it is clear that any linear combination of vectors in E_λ is also in E_λ). In fact, E_λ is an α-invariant subspace for suppose that $v \in E_\lambda$, and let $w = \alpha(v)$. Then $w = \lambda v$, so that $\alpha(w) = \lambda \alpha(v) = \lambda w$, and hence $w \in E_\lambda$.

If λ is not an eigenvalue of α, then $E_\lambda = \{0\}$. If λ is an eigenvalue of α, then E_λ consists of the zero vector together with all of the (non-zero) eigenvectors associated with λ. If v is an eigenvector of α then, as $v \neq 0$, we can find the *unique* eigenvalue λ associated with v from the equation $\alpha(v) = \lambda v$. Notice that if v is an eigenvector of α, then the set of scalar multiples of v is the *line* L_v (a one-dimensional subspace of V) consisting of all scalar multiples of v, and $\alpha(x) = \lambda x$ for every x in L_v. Finally, we remark that although eigenvectors are non-zero, eigenvalues can be zero; indeed, 0 is an eigenvalue of α if and only if $\ker(\alpha) \neq \{0\}$, and this is so if and only if α is not injective. The next result summarizes the discussion so far.

Theorem 10.1.3 *Suppose that $\alpha : V \to V$ is linear, and $I : V \to V$ is the identity map. Then, for any scalar λ, the following are equivalent:*

(1) λ *is an eigenvalue of α;*
(2) $\alpha - \lambda I$ *is not injective;*
(3) $\ker(\alpha - \lambda I)$ *has positive dimension.*

We now give three examples to show how eigenvalues arise naturally in geometry; in all of these examples, V is $\mathbb{R}^2$ or $\mathbb{R}^3$, and the scalars are real numbers.

Example 10.1.4 Let $\alpha : \mathbb{R}^3 \to \mathbb{R}^3$ be the reflection in a plane Π that contains the origin. As α preserves the length of a vector, any eigenvalue of α must be ± 1. It is clear from the geometry that $E_1 = \Pi$, and that E_{-1} is the line through the origin that is normal to Π. □

Example 10.1.5 Let $V = \mathbb{R}^2$, and let α be the rotation of angle θ about the vertical line (in $\mathbb{R}^3$) through the origin. If $\theta \neq 0, \pi$ then α has no eigenvalues. If $\theta = 0$ then $\alpha = I$, the only eigenvalue of α is 1 and $E_1 = \mathbb{R}^2$. If $\theta = \pi$, then $\alpha = -I$, the only eigenvalue of α is -1 and $E_{-1} = \mathbb{R}^2$. If we view α as a map of $\mathbb{R}^3$ onto itself, and if $\theta \neq 0, \pi$, then 1 is the only eigenvalue of α, and E_1 is the vertical axis. Notice that this example shows that *the eigenvectors of α need not span V*. □

Example 10.1.6 $\alpha : \mathbb{R}^2 \to \mathbb{R}^2$ be the linear map $(x, y) \mapsto (x + y, y)$; this is a *shear*. Now α acts like a non-zero horizontal translation of each horizontal line into itself, except on the x-axis where α acts as the identity map (or the zero translation). Thus, from the geometry, we see that the only eigenvalue is 1, and that E_1 is the real axis. It is easy to verify analytically that this is so for we simply have to find those real λ such that $(x + y, y) = \lambda(x, y)$ for some non-zero (x, y). The only solutions are $y = 0$ and $\lambda = 1$. □

We turn now to the question of the *existence* of eigenvalues and eigenvectors. We have seen (in the examples above) that not every linear map has eigenvalues, and that even if eigenvalues exist, there may not be enough eigenvectors to span V. However, it is true that every linear map of a *complex* vector space into itself has an eigenvalue (and corresponding eigenvectors). This is yet another consequence of the Fundamental Theorem of Algebra, and the proof of the existence of eigenvalues and eigenvectors that we give here is elementary in the sense that it does not use determinants (eigenvalues are often defined in terms of a certain determinant, and we shall discuss this approach later).

10.1 Eigenvalues and eigenvectors

Theorem 10.1.7 *Let V be a finite-dimensional complex vector space. Then each linear map $\alpha : V \to V$ has an eigenvalue and an eigenvector.*

Proof Let $\dim(V) = n$. Take any non-zero vector v in V, and consider the $n+1$ vectors $v, \alpha(v), \ldots, \alpha^n(v)$. As these vectors must be linearly dependent, there are scalars $a_0, \ldots, a_n$, not all zero, such that

$$a_0 v + a_1 \alpha(v) + \cdots + a_n \alpha^n(v) = 0. \tag{10.1.1}$$

Now we cannot have $a_1 = \cdots = a_n = 0$ for then $a_0 v = 0$ and hence (as $v \neq 0$) $a_0 = 0$ contrary to our assumption. It follows that there is some j with $j \geq 1$ and $a_j \neq 0$, so that if k is the largest such j, then we can create the non-constant polynomial $P(t) = a_0 + a_1 t + \cdots + a_k t^k$, where $1 \leq k \leq n$ and $a_k \neq 0$. The Fundamental Theorem of Algebra implies that there are complex numbers λ_j such that

$$P(t) = a_k(t - \lambda_1) \cdots (t - \lambda_k), \tag{10.1.2}$$

and (10.1.1) now implies that

$$\begin{aligned} a_k(\alpha - \lambda_1 I) \cdots (\alpha - \lambda_k I)(v) &= a_0 v + a_1 \alpha(v) + \cdots + a_k \alpha^k(v) \\ &= a_0 v + a_1 \alpha(v) + \cdots + a_n \alpha^n(v) \\ &= 0. \end{aligned}$$

This shows that not every map $\alpha - \lambda_j I$ is injective, for if they were then their composition would also be injective and then v would be 0. Thus for some j, $\alpha - \lambda_j I$ is not injective and then λ_j is an eigenvalue. By definition, every eigenvalue has a corresponding eigenvector. □

Another way to interpret Theorem 10.1.7 is that, in these circumstances, V has an α-invariant subspace of dimension one. There is an analogous result for real vector spaces, although the result is about invariant subspaces rather than eigenvectors. However, we shall see shortly that if the dimension of V is odd, then α has an eigenvector.

Theorem 10.1.8 *Let V be a finite-dimensional real vector space. Given a linear map $\alpha : V \to V$, then V has an α-invariant subspace of dimension one or two.*

Proof We repeat the proof of Theorem 10.1.7 up to (10.1.2) which must now be replaced by

$$P(t) = a_k(t - \lambda_1) \cdots (t - \lambda_\ell) q_1(t) \cdots q_m(t),$$

where the λ_j are the real eigenvalues, and the $q_j(t)$ are real quadratic polynomials that do not factorize into a product of real linear factors. As before, we now see that this and (10.1.1) implies that

$$c(\alpha - \lambda_1 I) \cdots (\alpha - \lambda_k I) q_1(\alpha) \cdots q_m(\alpha)$$

maps v to 0, so that one of the factors is not injective. If one of the linear factors is not injective, then α has a real eigenvalue and a real eigenvector, and the line through this eigenvector is a one-dimensional invariant subspace. If one of the quadratic factors, say $\alpha^2 + a\alpha + bI$, is not injective, there is a vector w such that $\alpha^2(w) = -a\alpha(w) - bw$. In this case the subspace U spanned by w and $\alpha(w)$ is invariant. Indeed, the general vector in U is, say $x = sw + t\alpha(w)$, and $\alpha(x) = s\alpha(w) + t\alpha^2(w)$ which is again in U. □

The next result gives important information about eigenvalues and eigenvectors, regardless of whether the scalars are real or complex.

Theorem 10.1.9 *Let V be an n-dimensional vector space over $\mathbb{F}$, and suppose that the linear map $\alpha : V \to V$ has distinct eigenvalues $\lambda_1, \ldots, \lambda_r$ and corresponding eigenvectors $v_1, \ldots, v_r$. Then $v_1, \ldots, v_r$ are linearly independent. In particular, α has at most n eigenvalues.*

Proof First, suppose that $a_1 v_1 + a_2 v_2 = 0$ for some scalars a_1 and a_2. As v_1 is in the kernel of $\alpha - \lambda_1 I$, we see that

$$0 = (\alpha - \lambda_1 I)(a_1 v_1 + a_2 v_2) = a_2(\lambda_2 - \lambda_1) v_2.$$

As $(\lambda_2 - \lambda_1) v_2 \neq 0$ we see that $a_2 = 0$ and, similarly, $a_1 = 0$; thus v_1, v_2 are linearly independent. Clearly, the same argument holds for any pair v_i, v_j. Now suppose that $a_1 v_1 + a_2 v_2 + a_3 v_3 = 0$. By applying $\alpha - \lambda_1 I$ to both sides of this equation we see that

$$a_2(\lambda_2 - \lambda_1) v_2 + a_3(\lambda_3 - \lambda_1) v_3 = 0.$$

As v_2, v_3 are linearly independent, and the λ_j are distinct, we see that $a_2 = a_3 = 0$, and hence that $a_1 = 0$. This shows that v_1, v_2, v_3 are linearly independent and, more generally, that v_i, v_j, v_k are linearly independent (with i, j, k distinct). The argument continues in this way (formally, by induction) to show that $v_1, \ldots, v_r$ are linearly independent. The last statement in Theorem 10.1.9 is clear for as $\dim(V) = n$, we can have at most n linearly independent vectors in V. □

We have seen that the eigenvectors of a linear map $\alpha : V \to V$ need not span V. However, if α has n *distinct* eigenvalues, where $\dim(V) = n$, then, by Theorem 10.1.9, it has n linearly independent eigenvectors, and these must be a basis of V. Thus we have the following result.

Corollary 10.1.10 *Suppose that* $\dim(V) = n$, *and that* $\alpha : V \to V$ *has n distinct eigenvalues. Then the there exists a basis of V consisting entirely of eigenvectors of α.*

When the eigenvectors of $\alpha : V \to V$ span V (equivalently, when there is a basis of V consisting of eigenvectors) one can analyze the geometric action of α in terms of the eigenvectors, and we shall do this in Section 10.3. When the eigenvectors fail to span V, we need the following definition and theorem (which we shall not prove).

Definition 10.1.11 *Suppose that $\alpha : V \to V$ is linear, and $\dim(V) = n$. A vector v is a* generalized eigenvector *of α if there is some eigenvalue λ of α such that $(\alpha - \lambda I)^n(v) = 0$.*

Theorem 10.1.12 *Suppose that V is a finite-dimensional vector space over $\mathbb{C}$, and that $\alpha : V \to V$ is linear. Then the generalized eigenvectors of α span V.*

Exercise 10.1

1. Let $\alpha : \mathbb{C}^2 \to \mathbb{C}^2$ be defined by $\alpha(z, w) = (z + w, 4z + w)$. Find all eigenvalues and eigenvectors of α. Do the eigenvectors span $\mathbb{C}^2$?
2. Let α be the unique linear transformation of $\mathbb{R}^2$ into itself for which $\alpha(e_1) = 3e_1 - e_2$ and $\alpha(e_2) = 4e_1 - e_2$. Find a linear relation between e_1, $\alpha(e_1)$ and $\alpha^2(e_1)$, and hence show that 1 is an eigenvalue of α. Do the eigenvectors of α span $\mathbb{R}^2$?
3. Let P_n be the vector space of real polynomials of degree at most n, and let $\alpha : P_n \to P_n$ be defined by
$$\alpha(a_0 + a_1 x + \cdots + a_n x^n) = a_1 + 2a_2 x + \cdots + na_n x^{n-1}$$
(α is differentiation). Show that α is a linear map, and that 0 is the only eigenvalue of α. What is E_0? Show that the conclusion of Theorem 10.1.12 holds in this example.
4. Suppose that the linear map $\alpha : V \to V$ has the property that every non-zero vector in V is an eigenvector of α. Show that, for some constant μ, $\alpha = \mu I$.
5. Let α be the linear map of $\mathbb{R}^3$ into itself defined by $x \mapsto a \times x$, where $\times$ is the vector product. What are the eigenvalues, and eigenvectors of α? Show that $-||a||^2$ is an eigenvalue of α^2.
6. Let V be the real vector space of differentiable functions $f : \mathbb{R} \to \mathbb{R}$, and let $\alpha : V \to V$ be defined by $\alpha(f) = df/dx$. Show that every real number is an eigenvalue of α. Given a real number λ, what is E_λ?

7. Suppose that V is vector space of finite dimension, and that $\alpha : V \to V$ is a linear map. Show that α can have at most m distinct non-zero eigenvalues, where m is the dimension of $\alpha(V)$. [Use Theorem 10.1.9.]

10.2 Eigenvalues and matrices

In this section we shall show how to find all eigenvalues and all eigenvectors of a given linear transformation $\alpha : V \to V$.

Theorem 10.2.1 *Suppose that a linear map $\alpha : V \to V$ is represented by the matrix A with respect to some basis $\mathcal{B}$ of V. Then λ is an eigenvalue of α if and only if $\det(A - \lambda I) = 0$.*

Proof The linear map $\alpha - \lambda I$ is represented by the matrix $A - \lambda I$ with respect to $\mathcal{B}$. By definition, λ is an eigenvalue of α if and only if $(\alpha - \lambda I)^{-1}$ fails to exist and, from Theorem 9.4.2, this is so if and only if $\det(A - \lambda I) = 0$. □

Theorem 10.2.1 contains an algorithm for finding all eigenvalues and eigenvectors of a given linear map $\alpha : V \to V$. First, we find a matrix representation A for α; then we solve $\det(A - \lambda I) = 0$ (this is a polynomial equation in λ of degree equal to the dimension of V) to obtain a complete set of distinct eigenvalues, say $\lambda_1, \ldots, \lambda_r$ of α. Finally, for each α_j, we find the kernel of $\alpha - \lambda_j I$ to give the eigenspace E_{λ_j} (and hence all of the eigenvectors). It must be emphasized that the eigenvalues of α are scalars and so *they must lie in* $\mathbb{F}$. This means that if V is a real vector space, then the eigenvalues of α are the *real solutions* of $\det(A - \lambda I) = 0$. On the other hand, if V is a complex vector space, then the eigenvalues of α are the *complex solutions* of $\det(A - \lambda I) = 0$. By the Fundamental Theorem of Algebra, this equation will always have complex solutions, but not necessarily real solutions, and this signals a fundamental difference between the study of real vector spaces and complex vector spaces. Let us give an example to illustrate these ideas.

Example 10.2.2 Let
$$A = \begin{pmatrix} 1 & -3 & 4 \\ 4 & -7 & 8 \\ 6 & -7 & 7 \end{pmatrix}.$$

We view A as a linear map
$$\alpha : \begin{pmatrix} x \\ y \\ z \end{pmatrix} \mapsto \begin{pmatrix} 1 & -3 & 4 \\ 4 & -7 & 8 \\ 6 & -7 & 7 \end{pmatrix} \begin{pmatrix} x \\ y \\ z \end{pmatrix},$$

10.2 Eigenvalues and matrices

and the eigenvalues of α are the solutions of $\det(A - tI) = 0$; that is of

$$\begin{vmatrix} 1-t & -3 & 4 \\ 4 & -7-t & 8 \\ 6 & -7 & 7-t \end{vmatrix} = 0.$$

This equation simplifies to $-3 - 5t - t^2 + t^3 = 0$, and this has solutions $-1, -1, 3$; thus the eigenvalues of α are -1 and 3. To find E_{-1} and E_3 we have to solve the two systems of equations

$$\begin{pmatrix} 1 & -3 & 4 \\ 4 & -7 & 8 \\ 6 & -7 & 7 \end{pmatrix} \begin{pmatrix} x \\ y \\ z \end{pmatrix} = - \begin{pmatrix} x \\ y \\ z \end{pmatrix},$$

and

$$\begin{pmatrix} 1 & -3 & 4 \\ 4 & -7 & 8 \\ 6 & -7 & 7 \end{pmatrix} \begin{pmatrix} x \\ y \\ z \end{pmatrix} = 3 \begin{pmatrix} x \\ y \\ z \end{pmatrix}.$$

The first of these yields

$$2x - 3y + 4z = 0,$$
$$4x - 6y + 8z = 0,$$
$$6x - 7y + 8z = 0,$$

and the first and second of these equations are the same. The solutions of the first and third equations are $(x, y, z) = t(1, 2, 1)$ for any real t. We conclude that the eigenspace E_{-1} has dimension one, and is spanned by the single vector $(1, 2, 1)$, even though -1 appears as a double root of the characteristic equation of α (see Definition 10.2.3). A similar argument (which we omit) shows that E_3 is spanned by the single vector $(1, 2, 2)$. □

Suppose now that a linear map $\alpha : V \to V$ is represented by matrices A and A' with respect to the bases $\mathcal{B}$ and $\mathcal{B}'$, respectively, of V. Then there is a matrix P such that $A' = P^{-1}AP$ (Theorem 9.5.1), and

$$\begin{aligned} \det(A' - \lambda I) &= \det(P^{-1}AP - \lambda I) \\ &= \det(P^{-1}AP - \lambda P^{-1}P) \\ &= \det(P^{-1}[A - \lambda I]P) \\ &= \det(P^{-1})\det(A - \lambda I)\det(P) \\ &= \det(P^{-1}P)\det(A - \lambda I) \\ &= \det(I)\det(A - \lambda I) \\ &= \det(A - \lambda I), \end{aligned} \qquad (10.2.1)$$

and this shows (algebraically) that we obtain the same eigenvalues of α regardless of whether we use A or A' (or B or B'). Of course, this must be so as the eigenvalues and eigenvectors were defined geometrically, and without reference to matrices.

Next, given an $n \times n$ matrix A, we define the polynomial $\mathcal{P}_A$ by

$$\mathcal{P}_A(t) = \det(A - tI) = 0. \tag{10.2.2}$$

The theory of determinants implies that $\mathcal{P}_A(t)$ is a polynomial of degree n in t, with leading term $(-1)^n t^n$, and it is called the *characteristic polynomial* of A. Now (10.2.1) shows that two matrix representations A and A' of a linear map α have the same characteristic polynomial, and therefore this may be (unambiguously) considered to be the characteristic polynomial of α.

Definition 10.2.3 The *characteristic polynomial* $P_\alpha(t)$ of $\alpha : V \to V$ is the polynomial $\det(A - tI)$, where A is any matrix representation of α, and $n = \dim(V)$. The *characteristic equation* for α is the equation $P_\alpha(t) = 0$.

Theorem 10.2.1 shows that the eigenvalues of a map $\alpha : V \to V$ are precisely the zeros of the characteristic polynomial P_α, and it is here that we begin to see a significant difference between real vector spaces and complex vector spaces. According to the Fundamental Theorem of Algebra, any polynomial of degree n has exactly n complex roots, and it follows from this that if V is a complex vector space of dimension n, then every map $\alpha : V \to V$ has exactly n eigenvalues when we count each according to its multiplicity as a zero of the characteristic equation. In the case of real vector spaces we are looking for the *real roots* of the characteristic polynomial and, as we well know, there need not be any. There is one case of real vector spaces where we can say something positive. If n is odd, then the characteristic equation of $\alpha : V \to V$ is a real polynomial (the entries of the matrix A are real) of odd degree n and so it has at least one real root. Thus we have the following result.

Theorem 10.2.4 *Let V be a real vector space of odd dimension. Then any linear map $\alpha : V \to V$ has at least one (real) eigenvalue and one line of eigenvectors.*

To be more specific, a rotation of $\mathbb{R}^2$ has no real eigenvalues and no eigenvectors. However, any linear map of $\mathbb{R}^3$ into itself must have at least one eigenvector, and so there must be at least one line through the origin that is mapped into itself by α. A linear map of $\mathbb{R}^4$ into itself need not have any real eigenvalues;

an example of such a matrix is

$$\begin{pmatrix} \cos\theta & \sin\theta & 0 & 0 \\ -\sin\theta & \cos\theta & 0 & 0 \\ 0 & 0 & \cos\varphi & \sin\varphi \\ 0 & 0 & -\sin\varphi & \cos\varphi \end{pmatrix},$$

and the reader can easily generalize this to $\mathbb{R}^{2m}$ for any integer m.

Finally, there are two 'multiplicities' associated with any eigenvalue λ of a linear map α, namely

(i) the multiplicity of λ as a root of the characteristic equation, and
(ii) the dimension of the eigenspace $E(\lambda)$.

These need not be the same and, as one might expect, the two concepts play an important role in any deeper discussion of eigenvalues and eigenvectors of linear transformations.

Exercise 10.2

1. Let A be an $n \times n$ complex matrix and suppose that for some integer m, A^m is the zero matrix. Show that zero is the only eigenvalue of A.
2. Let

$$A = \begin{pmatrix} 3 & 1 & 1 \\ 1 & 2 & 0 \\ 1 & 0 & 2 \end{pmatrix},$$

and let A define a linear map $\alpha : \mathbb{R}^3 \to \mathbb{R}^3$. Show that the eigenvalues of α are 1, 2 and 4. In each case find a corresponding eigenvector.
3. Let

$$A = \begin{pmatrix} 4 & -5 & 7 \\ 1 & -4 & 9 \\ -4 & 0 & 5 \end{pmatrix},$$

and let $\alpha : \mathbb{C}^3 \to \mathbb{C}^3$ be the corresponding map. Show that the *complex* eigenvalues of α are 1 and $2 \pm 3i$, and find the associated eigenspaces.
4. Let A be a real 3×3 matrix, and let $\alpha : \mathbb{C}^3 \to \mathbb{C}^3$ be the corresponding map. Suppose that α has eigenvalues λ, μ and $\bar{\mu}$, where λ is real but μ is not. Let v be an eigenvector corresponding to μ; show that $\bar{v}$ is an eigenvector corresponding to $\bar{\mu}$.

Now suppose that $v = v_1 + iv_2$, where v_1 and v_2 are real vectors. Show that if we now view A as defining a map α of $\mathbb{R}^3$ into itself, then α leaves the subspace spanned by v_1 and v_2 invariant. Illustrate this by taking A to

be the matrix in Exercise 2. In this case the invariant plane is $2x - 2y + z = 0$, and you can verify this directly.
5. Construct a linear map of $\mathbb{R}^4$ into itself that has exactly two real eigenvalues.
6. (i) Let $\alpha : \mathbb{C}^2 \to \mathbb{C}^2$ be the linear map defined by $\alpha(e_1) = e_2$ and $\alpha(e_2) = e_1$. Find the eigenvalues of α.
   (ii) Let $\alpha : \mathbb{C}^3 \to \mathbb{C}^3$ be the linear map defined by $\alpha(e_1) = e_2$, $\alpha(e_2) = e_3$ and $\alpha(e_3) = e_1$. Find the eigenvalues of α.
   (iii) Generalise (i) and (ii) to a map $\alpha : \mathbb{C}^n \to \mathbb{C}^n$.
   (iv) Let ρ be the permutation of $\{1, \ldots, 9\}$ be given by
$$\rho = \begin{pmatrix} 1 & 2 & 3 & 4 & 5 & 6 & 7 & 8 & 9 \\ 2 & 3 & 4 & 1 & 6 & 7 & 5 & 9 & 8 \end{pmatrix}$$
and let $\alpha : \mathbb{C}^9 \to \mathbb{C}^9$ be the linear map defined by by $\alpha(e_j) = e_{\rho(j)}$, $j = 1, \ldots, 9$. What are the eignevalues of α?
7. Let A be the $n \times n$ matrix (a_{ij}), where a_{ij} is 1 if $i \neq j$ and 0 if $i = j$. Show that -1 and $n - 1$ are eigenvalues of A and find the dimension of the corresponding eigenspaces. [It is not necessary to evaluate any determinant.]

10.3 Diagonalizable matrices

Suppose that A is a square matrix. How can we compute the successive powers $A, A^2, A^3, \ldots$ of A? This question is easily answered for diagonal matrices. A matrix $D = (d_{ij})$ is a *diagonal matrix* if $d_{ij} = 0$ whenever $i \neq j$; that is, if

$$D = \begin{pmatrix} d_{11} & 0 & \cdots & 0 \\ 0 & d_{22} & \cdots & 0 \\ \vdots & \vdots & \ddots & \vdots \\ 0 & 0 & \cdots & d_{nn} \end{pmatrix}. \qquad (10.3.1)$$

It is obvious that

$$D^k = \begin{pmatrix} d_{11}^k & 0 & \cdots & 0 \\ 0 & d_{22}^k & \cdots & 0 \\ \vdots & \vdots & \ddots & \vdots \\ 0 & 0 & \cdots & d_{nn}^k \end{pmatrix},$$

and although this is not very exciting it does lead to a more useful result. For this we need the concept of a diagonalizable matrix.

10.3 Diagonalizable matrices

Definition 10.3.1 An $n \times n$ matrix A is *diagonalizable* if there is a non-singular matrix X, and a diagonal matrix D, such that $X^{-1}AX = D$ or, equivalently, $A = XDX^{-1}$.

It is easy to compute the powers of a diagonalizable matrix. If A diagonalizable with X and D as in Definition 10.3.1, then

$$A^k = (XDX^{-1})^k = (XDX^{-1})(XDX^{-1})\ldots(XDX^{-1}) = XD^kX^{-1}$$

and so providing that we can find X and D we can easily compute A^k. An example will illustrate this idea.

Example 10.3.2 Suppose that

$$A = \begin{pmatrix} 5 & -6 \\ 1 & 0 \end{pmatrix}. \tag{10.3.2}$$

We shall show how to find X shortly, but with

$$X = \begin{pmatrix} 2 & 3 \\ 1 & 1 \end{pmatrix}, \qquad X^{-1} = \begin{pmatrix} -1 & 3 \\ 1 & -2 \end{pmatrix},$$

we have

$$X^{-1}AX = \begin{pmatrix} 2 & 0 \\ 0 & 3 \end{pmatrix} = D,$$

say, and hence

$$A^k = X \begin{pmatrix} 2^k & 0 \\ 0 & 3^k \end{pmatrix} X^{-1} = \begin{pmatrix} -2^{k+1} + 3^{k+1} & 6.2^k - 6.3^k \\ -2^k + 3^k & 3.2^k - 2.3^k \end{pmatrix}.$$

$\square$

In order to use this technique we need a criterion for A to be diagonalizable and, when it is, we need to know how to find the matrices X and D. Unfortunately, not every square matrix is diagonalizable as the next example shows.

Example 10.3.3 The matrix

$$B = \begin{pmatrix} 0 & 1 \\ 0 & 0 \end{pmatrix} \tag{10.3.3}$$

is not diagonalizable. To see this, suppose that B is diagonalizable; then there is a non-singular matrix X and a diagonal matrix D, such that $XB = DX$; say

$$\begin{pmatrix} a & b \\ c & d \end{pmatrix} \begin{pmatrix} 0 & 1 \\ 0 & 0 \end{pmatrix} = \begin{pmatrix} \lambda & 0 \\ 0 & \mu \end{pmatrix} \begin{pmatrix} a & b \\ c & d \end{pmatrix}.$$

This implies that $\lambda a = 0$ and $\lambda b = a$. As X^{-1} exists, $ad - bc \neq 0$. Thus $\lambda = 0$, and similarly, $\mu = 0$, so that $D = 0$, and hence $B = X^{-1}DX = 0$ which is false. We deduce that B is not diagonalizable. □

The following result gives us a condition for A to be diagonalizable, and it also tells us how to find X and D.

Theorem 10.3.4 *A real $n \times n$ matrix A is diagonalizable if and only if $\mathbb{R}^n$ has a basis of eigenvectors of A. Moreover, if $A = XDX^{-1}$, where D is a diagonal matrix, then the diagonal elements of D are the eigenvalues of A and the columns of X are the corresponding eigenvectors of A. A similar statement holds for complex matrices and $\mathbb{C}^n$.*

The reader should now consider the matrix A in (10.3.2) and verify that the columns of X used in that example are eigenvectors of A which span $\mathbb{R}^2$. Likewise, the reader should verify that the eigenvectors of B in (10.3.3) do not span $\mathbb{R}^2$.

Before giving the proof of Theorem 10.3.4, we shall illustrate its use by explaining the standard method of solution of homogeneous second-order difference equations.

Example 10.3.5 The *difference equations*

$$a_{n+2} - 5a_{n+1} + 6a_n = 0,$$

where a_0 and a_1 are given values, can be written as

$$\begin{pmatrix} a_{n+2} \\ a_{n+1} \end{pmatrix} = \begin{pmatrix} 5 & -6 \\ 1 & 0 \end{pmatrix} \begin{pmatrix} a_{n+1} \\ a_n \end{pmatrix}.$$

Thus

$$\begin{pmatrix} a_{n+2} \\ a_{n+1} \end{pmatrix} = \begin{pmatrix} 5 & -6 \\ 1 & 0 \end{pmatrix}^n \begin{pmatrix} a_2 \\ a_1 \end{pmatrix}. \quad (10.3.4)$$

Theorem 10.3.4 suggests that we find the eigenvalues and eigenvectors of A, and a calculation shows that

$$\begin{pmatrix} 5 & -6 \\ 1 & 0 \end{pmatrix} \begin{pmatrix} 2 \\ 1 \end{pmatrix} = 2 \begin{pmatrix} 2 \\ 1 \end{pmatrix}, \quad \begin{pmatrix} 5 & -6 \\ 1 & 0 \end{pmatrix} \begin{pmatrix} 3 \\ 1 \end{pmatrix} = 3 \begin{pmatrix} 3 \\ 1 \end{pmatrix}.$$

Thus

$$\begin{pmatrix} 5 & -6 \\ 1 & 0 \end{pmatrix} \begin{pmatrix} 2 & 3 \\ 1 & 1 \end{pmatrix} = \begin{pmatrix} 4 & 9 \\ 2 & 3 \end{pmatrix} = \begin{pmatrix} 2 & 3 \\ 1 & 1 \end{pmatrix} \begin{pmatrix} 2 & 0 \\ 0 & 3 \end{pmatrix},$$

and so

$$\begin{pmatrix} 5 & -6 \\ 1 & 0 \end{pmatrix}^n = \begin{pmatrix} 2 & 3 \\ 1 & 1 \end{pmatrix} \begin{pmatrix} 2^n & 0 \\ 0 & 3^n \end{pmatrix} \begin{pmatrix} 2 & 3 \\ 1 & 1 \end{pmatrix}^{-1}. \quad (10.3.5)$$

10.3 Diagonalizable matrices

If we combine this with (10.3.4) we can calculate a_{n+1}, but in fact, one does not normally carry out this computation. Once we know that the eigenvalues of A are 2 and 3, it is immediate from (10.3.4) and (10.3.5) that we must have $a_n = \alpha 2^n + \beta 3^n$ for some constants α and β. As $a_0 = \alpha + \beta$ and $a_1 = 2\alpha + 3\beta$, we can now find α and β in terms of the given a_0 and a_1 and hence obtain an explicit formula for a_n. □

The proof of Theorem 10.3.4 The matrix A acts on $\mathbb{R}^n$ as a linear transformation α, and the matrix of α with respect to the standard basis $\{e_1, \ldots, e_n\}$ is A. First, suppose that $v_1, \ldots, v_r$ is a basis of eigenvectors of A. Then the matrix of α with respect to this basis is a diagonal matrix D, and by Theorem 9.5.1 (which shows how a matrix changes under a change of basis) there is some matrix X with $A = XDX^{-1}$. Thus A is diagonalizable.

Next, suppose that A is diagonalizable. Then there are matrices X and D (diagonal) such that $A = XDX^{-1}$. Thus $AX = XD$ and so matrix multiplication shows that the columns $X_1, \ldots, X_n$ of X are eigenvectors of A; indeed,

$$\begin{pmatrix} a_{11} & \cdots & a_{nn} \\ \vdots & \ddots & \vdots \\ a_{n1} & \cdots & a_{nn} \end{pmatrix} \begin{pmatrix} x_{11} & \cdots & x_{nn} \\ \vdots & \ddots & \vdots \\ x_{n1} & \cdots & x_{nn} \end{pmatrix}$$

$$= \begin{pmatrix} x_{11} & \cdots & x_{nn} \\ \vdots & \ddots & \vdots \\ x_{n1} & \cdots & x_{nn} \end{pmatrix} \begin{pmatrix} d_{11} & \cdots & 0 \\ \vdots & \ddots & \vdots \\ 0 & \cdots & d_{nn} \end{pmatrix}$$

$$= \begin{pmatrix} d_1 x_{11} & \cdots & d_n x_{nn} \\ \vdots & \ddots & \vdots \\ d_1 x_{n1} & \cdots & d_n x_{nn} \end{pmatrix}.$$

We shall show that the vectors $X_1, \ldots, X_n$ are a basis of $\mathbb{R}^n$ and for this, it is sufficient to show that they are linearly independent. However, as the matrix X^{-1} exists (by assumption), we see that $\det(X) \neq 0$ so the columns $X_1, \ldots, X_n$ of X must be linearly independent.

Next, suppose that D is given by (10.3.1). Then the characteristic equation for D is

$$(t - d_{11}) \cdots (t - d_{nn}) = 0. \qquad (10.3.6)$$

However, $XAX^{-1} = D$, so that A and D have the same characteristic polynomial. We conclude that the diagonal elements of D are the eigenvalues of A. □

Finally, recall Theorem 10.1.9 that if $\lambda_1, \ldots, \lambda_r$ are distinct eigenvalues of a matrix A with corresponding eigenvectors $v_1, \ldots, v_r$, then these vectors are linearly independent. It follows that if the $n \times n$ matrix A has n distinct eigenvalues, then the corresponding eigenvectors $v_1, \ldots, v_n$ form a linearly independent set of n vectors in an n-dimensional space and so are a basis of that space. Combining this with Theorem 10.3.4, we obtain the following sufficient condition for a matrix to be diagonalizable.

Theorem 10.3.6 *If a real $n \times n$ matrix A has n distinct real eigenvalues, then it is diagonalizable. If a complex $n \times n$ matrix A has n distinct complex eigenvalues, then it is diagonalizable.*

This condition is sufficient, but not necessary; for example, the identity matrix is diagonal, and hence diagonalizable, but it does not have distinct eigenvalues.

We end with an example to illustrate Theorem 10.3.6.

Example 10.3.7 Let

$$A = \begin{pmatrix} 2-i & 0 & i \\ 0 & 1+i & 0 \\ i & 0 & 2-i \end{pmatrix}.$$

A calculation (which the reader should do) shows that A has eigenvalues 2, $1+i$ and $2-2i$. As these are distinct, it follows that the map

$$\alpha : \begin{pmatrix} z_1 \\ z_2 \\ z_3 \end{pmatrix} \mapsto \begin{pmatrix} 2-i & 0 & i \\ 0 & 1+i & 0 \\ i & 0 & 2-i \end{pmatrix} \begin{pmatrix} z_1 \\ z_2 \\ z_3 \end{pmatrix}$$

of $\mathbb{C}^{3,t}$ into itself has a three eigenvectors that form a basis of $\mathbb{C}^{3,t}$. These eigenvectors are $w_1 = (1, 0, 1)^t$ for the eigenvalue 2, $w_2 = (2, -1, 0)^t$ for the eigenvalue $1+i$, and $w_3 = (1, 0, -1)^t$ for the eigenvalue $2-2i$. Finally, the matrix of α with respect to the basis $\{w_1, w_2, w_3\}$ is

$$\begin{pmatrix} 2 & 0 & 0 \\ 0 & 1+i & 0 \\ 0 & 0 & 2-2i \end{pmatrix},$$

for obviously this maps the coordinates $(\lambda, \mu, \nu)^t$ of $w = \lambda w_1 + \mu w_2 + \nu w_3$ to the coordinates $\big(2\lambda, (1+i)\mu, (2-2i)\nu\big)^t$ of $\alpha(w)$. $\square$

Exercise 10.3

1. Find the eigenvalues of the linear transformation

$$\alpha : \begin{pmatrix} x \\ y \\ z \end{pmatrix} \mapsto \begin{pmatrix} 2 & 2 & 1 \\ 1 & 3 & 1 \\ 1 & 2 & 2 \end{pmatrix} \begin{pmatrix} x \\ y \\ z \end{pmatrix}$$

(they are all small positive integers). Show that $\mathbb{R}^{3,t}$ has a basis of eigenvectors of α even though α does not have three distinct eigenvalues. Show also that the plane $x + 2y + z = 0$ is α-invariant.

2. Is the matrix
$$A = \begin{pmatrix} 1 & 0 & -1 \\ 1 & 2 & 1 \\ 2 & 2 & 4 \end{pmatrix}$$
diagonalizable?

3. Is the matrix
$$A = \begin{pmatrix} 2 & 1 & 1 \\ 1 & 2 & 1 \\ 1 & 1 & 2 \end{pmatrix}$$
diagonalizable? Regard A as acting as a linear map from $\mathbb{R}^3$ to itself in the usual way, and write $A = I + B$, where I is the 3×3 identity matrix. Try to describe the action of A in geometric terms.

4. Show that the matrix
$$\begin{pmatrix} 1 & 1 & 0 & 0 \\ 0 & 1 & 1 & 0 \\ 0 & 0 & 1 & 0 \\ 0 & 0 & 0 & 2 \end{pmatrix}$$
is not diagonalizable.

10.4 The Cayley–Hamilton theorem

Having understood how to compute powers of diagonalizable matrices, we now consider polynomials of a matrix. Given an $n \times n$ matrix A, and a real or complex polynomial
$$p(t) = a_0 + a_1 t + a_2 t^2 + \cdots + a_k t^k,$$
we define the $n \times n$ matrix $p(A)$ by
$$p(A) = a_0 I + a_1 A + a_2 A^2 + \cdots + a_k A^k.$$
We are accustomed to trying to find the zeros of a given polynomial but there is little prospect of us being able to solve the equation $p(A) = 0$, where 0 is the $n \times n$ zero matrix. We can, however, reverse the process and, starting with the matrix A, try to find a non-zero polynomial p for which $p(A) = 0$. It is not immediately obvious that any such p exists, but we have seen in Section 7.10 that there is at least one such p (of degree at most n^2).

Again, we start our investigation with diagonal matrices, then consider diagonalizable matrices and finally (though we shall not give all of the details) general matrices. First, for any polynomial p, and any diagonal matrix D with $D = (d_{ij})$, we have

$$D \begin{pmatrix} d_{11} & \cdots & 0 \\ \vdots & \ddots & \vdots \\ 0 & \cdots & d_{nn} \end{pmatrix}, \quad p(D) = \begin{pmatrix} p(d_{11}) & \cdots & 0 \\ \vdots & \ddots & \vdots \\ 0 & \cdots & p(d_{nn}) \end{pmatrix}.$$

Let us now denote the characteristic polynomial of a matrix A by p_A. As the diagonal entries of the diagonal matrix D are precisely its eigenvalues, we see that for each i, $p_D(d_{ii}) = 0$, and it follows immediately that $p_D(D) = 0$; that is, the diagonal matrix D satisfies its own characteristic equation. This is trivial, but it is a start and it suggests what might be true for non-diagonal matrices. In fact, it is true for *all matrices* and this is the celebrated

Cayley–Hamilton theorem. *Every complex square matrix satisfies its own characteristic equation.*

We have verified the conclusion for a diagonal matrix, and it is easy to extend this proof so as to include all diagonalizable matrices. We shall then indicate briefly how to extend the proof still further so as to include all matrices.

To prove the result for a diagonalizable matrix A, let $A = XDX^{-1}$, where D is a diagonal matrix. Then, for any integer k,

$$A^k = (XDX^{-1})^k = (XDX^{-1})\ldots(XDX^{-1}) = XD^kX^{-1}.$$

Now A and D have the same characteristic polynomial, say $p(z) = a_0 + \cdots + a_n z^n$, and $p(D) = 0$. We want to show that $p(A) = 0$, and this is so because

$$p(A) = \sum_{k=0}^{n} a_k A^k$$

$$= \sum_{k=0}^{n} a_k X D^k X^{-1}$$

$$= X \left(\sum_{k=0}^{n} a_k D^k \right) X^{-1}$$

$$= X p(D) X^{-1}$$

$$= X 0 X^{-1}$$

$$= 0.$$

We have now shown that *any diagonalizable matrix A satisfies its own characteristic equation.*

10.4 The Cayley–Hamilton theorem

We now sketch the proof for the general matrix. First, if A has n distinct eigenvalues then, by Theorem 10.3.6, it is diagonalizable and so it satisfies its own characteristic equation. The remaining cases are those in which A has a repeated eigenvalue, and this happens if and only if the coefficients of A satisfy some (known) algebraic relation (see Theorem 9.6.5). Perturbing the coefficients appropriately destroys the validity of this relation, and this means that every matrix A *is the limit of a sequence A_j of diagonalizable matrices.* Let p_j be the characteristic polynomial of A_j, so that $p_j(A_j) = 0$. As the coefficients of p_j depend continuously on the coefficients of A_j, we see that $p_j \to p$; thus

$$p(A) = \lim_{j \to \infty} p_j(A_j) = \lim_{j \to \infty} 0 = 0$$

as required. There is much to be done to validate these steps, but the general plan should be clear. There are, of course, purely algebraic proofs of the Cayley–Hamilton theorem; our proof is based on the idea that as the conclusion is true for 'almost all' matrices, it can be obtained for the remaining matrices by taking limits.

The Cayley–Hamilton theorem provides a useful method for computing the powers, and the inverse of a matrix, especially when the matrix is not too large. We consider the inverse first. Suppose that A is an invertible $n \times n$ matrix, with characteristic polynomial $a_0 + \cdots + a_{n-1}z^{n-1} + z^n$; then

$$A^n + a_{n-1}A^{n-1} + \cdots + a_1 A + a_0 I = 0.$$

If $a_0 = 0$ we can multiply both sides of this by A^{-1}, so that the new equation has constant term a_1. This process can be repeated until the constant term is non-zero, so we may assume now that $a_0 \neq 0$. Then

$$A^{-1} = \frac{-1}{a_0}\left[a_1 I + a_2 A + \cdots + a_{n-1}A^{n-2} + A^{n-1}\right],$$

and this gives a way of calculating A^{-1} simply from the positive powers of A. As an example, if

$$A = \begin{pmatrix} 1 & 0 & 1 \\ 2 & 1 & -1 \\ 0 & 1 & 2 \end{pmatrix},$$

then

$$A^2 = \begin{pmatrix} 1 & 1 & 3 \\ 4 & 0 & -1 \\ 2 & 3 & 3 \end{pmatrix},$$

and the characteristic equation of A is $t^3 - 4t^2 + 6t - 5$. Thus

$$A^{-1} = \frac{1}{5}(A^2 - 4A + 6I) = \frac{1}{5}\begin{pmatrix} 3 & 1 & -1 \\ -4 & 2 & 3 \\ 2 & -1 & 1 \end{pmatrix}.$$

We now illustrate how to use the Cayley–Hamilton theorem to compute powers of a matrix. Consider, for example, the matrix A given in (10.3.2). Its characteristic polynomial is $p(t) = t^2 - 5t + 6$ and so

$$A^2 = 5A - 6I,$$
$$A^3 = A(5A - 6I) = 5A^2 - 6A = 19A - 30I,$$
$$A^4 = A(19A - 30I) = 19(5A - 6I) - 30A = 65A - 114I,$$

and so on. The reader may care to check this against the result obtained (by essentially the same method) in Example 10.3.2.

Exercise 10.4

1. Show that the matrix

$$A = \begin{pmatrix} -1 & 2 \\ -3 & 4 \end{pmatrix}$$

satisfies its own characteristic equation. Use this to compute A^4.

2. The space $M^{2\times 2}$ of real 2×2 matrices has dimension four. Use the Cayley–Hamilton theorem to show that if A is any 2×2 matrix, then all of the matrices $A, A^2, A^3, \ldots$ lie in the two-dimensional subspace of $M^{2\times 2}$ spanned by I and A.

3. Let

$$A = \begin{pmatrix} a & * & * \\ 0 & b & * \\ 0 & 0 & c \end{pmatrix},$$

where $*$ denotes an unspecified entry of A. Show that if $a + b + c = 0$ and $1/a + 1/b + 1/c = 0$, then $A^3 = abcI$.

4. Let $a \times b$ be the standard vector product on $\mathbb{R}^3$, and define the linear map $\alpha : \mathbb{R}^3 \to \mathbb{R}^3$ by $\alpha(x) = a \times x$, where a is a given vector of unit length. Show that the matrix of α with respect to the standard basis is

$$A = \begin{pmatrix} 0 & -a_3 & a_2 \\ a_3 & 0 & -a_1 \\ -a_2 & a_1 & 0 \end{pmatrix}.$$

Find the characteristic equation of α, and verify that the conclusion of the Cayley–Hamilton theorem holds in the case. Derive the same result by

vector methods. Deduce that for all x,
$$a \times (a \times (a \times (a \times (a \times x)))) = a \times x.$$

5. Let $F : \mathbb{R}^2 \to \mathbb{R}^2$ be any map and, given a point (x_0, y_0) in $\mathbb{R}^2$ define x_n and y_n by $(x_{n+1}, y_{n+1}) = F(x_n, y_n)$. We study the *dynamics* of the map F by studying the limiting behaviour of the sequence (x_n, y_n) as $n \to \infty$ for different choices of the starting point (x_0, y_0). What is the limiting behaviour of (x_n, y_n) when F is defined by $F(x, y) = (x', y')$, where

$$\begin{pmatrix} 0 & 2 \\ 1 & 1 \end{pmatrix} \begin{pmatrix} x \\ y \end{pmatrix} = \begin{pmatrix} x' \\ y' \end{pmatrix}?$$

Sketch some sequences (x_n, y_n) for a few different starting points. [Find the eigenvalues of the matrix. The lines $y = x$ and $x + 2y = 0$ should figure prominently in your solution.]

10.5 Invariant planes

In this section we give some examples to illustrate the ideas discussed earlier in the chapter (and in these the reader is expected to provide some of the details).

Example 10.5.1 Consider the mapping α of $\mathbb{R}^{3,1}$ into itself given by

$$\begin{pmatrix} x \\ y \\ z \end{pmatrix} \mapsto \begin{pmatrix} 4 & -5 & 7 \\ 1 & -4 & 9 \\ -4 & 0 & 5 \end{pmatrix} \begin{pmatrix} x \\ y \\ z \end{pmatrix}.$$

The characteristic equation of α is $t^3 - 5t^2 + 17t - 13 = 0$, so the eigenvalues are 1 and $2 \pm 3i$. It is easy to see that $(x, y, z)^t$ an eigenvector corresponding to the eigenvalue 1 if and only if it is a scalar multiple of the vector $(1, 2, 1)^t$, so the eigenspace corresponding to the eigenvalue 1 is a line.

Let us now consider α to be acting on $\mathbb{C}^{3,1}$, and seek the eigenvectors in this space. A straightforward calculation shows that $(3 - 3i, 5 - 3i, 4)^t$ is an eigenvector for $2 + 3i$, and that (as the entries in the matrix are real) $(3 + 3i, 5 + 3i, 4)^t$ is an eigenvector for $2 - 3i$. Let us write these eigenvectors as $p + iq$ and $p - iq$, respectively, where p and q are real column vectors. Then $\alpha(p + iq) = (2 + 3i)(p + iq)$, so that $\alpha(p) = 2p - 3q$ and $A(q) = 3p + 2q$. Of course this can be verified directly (once one knows which vectors to look at). It follows that the real subspace of $\mathbb{R}^{3,1}$ spanned by p and q is invariant under α. Of course, this is the plane whose normal is orthogonal to both p and q. Now $p^t = (3, 5, 4)$ and $q^t = (-3, -3, 0)$ so that (using the vector product in $\mathbb{R}^3$ to compute this normal), we see that plane given by $2x - 2y + z = 0$ is invariant under A. This can now be checked directly for if we write

$$\begin{pmatrix} x' \\ y' \\ z' \end{pmatrix} = A \begin{pmatrix} x \\ y \\ z \end{pmatrix},$$

then (as is easily checked) $2x' - 2y' + z' = 2x - 2y + z$. Finally, the vectors

$$r = \begin{pmatrix} 1 \\ 2 \\ 2 \end{pmatrix}, \quad p = \begin{pmatrix} 3 \\ 5 \\ 4 \end{pmatrix}, \quad q = \begin{pmatrix} -3 \\ -3 \\ 0 \end{pmatrix}$$

form a basis of $\mathbb{R}^{3,t}$, and the matrix of α with respect to this basis (in this order) is

$$\begin{pmatrix} 1 & 0 & 0 \\ 0 & 2 & -3 \\ 0 & 3 & 2 \end{pmatrix}.$$

Here, the single diagonal element corresponds to the single line of eigenvectors, and the 'obvious' 2×2 part of the matrix represents the action of A on the invariant plane. □

Example 10.5.2 Let α be the map

$$\begin{pmatrix} x \\ y \\ z \end{pmatrix} \mapsto \begin{pmatrix} 0 & 1 & 0 \\ 0 & 0 & 1 \\ 1 & -3 & 3 \end{pmatrix} \begin{pmatrix} x \\ y \\ z \end{pmatrix} = \begin{pmatrix} x' \\ y' \\ z' \end{pmatrix}.$$

Then the characteristic equation of α is $(t-1)^3 = 0$, so 1 is the only eigenvalue of α. Moreover, $(x, y, z)^t$ is an eigenvector if and only if it is a scalar multiple of the vector $(1, 1, 1)^t$. We now ask whether α has an invariant plane (containing the origin) or not. The plane $px + qy + rz = 0$ is mapped to the plane $(3p + q)x' + (r - 3p)y' + pz' = 0$ and these planes will coincide if their normals are in the same direction; that is, if there is a scalar λ such that $\lambda(p, q, r) = (3p + q, r - 3p, p)$. A solution of this is $\lambda = 1$, and $(p, q, r) = (1, -2, 1)$ so we conclude that the plane $x - 2y + z = 0$ is invariant under α. This can be confirmed by noting that $x - 2y + z = x' - 2y' + z'$. □

Example 10.5.3 Let α be the map

$$\begin{pmatrix} x \\ y \\ z \end{pmatrix} \mapsto \begin{pmatrix} -3 & -9 & -12 \\ 1 & 3 & 4 \\ 0 & 0 & 1 \end{pmatrix} \begin{pmatrix} x \\ y \\ z \end{pmatrix} = \begin{pmatrix} x' \\ y' \\ z' \end{pmatrix}.$$

This has eigenvalues 1, 0 and 0, where the eigenspace E_1 is the line through $(-12, 4, 1)^t$, and the eigenspace E_0 is the line through $(3, -1, 0)^t$. We can analyse the action of α in more detail if we note that with the notation

$$e_1^t = \begin{pmatrix} 1 \\ 0 \\ 0 \end{pmatrix}, \ e_2^t = \begin{pmatrix} 0 \\ 1 \\ 0 \end{pmatrix}, \ e_3^t = \begin{pmatrix} 0 \\ 0 \\ 1 \end{pmatrix}, \ u = \begin{pmatrix} -3 \\ 1 \\ 0 \end{pmatrix}, \ v = \begin{pmatrix} -12 \\ 4 \\ 1 \end{pmatrix},$$

we have $\alpha(e_1) = u, \alpha(e_2) = 3u, \alpha(e_3) = v = 4u + e_3,$ and $\alpha(u) = 0, \alpha(v) = v$. Thus, for example, the plane containing u and v (namely $x + 3y = 0$), and also the plane containing e_1 and e_2 (namely $z = 0$), is invariant under α. □

Example 10.5.4 Consider the mapping α of $\mathbb{R}^{3,t}$ into itself given by

$$\begin{pmatrix} x \\ y \\ z \end{pmatrix} \mapsto \begin{pmatrix} 3 & 1 & -1 \\ 2 & 2 & -1 \\ 2 & 2 & 0 \end{pmatrix} \begin{pmatrix} x \\ y \\ z \end{pmatrix} = \begin{pmatrix} x' \\ y' \\ z' \end{pmatrix}.$$

The characteristic equation of α is $(t-1)(t-2)^2 = 0$, and it is easily seen that E_1 is the line through $(1, 0, 2)^t$, while E_2 is the line through $(1, 1, 2)^t$.

Let A be the matrix in the definition of α. Then, by the Cayley–Hamilton theorem, $(A - I)(A^2 - 4A + 4I) = 0$. If $A^2 - 4A + 4I$ is non-singular, this would imply that $A - I = 0$ which is false. It follows that there is some non-zero vector v such that $(A^2 - 4A + 4I)(v) = 0$, and hence that $A^2(v) = 4A(v) - 4v$. This implies that the plane spanned by v and $A(v)$ is invariant under α, so we need to find a suitable v. Now we can take any v in the kernel of $A^2 - 4A + 4I$, and as

$$A^2 - 4A + 4I = \begin{pmatrix} 1 & -1 & 0 \\ 0 & 0 & 0 \\ 2 & -2 & 0 \end{pmatrix},$$

and it is easy to check that the kernel of $A^2 - 4A + 4I$ is the plane given by $x = y$. It follows that we can take $v = (1, 1, 0)^t$, and then $\alpha(v) = (4, 4, 4)^t$. The normal to the two vectors v and $\alpha(v)$ can be found by computing their vector product (consider them as row vectors), and we find that α leaves the plane $x - y = 0$ invariant. To check this directly, observe that $x' - y' = x - y$; thus if $x = y$ then $x' = y'$. □

Exercise 10.5

1. Find an invariant line, and an invariant plane, for the map α given by

$$\begin{pmatrix} x \\ y \\ z \end{pmatrix} \mapsto \begin{pmatrix} 6 & -3 & -2 \\ 4 & -1 & -2 \\ 10 & -5 & -3 \end{pmatrix} \begin{pmatrix} x \\ y \\ z \end{pmatrix}.$$

2. Let α be the linear map
$$\begin{pmatrix} x \\ y \\ z \end{pmatrix} \mapsto \begin{pmatrix} 1 & -1 & -1 \\ 1 & -1 & 0 \\ 1 & 0 & -1 \end{pmatrix} \begin{pmatrix} x \\ y \\ z \end{pmatrix}.$$
Find the (complex) eigenvalues of α and show that α has an invariant plane that does not contain any (real) eigenvector of α. By considering the complex eigenvectors, deduce that α^4 is the identity map.

3. Find an invariant plane for the linear map α given by
$$\begin{pmatrix} x \\ y \\ z \end{pmatrix} \mapsto \begin{pmatrix} 3 & 0 & 2 \\ -5 & 2 & -5 \\ -5 & 1 & -4 \end{pmatrix} \begin{pmatrix} x \\ y \\ z \end{pmatrix}.$$

11
Linear maps of Euclidean space

11.1 Distance in Euclidean space

We recall the standard basis $e_1, \ldots, e_n$ of $\mathbb{R}^n$. If $x = \sum_j x_j e_j$, and similarly for y, we write

$$x \cdot y = \sum_{j=1}^n x_j y_j \quad, \quad ||x||^2 = x \cdot x = \sum_{j=1}^n x_j^2,$$

and $x \perp y$ when $x \cdot y = 0$. The distance $||x - y||$ between the points x and y is given by the natural extension of Pythagoras' theorem, and it is important to know that it satisfies the *triangle inequality*: for all x, y and z in $\mathbb{R}^n$,

$$||x - z|| \leq ||x - y|| + ||y - z||. \tag{11.1.1}$$

To prove this it is sufficient to show that $|x \cdot y| \leq ||x|| \, ||y||$ for this leads to $||x + y|| \leq ||x|| + ||y||$, and then a change of variables gives the triangle inequality exactly as in the case of $\mathbb{R}^3$. It is sufficient, therefore, to prove the following important inequality.

Theorem 11.1.1: the Cauchy–Schwarz inequality *For all x and y,*

$$|x \cdot y| \leq ||x|| \, ||y||. \tag{11.1.2}$$

Further, equality holds if and only if $||x||y = \pm ||y||x$.

Proof Let $x = (x_1, \ldots, x_n)$ and $y = (y_1, \ldots, y_n)$. Now (11.1.2) is true when $x = 0$, and also when $y = 0$; thus we may assume that $||x|| > 0$ and $||y|| > 0$. Now simple algebra gives

$$0 \leq \sum_{j=1}^n \bigl(||y||x_j - ||x||y_j\bigr)^2 = 2||x|| \, ||y|| \bigl(||x|| \, ||y|| - x \cdot y\bigr),$$

so that $x \cdot y \leq ||x||\,||y||$. If we replace x by $-x$, we see that $-x \cdot y \leq ||x||\,||y||$, and (11.1.2) follows. Finally, equality holds in (11.1.2) if and only if $x \cdot y = \pm ||x||\,||y||$. However, the argument above shows that $x \cdot y = ||x||\,||y||$ if and only if $||x||y = ||y||x$, and a similar argument shows that $x \cdot y = -||x||\,||y||$ if and only if $||x||y = -||y||x$. Thus equality holds in (11.1.2) if and only if $||x||y = \pm||y||x$. □

The Cauchy–Schwarz inequality implies that, for any vectors x and y in $\mathbb{R}^n$, there is some unique real number β in $[0, \pi]$ that satisfies $x \cdot y = ||x||\,||y|| \cos \beta$. As the Cauchy–Schwarz inequality has been established without the use of geometry, we can now *use this equation to define the angle β between the segments $[0, x]$ and $[0, y]$* in $\mathbb{R}^n$. For example, the angle β between these two segments in $\mathbb{R}^4$ when $x = (1, 1, 1, 1)$ and $y = (1, 2, 2, 0)$ is given by $\cos = 5/6$.

Exercise 11.1

1. Let $z_1, \ldots, z_n, w_1, \ldots, w_n$ be complex numbers. Show that
$$|z_1 w_1 + \cdots + z_n w_n|^2 \leq \sum_{j=1}^{n} |z_j|^2 \sum_{j=1}^{n} |w_j|^2.$$
This is the Cauchy–Schwarz inequality for complex numbers.
2. The unit cube in $\mathbb{R}^n$ has vertices $(\varepsilon_1, \ldots, \varepsilon_n)$, where $\varepsilon_1, \ldots, \varepsilon_n = 0, 1$. What is the angle θ_n between the segments $[0, e_1]$ and $[0, e_1 + \cdots + e_n]$? How does θ_n behave as $n \to \infty$?
3. Consider the cube in $\mathbb{R}^n$ with centre 0 and vertices $(a_1, \ldots, a_n)$, where $a_j = \pm 1$. Show that if n is even then the cube has two diagonals that are orthogonal to each other (a diagonal of the cube is a line through 0 and a vertex). Show that no such diagonals exist when n is odd.

11.2 Orthogonal maps

The most useful feature of the basis $e_1, \ldots, e_n$ is that it is a basis of mutually orthogonal unit vectors; that is, for all i, $||e_i|| = 1$, and for all i and j with $i \neq j$, $e_i \perp e_j$. This property is so important that we enshrine it in a definition.

Definition 11.2.1 A basis $v_1, \ldots, v_n$ of $\mathbb{R}^n$ is an *orthonormal basis* if $||v_j|| = 1$ for every j, and $v_i \perp v_j$ whenever $i \neq j$.

The following lemma is useful.

11.2 Orthogonal maps

Lemma 11.2.2 *If $v_1, \ldots, v_n$ are mutually orthogonal unit vectors in $\mathbb{R}^n$ then they form an orthonormal basis of $\mathbb{R}^n$.*

Proof We must show that the v_j are linearly independent, so suppose that $\sum_j \lambda_j v_j = 0$. If we take the scalar product with each side of the equation with v_k, we find that $\lambda_k = 0$. As this holds for all k, the conclusion follows. □

It makes sense to speak of an orthonormal basis of a subspace of $\mathbb{R}^n$, and the next result guarantees that these always exist.

Theorem 11.2.3 *Any subspace of $\mathbb{R}^n$ has an orthonormal basis.*

Proof Let W be a non-trivial subspace of $\mathbb{R}^n$, and let $w_1, \ldots, w_k$ be any basis of W. By replacing w_1 by a scalar multiple of itself we may suppose that $||w_1|| = 1$. Now let $w_2' = w_2 - (w_1 \cdot w_2)w_1$, and note that $w_1, w_2', w_3, \ldots, w_k$ is a basis of W with $||w_1|| = 1$ and $w_1 \perp w_2'$. This orthogonality is preserved if we replace w_2' by a scalar multiple of itself; thus, after relabelling, we may now assume that $||w_1|| = ||w_2|| = 1$ and $w_1 \perp w_2$. Now let $w_3' = w_3 - (w_1 \cdot w_3)w_1 - (w_2 \cdot w_3)w_2$. It is clear that $w_1, w_2, w_3', w_4, \ldots, w_k$ is a basis of W, and that w_1, w_2 and w_3' are orthogonal to each other. Again, we may replace w_3' by a scalar multiple of itself and so assume (after relabelling) that w_1, w_2 and w_3 are now mutually orthogonal unit vectors. The process can clearly be continued until it produces an orthonormal basis for W. □

We now define orthogonal maps, and give equivalent definitions.

Definition 11.2.4 A linear map $\alpha : \mathbb{R}^n \to \mathbb{R}^n$ is an *orthogonal map* if $\alpha(e_1), \ldots, \alpha(e_n)$ is an orthonormal basis of $\mathbb{R}^n$.

Theorem 11.2.5 *Let $\alpha : \mathbb{R}^n \to \mathbb{R}^n$ be a linear map. Then the following are equivalent:*

(1) *α is an orthogonal map;*
(2) *α preserves scalar products (for all x and y, $\alpha(x) \cdot \alpha(y) = x \cdot y$);*
(3) *α preserves lengths of vectors (for all x, $||\alpha(x)|| = ||x||$);*
(4) *if $v_1, \ldots, v_n$ is an orthonormal basis, then so is $\alpha(v_1), \ldots, \alpha(v_n)$.*

Proof Assume that (1) holds. As α is linear, $\alpha(x) = \alpha\bigl(\sum_j x_j e_j\bigr) = \sum_j x_j \alpha(e_j)$, and similarly for y. As $\alpha(e_1), \ldots, \alpha(e_n)$ is an orthonormal basis, we see that

$$\alpha(x) \cdot \alpha(y) = \sum_{i,j} x_i y_j \alpha(e_i) \cdot \alpha(e_j) = \sum_j x_j y_j = x \cdot y;$$

thus (2) holds. Clearly (2) implies (3) for $||x||^2 = x \cdot x$ and similarly for $\alpha(x)$.

Conversely, if (3) holds, then the identity

$$x \cdot y = \frac{1}{2}(\|x\|^2 + \|y\|^2 - \|x-y\|^2),$$

and the corresponding identity for $\alpha(x)$ and $\alpha(y)$, shows that (2) holds. Thus (2) and (3) are equivalent. Next, if (3) holds then both (2) and (3) hold, and (4) follows from these. Finally, if (4) holds we let $v_j = e_j$ and then (1) follows. □

We can describe orthogonal maps in terms of their matrices.

Theorem 11.2.6 *Let $\alpha : \mathbb{R}^n \to \mathbb{R}^n$ be a linear map, and let A be the matrix representation of α with respect to the standard basis $e_1, \ldots, e_n$ of $\mathbb{R}^n$. Then the following are equivalent:*

(1) *α is an orthogonal map;*
(2) *the columns of A form an orthonormal basis of $(\mathbb{R}^n)^t$;*
(3) *$A^t A = I$;*
(4) *$A A^t = I$;*
(5) *the rows of A form an orthonormal basis of $\mathbb{R}^n$.*

Proof Let c_j be the column vectors of A. As $\alpha(e_j) = c_j{}^t$, it is immediate that (1) and (2) are equivalent. Next, the definition of a matrix product shows that (2) is equivalent to (3), and that (4) is equivalent to (5). Finally, (3) and (4) are equivalent by Theorem 9.4.4. □

Theorem 11.2.6 suggests the following definition.

Definition 11.2.7 A real square matrix A is *orthogonal* if it is non-singular (that is, invertible), and $A^{-1} = A^t$.

Example 11.2.8 We show that a 2×2 orthogonal matrix is the matrix of either a rotation or a reflection in $\mathbb{R}^2$. Suppose that

$$A = \begin{pmatrix} a & b \\ c & d \end{pmatrix}$$

is orthogonal. Then, by Theorem 11.2.6,

$$a^2 + b^2 = a^2 + c^2 = c^2 + d^2 = b^2 + d^2 = 1, \quad ab + cd = ac + bd = 0.$$

We may choose θ such that $a = \cos\theta$ and $c = \sin\theta$, and then it is easy to see that A is one of the matrices

$$\begin{pmatrix} \cos\theta & -\sin\theta \\ \sin\theta & \cos\theta \end{pmatrix}, \quad \begin{pmatrix} \cos\theta & \sin\theta \\ \sin\theta & -\cos\theta \end{pmatrix}.$$

11.2 Orthogonal maps

These matrices represent a rotation of angle θ, and the reflection in the line $y\cos(\theta/2) = x\sin(\theta/2)$, respectively. □

We shall need the next result in the next section.

Lemma 11.2.9 *Let $\mathcal{B} = \{u_1, \ldots, u_n\}$ and $\mathcal{B}' = \{v_1, \ldots, v_n\}$ be orthonormal bases of $\mathbb{R}^n$, and suppose that the identity map $I : \mathbb{R}^n_{\mathcal{B}} \to \mathbb{R}^n_{\mathcal{B}'}$ has matrix A. Then A is an orthogonal matrix.*

Proof In general, a map $\alpha : \mathbb{R}^n_{\mathcal{B}} \to \mathbb{R}^n_{\mathcal{B}'}$ has matrix $A = (a_{ij})$, where

$$\alpha(u_j) = \sum_{k=1}^{n} a_{kj} v_k, \quad j = 1, \ldots, n.$$

If $\alpha = I$ then $u_j = \sum_k a_{kj} v_k$ and, as

$$u_i \cdot u_j = v_i \cdot v_j = \begin{cases} 1 & \text{if } i = j; \\ 0 & \text{if } i \neq j, \end{cases}$$

this implies that

$$\sum_{k=1}^{n} a_{ki} a_{kj} = \begin{cases} 1 & \text{if } i = j; \\ 0 & \text{if } i \neq j. \end{cases}$$

As this sum is the scalar product of the i-th and j-th columns of A, we see from Theorem 11.2.6 that A is orthogonal. □

The next result gives information about the determinant and eigenvalues of an orthogonal matrix.

Theorem 11.2.10 *Suppose that A is a real $n \times n$ orthogonal matrix. Then*

(1) $\det(A) = \pm 1$;
(2) *any (complex) eigenvalue λ of A satisfies $|\lambda| = 1$;*
(3) *if n is odd, then 1 or -1 is an eigenvalue of A;*
(4) *if n is even, then A need not have any real eigenvalues.*

Proof As $AA^t = I$, $\det(AB) = \det(A)\det(B)$ and $\det(A^t) = \det(A)$, (1) follows. To prove (2), we note that the orthogonal matrix A defines a linear map α of the complex space $\mathbb{C}^{n,1}$ of column vectors into itself by $\alpha(z) = Az$. Now let λ be any eigenvalue of α, and let v be a corresponding (non-zero) eigenvector; thus $Av = \lambda v$. For any complex column vector z, we let $\bar{z}$ be the vector formed by taking the complex conjugate of each of its elements. As A is a real matrix, this gives $A\bar{v} = \bar{A}\bar{v} = \overline{Av} = \overline{\lambda v} = \bar{\lambda}\bar{v}$. Thus

$$|\lambda|^2 v^t \bar{v} = (\lambda v)^t (\bar{\lambda}\bar{v}) = (Av)^t(A\bar{v}) = v^t(A^t A)\bar{v} = v^t(\bar{v}).$$

Now $v^t \bar{v} = \sum_j |v_j|^2$, where $v = (v_1, \ldots, v_n)^t$, and this is non-zero (as every eigenvector is non-zero). Thus $|\lambda| = 1$ and (2) follows. Now (3) follows immediately because if n is odd, then the characteristic equation for A is a real equation of odd degree and so A has at least one real eigenvalue which, by (2), has unit modulus. Finally, the matrix

$$\begin{pmatrix} \cos\theta & -\sin\theta & 0 & 0 \\ \sin\theta & \cos\theta & 0 & 0 \\ 0 & 0 & \cos\phi & -\sin\phi \\ 0 & 0 & \sin\phi & \cos\phi \end{pmatrix}.$$

is an orthogonal 4×4 matrix with no real eigenvalue. A similar matrix exists whenever n is even and this proves (4). □

It is clear from Theorem 11.2.6 that if A and B are orthogonal matrices, then so too are AB and A^{-1}; indeed, $(AB)^t(AB) = B^t(A^tA)B = B^tB = I$, and $(A^{-1})^tA^{-1} = (A^{-1})^tA^t = (AA^{-1})^t = I$. Thus we have the next result (which justifies the definition that follows it).

Theorem 11.2.11 *The class of $n \times n$ orthogonal matrices is a group under matrix multiplication.*

Definition 11.2.12 The group of real orthogonal $n \times n$ matrices is called the *orthogonal group*, and it is denoted by $O(n)$. The subgroup of matrices with determinant $+1$ (see Theorem 11.2.10) is called the *special orthogonal group*, and is denoted by $SO(n)$.

Example 11.2.13 Let A be a 3×3 orthogonal matrix, and let α be the linear map of $\mathbb{R}^3$ into itself whose matrix is A relative to the standard basis. By Theorem 11.2.10, α has an eigenvector v corresponding to an eigenvalue of ± 1. As α preserves scalar products, the plane Π orthogonal to v is mapped into itself by α. As α preserves lengths and scalar products, its action on the plane Π is orthogonal, and so it either acts as a rotation of Π, or as the reflection across a line in Π (see Example 11.2.8). It follows that the action of α relative to the basis v_1, v_2, v_3 (where $v_1 = v$, and v_2, v_3 is a basis of Π) is of one of the forms

$$\begin{pmatrix} \pm 1 & 0 & 0 \\ 0 & \cos\theta & -\sin\theta \\ 0 & \sin\theta & \cos\theta \end{pmatrix}, \quad \begin{pmatrix} \pm 1 & 0 & 0 \\ 0 & \cos\theta & \sin\theta \\ 0 & -\sin\theta & \cos\theta \end{pmatrix}.$$

□

The ideas in Example 11.2.13 can be generalized to higher dimensions.

Theorem 11.2.14 *Let A be a real $n \times n$ orthogonal matrix. Then there exists a real orthogonal matrix Q such that (with the obvious meaning)*

$$QAQ^{-1} = \begin{pmatrix} A_1 & & & & \\ & \ddots & & & \\ & & A_r & & \\ & & & I_s & \\ & & & & -I_t \end{pmatrix}, \quad A_k = \begin{pmatrix} \cos\theta_k & -\sin\theta_k \\ \sin\theta_k & \cos\theta_k \end{pmatrix},$$

where all of the unspecified entries are zero, and where I_m denotes the unit $m \times m$ matrix.

The significance of Theorem 11.2.14 is that given an orthogonal matrix A acting on $\mathbb{R}^n$, we can decompose $\mathbb{R}^n$ into a number of two-dimensional subspaces, each of which is invariant by A, and on each of which A acts as a rotation or a reflection, and two other subspaces on which A acts as I and $-I$, respectively. Moreover, any two of these subspaces are orthogonal to each other.

Briefly, the proof is as follows. We know that any real linear map (given by the matrix A) of a real vector space into itself has an invariant subspace, say U, of dimension one or two (Theorem 10.1.8). As A preserves scalar products, the set of vectors that are orthogonal to all vectors in U is also a subspace that is invariant under A, and as A preserves lengths, the action of A that takes U into itself is also orthogonal. We can continue in this way to obtain the desired decomposition of the whole space into mutually orthogonal subspaces of dimension one or two, and the result now follows from the ideas discussed above. □

Exercise 11.2

1. Find an orthonormal basis of $\mathbb{R}^3$ that contains the vector $\frac{1}{9}(1, 4, 8)$.
2. Show that the vectors $(1, 0, 1, 0)$, $(1, 1, -1, 0)$ and $(1, -2, -1, 1)$ in $\mathbb{R}^4$ are mutually orthogonal. Find a fourth vector that is orthogonal to each of these, and hence find an orthonormal basis of $\mathbb{R}^4$ that contains a scalar multiple of each of these vectors.
3. Show that if there is an orthonormal basis of $\mathbb{R}^n$ that consists of eigenvectors of both of the $n \times n$ matrices A and B, then $AB = BA$.
4. Show that for suitable p, q, r, s the matrix
$$\begin{pmatrix} p & p & p & p \\ q & -q & q & -q \\ r & 0 & -r & 0 \\ 0 & s & 0 & -s \end{pmatrix}$$
is orthogonal.

5. Show that if B is a square invertible matrix then $(B^t)^{-1} = (B^{-1})^t$. Now suppose that A is a square matrix, and that $I + A$ is invertible. Show
   (a) if A is orthogonal then $(I - A)(I + A)^{-1}$ is skew-symmetric;
   (b) if A is skew-symmetric then $(I - A)(I + A)^{-1}$ is orthogonal.

11.3 Isometries of Euclidean n-space

We shall now examine the group of isometries of $\mathbb{R}^n$.

Definition 11.3.1 A map $f : \mathbb{R}^n \to \mathbb{R}^n$ is an *isometry* if it preserves distances; that is, if $||f(x) - f(y)|| = ||x - y||$ for every x and y in $\mathbb{R}^n$.

Each translation is an isometry, and if A is an $n \times n$ orthogonal matrix then g defined by $g(x) = xA$ is an isometry (note: we must write xA for a row vector x, and Ax for a column vector x). Indeed, g is linear, and it preserves lengths as

$$||g(x) - g(y)|| = ||g(x - y)|| = ||(x - y)A|| = ||x - y||.$$

It follows that each map of the form $x \mapsto xA + a$, where A is orthogonal, is an isometry, and we shall now show that every isometry is of this form.

Theorem 11.3.2 *Every isometry of $\mathbb{R}^n$ is of the form $x \mapsto a + xA$ for some a in $\mathbb{R}^n$ and some orthogonal matrix A. In particular, f is an isometry with $f(0) = 0$ if and only if $f(x) = xA$ for some orthogonal matrix A.*

Proof Let f be any isometry, and let $g(x) = f(x) - f(0)$; then it suffices to show that $g(x) = xA$ for some orthogonal matrix A. Now g is clearly an isometry (it is the composition of the isometry f with a translation), and $g(0) = 0$. Thus, for all x and y,

$$||g(x)|| = ||g(x) - g(0)|| = ||x - 0|| = ||x||, \quad ||g(y)|| = ||y||;$$

that is, g preserves the length of each vector. Also, as g is an isometry, $||g(x) - g(y)|| = ||x - y||$, and as

$$2x \cdot y = ||x||^2 + ||y||^2 - ||x - y||^2,$$
$$2g(x) \cdot g(y) = ||g(x)||^2 + ||g(y)||^2 - ||g(x) - g(y)||^2,$$

we see that $g(x) \cdot g(y) = x \cdot y$; that is, g preserves scalar products. Thus, by Lemma 11.2.2, the vectors $g(e_1), \ldots, g(e_n)$ form an orthonormal basis of $\mathbb{R}^n$. Now write

$$x = \sum x_j e_j, \quad g(x) = \sum x'_j g(e_j).$$

11.3 Isometries of Euclidean n-space

As g preserves scalar products, for each k, $x_k = x \cdot e_k = g(x) \cdot g(e_k) = x'_k$, and hence

$$g\left(\sum_j x_j e_j\right) = \sum_j x_j g(e_j).$$

This implies that g is a linear map and so, from Theorems 11.2.5 and 11.2.6, $g(x) = xA$ for some orthogonal matrix A. □

We have already seen that each isometry of $\mathbb{R}^3$ is the composition of at most four reflections (Theorem 6.1.4), and we shall now extend this by showing that every isometry of $\mathbb{R}^n$ is a composition of at most $n + 1$ reflections.

Theorem 11.3.3 *Every isometry of $\mathbb{R}^n$ can be written as the composition of at most $n + 1$ reflections.*

We must comment on the idea of a reflection in $\mathbb{R}^n$. In Section 8.1 we defined a hyperplane to be an $(n - 1)$-dimensional subspace of $\mathbb{R}^n$ and, consequently, each hyperplane was given by an equation of the form $x \cdot a = 0$. However, we now want to discuss 'hyperplanes' that do not contain the origin, and so we must now modify the earlier definition. In this discussion a hyperplane will be the set of vectors x that satisfy $x \cdot a = d$, where $a \in \mathbb{R}^n$ and $d \in \mathbb{R}$. The ideas in Chapter 3 generalize immediately, and the reflection across the hyperplane with equation $x \cdot a = d$, where $||a|| = 1$, is defined to be the map $R(x) = x + 2(d - (x \cdot a))a$.

The proof of Theorem 11.3.3 It is sufficient to show that if f is an isometry with $f(0) = 0$, then f can be expressed as the composition of at most n reflections, and we shall prove this by induction on n. First, we know the result to be true when n is 1, 2 or 3. We now suppose that the conclusion holds for any isometry of $\mathbb{R}^k$, where $k = 1, \ldots, n - 1$, and we consider an isometry f of $\mathbb{R}^n$ that fixes the origin. As $||f(e_n)|| = ||f(e_n) - f(0)|| = ||e_n - 0|| = 1$, we see that there is a reflection, say R, in some hyperplane through the origin so that Rf is an isometry that fixes both 0 and e_n. Now let $g = Rf$. It is easy to see that g fixes every point of the line L through 0 and e_n (consider a point $(0, \ldots, 0, t)$ and require that its image is a distance $|t|$ from 0, and $|t - 1|$ from e_n). Next, let

$$W = \{(x_1, \ldots, x_{n-1}, 0) : x_1, \ldots, x_{n-1} \in \mathbb{R}\}.$$

Then, as W is the set of points that are orthogonal to L, and as g preserves scalar products, we see that g maps W into itself. More generally, if $g(x) = y$, then $y_n = x_n$ (because $||y|| = ||x||$, and $||y - te_n||^2 = ||x - te_n||^2$ for every real t).

Now we can regard g as a map of the $(n - 1)$-dimensional space W into itself; formally, if $g(x_1, \ldots, x_n) = (y_1, \ldots, y_n)$, then $y_n = x_n$, and we can define

$g^*: \mathbb{R}^{n-1} \to \mathbb{R}^{n-1}$ by

$$g^*(x_1, \ldots, x_{n-1}) = (y_1, \ldots, y_{n-1}).$$

We can identify W with $\mathbb{R}^{n-1}$ and, by the induction hypothesis, we can now write $g^* = R_1^* \cdots R_k^*$, where each R_j^* is a reflection across some hyperplane in $\mathbb{R}^{n-1}$ that contains the origin, and $k \leq n-1$. Now each R_j^* is given by some equation

$$R_j^*(x^*) = x^* - 2(x^* \cdot a_j^*) a_j^*,$$

where $x^* = (x_1, \ldots, x_{n-1})$, and $a_j^* = (a_1, \ldots, a_{n-1})$ with $||a_j^*|| = 1$. This means that we can define a reflection R_j, acting on $\mathbb{R}^n$, by $R_j(x) = x - 2(a_j \cdot x) a_j$, where $a_j = (a_1, \ldots, a_{n-1}, 0)$. It is clear that each R_j preserves the n-th coordinate of any vector in $\mathbb{R}^n$, and hence that the first $n-1$ coordinates of $R_1 \cdots R_k(x_1, \ldots, x_n)$ agree with the corresponding coordinates of $R_1^* \cdots R_k^*(x_1, \ldots, x_{n-1})$. It follows that $g = R_1 \cdots R_k$, and hence that $f = R R_1 \cdots R_k$ as required. □

Theorem 11.3.3 enables us to divide the isometries of $\mathbb{R}^n$ into two classes, namely the direct and indirect isometries. We say that the isometry $x \mapsto a + xA$ is a *direct isometry* if $\det(A) = 1$, and it is an *indirect isometry* if $\det(A) = -1$ (recall that as A is orthogonal, $\det(A) = \pm 1$).

Exercise 11.3

1. Find an isometry of $\mathbb{C}$ ($= \mathbb{R}^2$) that requires three reflections in the sense of Theorem 11.3.3 (that is, that cannot be written as the product of one or two reflections).
2. Show that every rotation of $\mathbb{R}^3$ can be written as the product of two rotations of order two. [Hint: a rotation of $\mathbb{R}^3$ is a product of two reflections.] Is a similar statement true in $\mathbb{R}^2$?

11.4 Symmetric matrices

The study of homogeneous quadratic forms, that is, expressions of the form $\sum_{i,j} a_{ij} x_i x_j$, is important because these forms arise naturally in many parts of mathematics. If we make a linear change of variables, then the new form will have different coefficients, and may be simpler to analyse. For example, if $u = (x+y)/\sqrt{2}$ and $v = (x-y)/\sqrt{2}$, then $x^2 + 4xy + y^2 = 3u^2 - v^2$, and it is now obvious that the form $x^2 + 4xy + y^2$ can take negative as well as

11.4 Symmetric matrices

positive values. Our aim in this section is to show that *any real quadratic form can, by a suitable change of variables, be transformed into a form of the type* $\sum_j a_j x_j^2$. In order to achieve this we shall need to use orthogonal matrices, and orthonormal bases of eigenvectors, and we begin with a simple example to illustrate why these may be relevant.

Example 11.4.1 We can express the quadratic form $x^2 + 4xy + y^2$ in terms of a matrix in many ways, namely

$$x^2 + 4xy + y^2 = (x, y)\begin{pmatrix} 1 & s \\ t & 1 \end{pmatrix}\begin{pmatrix} x \\ y \end{pmatrix},$$

where $s + t = 4$. Let A be the matrix in this expression. Then A has real, distinct eigenvalues if and only if $st > 0$, so let us assume that this is so. Then the eigenvalues of A are $1 + \sqrt{st}$ and $1 - \sqrt{st}$, and the corresponding eigenvectors are $(\sqrt{t}, \sqrt{s})$ and $(\sqrt{t}, -\sqrt{s})$, respectively. Now these eigenvectors are orthogonal if and only if $s = t$, so that (in this example, at least), *the eigenvectors exist and are orthogonal if and only if A is symmetric*. □

There is another simple argument that suggests that symmetry and orthogonality might be related. Suppose we are given a linear map $\alpha : \mathbb{R}^n \to \mathbb{R}^n$ that is represented by a symmetric matrix A with respect to some basis. If we now change this basis in at attempt to find a diagonal matrix that represents α, the new matrix will be, say, PAP^{-1}. Now, in general, PAP^{-1} will not be symmetric, and so certainly not diagonal. It seems reasonable, then, to ask which matrices X at least have the property that XAX^{-1} is symmetric whenever A is symmetric. In fact, *this is true for all orthogonal matrices X*, and the proof is easy. If A is symmetric and X is orthogonal, then

$$(XAX^{-1})^t = (XAX^t)^t = (X^t)^t A^t X^t = XAX^{-1},$$

so that XAX^{-1} is symmetric. We record this for future use.

Lemma 11.4.2 *If A is symmetric and X is orthogonal, then XAX^{-1} is also symmetric.*

It is usually better to think about linear maps rather than matrices, so we need to understand which properties of linear maps are related to symmetric matrices. We shall now define what it means for a linear *map* to be symmetric, and then relate this to symmetric matrices.

Definition 11.4.3 A linear map $\alpha : \mathbb{R}^{n,t} \to \mathbb{R}^{n,t}$ is *symmetric* if and only if given any orthonormal basis $\mathcal{B}$ of $\mathbb{R}^{n,t}$, the matrix representation of α relative to $\mathcal{B}$ is symmetric.

Although this definition requires that the matrix representation of α relative to *every* orthonormal basis is symmetric, it is sufficient to know this for just one orthonormal basis.

Lemma 11.4.4 *A linear map $\alpha : \mathbb{R}^{n,t} \to \mathbb{R}^{n,t}$ is symmetric if there is one orthonormal basis of $\mathbb{R}^{n,t}$ for which the matrix representation of α is symmetric.*

Proof Suppose that there is some orthonormal basis $\mathcal{B}$ with respect to which α has a symmetric matrix representation A. Let $\mathcal{B}'$ be any orthonormal basis, and let X be the matrix of the identity map I from the basis $\mathcal{B}$ to the basis $\mathcal{B}'$. Then, by Theorem 9.5.1, the matrix representation of α relative to the basis $\mathcal{B}'$ is XAX^{-1}. By Lemma 11.2.9, X is orthogonal, and by Lemma 11.4.2, XAX^{-1} is symmetric. □

There is another characterization of symmetric linear maps.

Theorem 11.4.5 *A linear map $\alpha : \mathbb{R}^{n,t} \to \mathbb{R}^{n,t}$ is symmetric if and only if for all vectors x and y, $\alpha(x) \cdot y = x \cdot \alpha(y)$.*

Proof Suppose that α is symmetric; then the matrix, say A, of α with respect to the standard basis $e_1^t, \ldots, e_n^t$ is symmetric. Take any x and y in $\mathbb{R}^{n,t}$; then

$$\alpha(x) \cdot y = (Ax)^t y = (x^t A) y = x^t (Ay) = x \cdot \alpha(y).$$

Now suppose that for all x and y, $\alpha(x) \cdot y = x \cdot \alpha(y)$, and let A be the matrix representation of α relative to the standard basis of $\mathbb{R}^{n,t}$. Then (as above) $x^t A^t y = xAy$. If we now let $x = e_i^t$ and $y = e_j^t$, then we see that $a_{ij} = a_{ji}$, so that A is symmetric. Lemma 11.4.4 now implies that α is symmetric. □

We need one more result before we can achieve our objective. The characteristic polynomial of an $n \times n$ real symmetric matrix A is a real polynomial of degree n and, by the Fundamental Theorem of Algebra, this will always have n complex roots. We need to show that all of these roots all real.

Theorem 11.4.6 *Every root of the characteristic equation of a real symmetric matrix is real.*

Proof Let A be a real $n \times n$ symmetric matrix (a_{ij}). Then A provides a linear map of the complex vector space $\mathbb{C}^{n,t}$ into itself by the rule

$$\begin{pmatrix} z_1 \\ \vdots \\ z_n \end{pmatrix} \mapsto \begin{pmatrix} a_{11} & \cdots & a_{1n} \\ \vdots & \ddots & \vdots \\ a_{n1} & \cdots & a_{nn} \end{pmatrix} \begin{pmatrix} z_1 \\ \vdots \\ z_n \end{pmatrix},$$

which we shall write as $z^t \mapsto Az^t$, where $z = (z_1, \ldots, z_n)$. We shall also use the notation $\bar{z} = (\bar{z}_1, \ldots, \bar{z}_n)$, where $\bar{z}_j$ is the usual complex conjugate of z_j.

11.4 Symmetric matrices

Now take any (possibly complex) eigenvalue λ of A, and a corresponding (possibly complex) eigenvector w^t; thus $Aw^t = \lambda w^t$. We now take the complex conjugate, and then the transpose, of each side of this equation. As A is real and symmetric, $\bar{A} = A = A^t$, and this gives $\bar{w}A = \bar{\lambda}\bar{w}$. Thus

$$\bar{\lambda}\bar{w}w^t = (\bar{w}A)w^t = \bar{w}(Aw^t) = \lambda\bar{w}w^t,$$

and hence $(\bar{\lambda} - \lambda)(|w_1|^2 + \cdots + |w_w|^2) = 0$. As w is an eigenvector, $w \neq 0$ and so some w_j is non-zero; thus λ is real. □

Finally, we are now in a position to state and prove our main result.

Theorem 11.4.7 *Given a linear symmetric map $\alpha : \mathbb{R}^{n,t} \to \mathbb{R}^{n,t}$, there is an orthonormal basis of $\mathbb{R}^{n,t}$ such that each basis element is an eigenvector of α.*

If we rephrase Theorem 11.4.7 in terms of matrices rather than linear maps, we obtain the following result.

Corollary 11.4.8 *A real symmetric $n \times n$ matrix A has n mutually orthogonal eigenvectors. In particular, there exists an orthogonal matrix P such that $PAP^t (= PAP^{-1})$ is diagonal.*

Proof of Theorem 11.4.7 We prove this by induction on n, and the conclusion is true when $n = 1$. Now suppose the result is true for $n = 1, \ldots m - 1$ and consider the case when $n = m$.

By Theorem 11.4.6, α has a real eigenvalue, say λ_1 and a corresponding eigenvector w_1, with $||w_1|| = 1$. Let $W_1 = \{x \in \mathbb{R}^{m,t} : x \perp w_1\}$. Then, by Theorem 8.1.2, W_1 has dimension $m - 1$. Moreover, α maps W_1 into itself because if $x \in W_1$, then $\alpha(x) \cdot w_1 = x \cdot \alpha(w_1) = x \cdot (\lambda_1 w_1) = 0$. Now the action of α on W_1 is symmetric (because $\alpha(x) \cdot y = x \cdot \alpha(y)$ for all x and y in $\mathbb{R}^{m,t}$, and hence certainly for x and y in W_1). Thus, by the induction hypothesis, W_1 has an orthonormal basis consisting entirely of eigenvectors of α, and the proof is complete. □

We give one example to illustrate these ideas.

Example 11.4.9 Let

$$A = \begin{pmatrix} 4 & 2 \\ 2 & 1 \end{pmatrix}.$$

Then the eigenvalues of A are 0, with eigenvector $(1, -2)^t$, and 5, with eigenvector $(2, 1)^t$. These vectors are orthogonal to each other, and if we scale these so that they have unit length, we obtain an orthonormal basis of $\mathbb{R}^{2,t}$, namely

$(\mu, -2\mu)^t$ and $(2\mu, \mu)^t$, where $\mu = 1/\sqrt{5}$. Thus

$$\begin{pmatrix} 4 & 2 \\ 2 & 1 \end{pmatrix}\begin{pmatrix} \mu \\ -2\mu \end{pmatrix} = 0\begin{pmatrix} \mu \\ -2\mu \end{pmatrix}, \quad \begin{pmatrix} 4 & 2 \\ 2 & 1 \end{pmatrix}\begin{pmatrix} 2\mu \\ \mu \end{pmatrix} = 5\begin{pmatrix} 2\mu \\ \mu \end{pmatrix}.$$

Together, these give

$$\begin{pmatrix} 4 & 2 \\ 2 & 1 \end{pmatrix}\begin{pmatrix} \mu & 2\mu \\ -2\mu & \mu \end{pmatrix} = \begin{pmatrix} \mu & 2\mu \\ -2\mu & \mu \end{pmatrix}\begin{pmatrix} 0 & 0 \\ 0 & 5 \end{pmatrix},$$

which can be writen in the form

$$\begin{pmatrix} \mu & -2\mu \\ 2\mu & \mu \end{pmatrix}\begin{pmatrix} 4 & 2 \\ 2 & 1 \end{pmatrix}\begin{pmatrix} \mu & 2\mu \\ -2\mu & \mu \end{pmatrix} = \begin{pmatrix} 0 & 0 \\ 0 & 5 \end{pmatrix};$$

that is, in the form $P^{-1}AP = D$, where the diagonal matrix D has the eigenvalues of A as its diagonal entries. □

It is perhaps worth stressing that it is not just the symmetry of the matrix that is important in this discussion; it is the symmetry of the matrix *together with the existence of a scalar product*. We are so familiar with the scalar product (at least in $\mathbb{R}^3$) that we tend to take its existence for granted; however without it, symmetry is of no consequence.

It is also important to note that the conclusion of Corollary 11.4.8 do not generally hold for symmetric *complex* matrices : for example, the symmetric matrix A given by

$$A = \begin{pmatrix} 3 & i \\ i & 1 \end{pmatrix}$$

is not diagonalizable. Indeed, if there is a complex non-singular matrix X such that XAX^{-1} is a diagonal matrix D, then the diagonal entries in D are the eigenvalues of A, and these are 2, 2. Thus $D = 2I$, and hence $A = X^{-1}(2I)X = 2I$, which is not so. In fact, a similar theory holds for complex matrices if we use the scalar product $z \cdot w = \sum_j z_j \bar{w}_j$ and require that the matrices A satisfy $\bar{A}^t = A$ instead of $A^t = A$.

Exercise 11.4

1. Verify the details regarding the eigenvalues and eigenvectors given in Example 11.4.1. In particular, show that if $s = t = 2$ then A is symmetric, and has eigenvalues 3 and -1 with corresponding (orthogonal) eigenvectors $(1, 1)$ and $(1, -1)$. Use this to find an orthogonal matrix P such that

$$P\begin{pmatrix} 1 & 2 \\ 2 & 1 \end{pmatrix}P^{-1} = \begin{pmatrix} 3 & 0 \\ 0 & -1 \end{pmatrix}.$$

2. Let
$$A = \begin{pmatrix} 1 & 1 \\ 1 & 1 \end{pmatrix}, \quad X = \begin{pmatrix} 1 & 2 \\ 0 & 1 \end{pmatrix}.$$
Show that A is symmetric but XAX^{-1} is not. Is X orthogonal?

3. For each of the following matrices, find an orthonormal basis of $\mathbb{R}^{3,t}$ that consists of eigenvectors of the matrix:
$$\begin{pmatrix} 0 & 1 & 1 \\ 1 & 0 & 1 \\ 1 & 1 & 0 \end{pmatrix}, \quad \begin{pmatrix} 0 & 0 & 1 \\ 0 & 0 & 1 \\ 1 & 1 & 0 \end{pmatrix}, \quad \begin{pmatrix} 2 & 1 & 1 \\ 1 & 2 & 1 \\ 1 & 1 & 2 \end{pmatrix}.$$

4. Show, without the use of matrices, that under a suitable linear change of variables, $2yz + 2xz$ can be expressed in the form $X^2 - Y^2$. Now use the theory of symmetric matrices to achieve a similar result.

5. Find a linear change of variables that transforms the quadratic form $\sum_{1 \leq i < j \leq 4} 2x_i x_j$ into $-y_1^2 - y_2^2 - y_3^2 + 3y_4^2$.

11.5 The field axioms

In this and the next section we briefly discuss Euclidean n-dimensional space $\mathbb{R}^n$, and ask to what extent it is possible to define some kind of product on this space. The single result in this section says that if $n \geq 3$ then it is impossible to create a multiplication on $\mathbb{R}^n$ which, when combined with the natural addition on $\mathbb{R}^n$, makes $\mathbb{R}^n$ into a field. Thus it is only in the case of real and complex numbers (that is, for $n = 1, 2$) can we obtain a field (and some suggestion that this might be so was given in Section 4.3).

Theorem 11.5.1 *Suppose that there is a multiplication defined on $\mathbb{R}^n$ which, with vector addition, gives a field. Then n is 1 or 2.*

Proof The proof is by contradiction, so we suppose that $n \geq 3$, and that we have a multiplication in $\mathbb{R}^n$ which, with vector addition, makes $\mathbb{R}^n$ into a field. We denote the product of x and y by xy. Now the field axioms guarantee the existence of a vector e such that $ex = x = xe$ for every x. The vector 0 is, of course, the additive identity. Now in any field, $0x = 0 = x0$, and this implies that $0 \neq e$ (as otherwise $x = xe = x0 = 0$ and so 0 would be the only vector in the space).

Now choose a vector x that is not a multiple of e, and consider the vectors $e, x, x^2, \ldots, x^n$. As these vectors are linearly dependent, there are real numbers λ_j, not all zero, such that $\lambda_0 e + \lambda_1 x + \cdots + \lambda_n x^n = 0$. Notice that we cannot have $\lambda_1 = \cdots = \lambda_n = 0$ else $\lambda_0 e = 0$ and then $e = 0$. Now define the real polynomial P by $P(t) = \sum_j \lambda_j t^j$. As some λ_j, where $j > 0$, is non-zero, P is not constant.

As P has real coefficients, it is a product of real linear factors and real quadratic factors, where the quadratic factors (if any) do not factorize into a product of linear factors. Now in any field, $uv = 0$ implies that $u = 0$ or $v = 0$ so, as $P(x) = 0$, we see that either there is some linear factor that vanishes at x, or there is some quadratic factor that vanishes at x. Now a linear factor cannot be zero for if $ax + be = 0$, then x would be a multiple of e, contrary to our original choice of x. It follows that there is some quadratic factor that is zero, say $Ax^2 + Bx + Ce = 0$, where $A \neq 0$ and $B^2 - 4AC < 0$ (because we know that this quadratic expression is not a product of linear factors). It follows that

$$(x + B(2A)^{-1}e)^2 = (B^2 - 4AC)(4A^2)^{-1}e = -\mu^2 e,$$

where $\mu > 0$, and so there are real numbers α and β such that, for the given x, $(\alpha x + \beta e)^2 = -e$.

Now take any other y that is not a multiple of e. Then, exactly as for x, there are real numbers γ and δ such that $(\gamma y + \delta e)^2 = -e$. We deduce that

$$0 = (\alpha x + \beta e)^2 - (\gamma y + \delta e)^2$$
$$= (\alpha x + \beta e + \gamma y + \delta e)(\alpha x + \beta e - \gamma y - \delta e).$$

As one of these factors must be 0, this means that x, y and e are linearly dependent. As $n \geq 3$, we could have chosen y so that it was not linearly dependent on x and e; thus we have a contradiction. □

11.6 Vector products in higher dimensions

This section is intended to give the reader more insight into the very special nature of the vector product in $\mathbb{R}^3$. We begin by showing how the vector product in $\mathbb{R}^3$ is intimately connected to rotations. We recall from (6.1.5) that if A is a rotation of $\mathbb{R}^3$ about the origin then $A(x) \times A(y) = A(x \times y)$. The next result is a type of converse of this in the sense that it shows that the vector product is characterized by linearity and its invariance under rotations.

Theorem 11.6.1 *Suppose that there is some product of vectors in $\mathbb{R}^3$, say $x \star y$, that is itself a vector in $\mathbb{R}^3$, with the properties*

(a) $x \star y$ *is linear in x, and linear in y, and*
(b) $x \star y$ *is invariant under all rotations; that is, $A(x) \star A(y) = A(x \star y)$.*

Then $x \star y$ is a constant multiple of the vector product $x \times y$.

11.6 Vector products in higher dimensions

Proof First, consider a non-zero vector x, take any vector y orthogonal to x, and let A be the rotation of angle π about the axis along y. As $A(x) = -x$, (a) and (b) imply that

$$A(x \star x) = A(x) \star A(x) = (-x) \star (-x) = x \star x.$$

As the only solutions of $A(x) = x$ are scalar multiples of y, we see that $x \star x$ is a scalar multiple of y. However, the direction of y is arbitrary (subject to being orthogonal to x); thus we must have $x \star x = 0$. If we now expand $(x + y) \star (x + y)$ we see that for all x and y, $x \star y = -y \star x$.

Take any two non-zero vectors x and y that do not lie along the same line. Then they determine a plane Π, say, with normal along the vector n. Let B be the rotation of angle π about the axis n. Then

$$B(x \star y) = B(x) \star B(y) = (-x) \star (-y) = x \star y,$$

so that $x \star y$ must be a scalar multiple of n for these are the only solutions of $B(x) = x$. This shows that $x \star y$ is orthogonal to x and to y. In particular, $\mathbf{i} \star \mathbf{j}$ is a multiple of $\mathbf{k}$ and so on. Now any fixed scalar multiple of a product satisfying (a) and (b) also satisfies (a) and (b), so these can only determine such a product to within a scalar multiple. In such situations one chooses an arbitrary normalization and, as $\mathbf{i} \star \mathbf{j}$ is a multiple of $\mathbf{k}$, we may choose the normalization $\mathbf{i} \star \mathbf{j} = \mathbf{k}$.

Finally, let C be the rotation of angle $2\pi/3$ about the axis given by $x_1 = x_2 = x_3$. Then A cyclically permutes $\mathbf{i}$, $\mathbf{j}$ and $\mathbf{k}$ and so, using (b), we see that $\mathbf{j} \star \mathbf{k} = \mathbf{i}$ and $\mathbf{k} \star \mathbf{i} = \mathbf{j}$. We now know $x \star y$ whenever x and y are any of $\mathbf{i}, \mathbf{j}$ and $\mathbf{k}$, and as these last three vectors are a basis of $\mathbb{R}^3$, the linearity of $\star$ now implies that the product, $x \star y$, is completely determined. Moreover, we can compute what it must be and when we do this we find that $x \star y = x \times y$. □

We have defined a multiplication in $\mathbb{R}^3$, namely the vector product, and although this does not make $\mathbb{R}^3$ into a field, it does have some very useful properties. We have also seen that we cannot make $\mathbb{R}^n$ into a field. Might it be possible, then, to define a vector product in other spaces $\mathbb{R}^n$? Of course, this depends on what we mean by a 'vector product' but, as the (very) brief sketch below shows, if we extract some of the fundamental properties of the vector product and then ask in which spaces does such a product exist, the answer, rather surprisingly, is 'only in $\mathbb{R}^3$ and $\mathbb{R}^7$'.

Theorem 11.6.2 *Suppose that $\mathbb{R}^n$ supports a non-zero vector product, which we denote by $x \star y$, with the following properties:*

(1) $x \star y$ *is linear in x and in y;*

(2) $x \star y$ is orthogonal to x and to y;
(3) $||x \star y||^2 + (x \cdot y)^2 = ||x||^2 ||y||^2$.

Then n is 3 or 7.

We can only comment briefly on this result. First, (3) implies that $x \star x = 0$. Thus if we expand $(x+y) \star (x+y)$ we see that $x \star y = -y \star x$. If $n = 1$ then we have $x \star y = xy(1 \star 1) = 0$, so no product of this type exists on $\mathbb{R}$ (except the zero product). Likewise, none can exist on $\mathbb{R}^2$, since (2) clearly fails in this case. If $n = 3$ the vector product $x \times y$ satisfies the requirements given in Theorem 11.6.1 (and this is the only such product up to multiplication by a scalar). How does one handle the cases $n \geq 4$? Briefly, one shows that if such a vector product exists on $\mathbb{R}^n$, then one can construct a product, say $\otimes$, on $\mathbb{R}^{n+1}$ exactly as one did for quaternions, namely

$$(a, x) \otimes (b, y) = (ab - x \cdot y, ay + bx + (x \star y)),$$

which

(1) is linear in x and in y,
(2) has an identity element e, and
(3) is such that the norm of the product is the product of the norms.

A result of Hurwitz (in 1898) is that such a product can only occur if $n+1$ is 1, 2, 4 or 8. Thus n is 3 or 7.

Exercise 11.6

1. Suppose that $\Phi(\mathbf{a}, \mathbf{b}, \mathbf{c})$ is a real-valued function of the vectors $\mathbf{a}$, $\mathbf{b}$ and $\mathbf{c}$ in $\mathbb{R}^3$ with the following properties:
   (1) $\Phi(\mathbf{a}, \mathbf{b}, \mathbf{c})$ is linear in $\mathbf{a}$, $\mathbf{b}$ and $\mathbf{c}$;
   (2) $\Phi(\mathbf{a}, \mathbf{b}, \mathbf{c}) = 0$ if $\mathbf{a} = \mathbf{b}$, or $\mathbf{b} = \mathbf{c}$, or $\mathbf{c} = \mathbf{a}$;
   (3) $\Phi(\mathbf{i}, \mathbf{j}, \mathbf{k}) = 1$.
   Show that $\Phi(\mathbf{a}, \mathbf{b}, \mathbf{c}) = [\mathbf{a}, \mathbf{b}, \mathbf{c}]$.

12
Groups

12.1 Groups

In Chapter 1 we defined what it means to say that a set G is a group with respect to an operation $*$, and we studied groups of permutations of a set. In this chapter we shall study 'abstract' groups. Although this may seem like a more difficult task, every 'abstract' group is the permutation group of some set so that, in some sense, this apparent change of direction is only an illusion.

We have already seen many examples of groups: the sets $\mathbb{Z}$, $\mathbb{R}$ and $\mathbb{C}$ of integers, real numbers, and complex numbers, respectively, the spaces $\mathbb{R}^n$ and $\mathbb{C}^n$, the set of matrices of a fixed size, and indeed any vector space, all form a group with respect to addition. Likewise, the set $\mathbb{R}^+$ of positive real numbers, the set $\mathbb{C}^*$ of non-zero complex numbers, the set of complex numbers of modulus one, the set of non-singular $n \times n$ matrices, and the set of n-th roots of unity, all form a group with respect to multiplication. Finally, the set of bijections of a given set onto itself, the set of isometries of $\mathbb{C}$, and of $\mathbb{R}^n$, the permutations of $\{1, \ldots, n\}$, and the set of non-singular (invertible) linear transformations of a vector space onto itself, all form a group with respect to the usual composition of functions (that is, if $f * g$ is the function defined by $(f * g)(x) = f(g(x))$). It is hardly surprising, then, that the study of general groups is so useful and important.

Often a group is a familiar object with an accepted symbol instead of $*$; for example, where there is an operation of addition we use $x + y$ instead of $x * y$, and 0 instead of e (the identity element). Likewise, if $*$ is multiplication we usually write xy instead of $x * y$, and 1 instead of e. In these and similar cases we shall retain the use of the accepted symbols. From now on we shall, for the general group, adopt the notation gh for $g * h$; however, it must always be remembered that underlying the notation gh there is some (perhaps unspecified) operation $*$.

We recall that a group G contains a unique identity element e such that $ge = g = eg$ for every g, and that every g has a unique inverse g^{-1} such that $gg^{-1} = e = g^{-1}g$. Note also that $(gh)^{-1} = h^{-1}g^{-1}$ for obviously,

$$(gh)(h^{-1}g^{-1}) = e = (h^{-1}g^{-1})(gh).$$

Next, suppose that g is an element of a group G. We denote the composition of g with itself n times by g^n; formally, this is defined for $n = 1, 2, \ldots$ by $g^0 = e$ and $g^{n+1} = g^n g$. We also define g^{-n}, where $n \geq 1$, to be $(g^{-1})^n$. With these definitions it is straightforward, though tedious, to show that for all integers m and n, $g^m g^n = g^{m+n}$. We omit the proof (which is by induction).

We now define the order of a group, and the order of an element of a group. If E is a finite set we denote the number of elements in E by $|E|$; this is called the *cardinality* of E. When E is a group, the word *order* is preferred to cardinality.

Definition 12.1.1 Let G be a group. If G is a finite set, the *order* of G is $|G|$. If G is an infinite set then G is said to have *infinite order*.

Definition 12.1.2 Let G be a group, and suppose that $g \in G$. The element g is said to have (or be of) *finite order* if there is some integer m such that $g^m = e$. If g ($\neq e$) has finite order, then the *order* of g is the smallest positive integer k such that $g^k = e$. If g is not of finite order then it has (or is of) *infinite order*.

For example, the group $\{1, i, -1, -i\}$ of complex numbers under multiplication has order four, and the elements $1, i, -1$ and $-i$ have orders 1, 4, 2 and 4, respectively. The permutation group acting on $\{1, \ldots, n\}$ has order $n!$, and a cycle $(n_1, \ldots, n_t)$, has order t. The permutation $(1\,2)(3\,4\,5)$ has order six; the permutation $(1\,2)(1\,3)$ has order three.

Suppose that g in G has order t; then $e, g, \ldots, g^{t-1}$ are distinct elements of G (for if $g^i = g^j$, where $0 \leq i < j < t$, then $g^{j-i} = e$ and $0 < j - i < t$ which contradicts the fact that g has order t). It follows that $|G| \geq t$ so that if $g \in G$ then the order of g is not greater than the order of G. In Section 12.3 we shall prove that *the order of g divides the order of G*. In any event, *every element of a finite group has finite order*.

All groups G have the property that for certain g and h in G, $gh = hg$. However, this need not be so for every pair of elements in G; for example, $(1\,2\,3)(1\,2) \neq (1\,2)(1\,2\,3)$ in the permutation group on $\{1, 2, 3\}$. For convenience, we repeat Definition 1.2.6 here.

Definition 12.1.3 The elements g and h in a group G are said to *commute* if $gh = hg$. The group G is *abelian*, or *commutative*, if every pair of elements in G commute. The word 'abelian' is in honour of the Norwegian mathematician Abel.

12.1 Groups

We pause to illustrate these ideas with two results about elements of order two. First, suppose that G is a group in which every element has order two, and consider any x and y in G. Then x^2, $(xy)^2$ and y^2 are all e, so that

$$xy = xey = x(xyxy)y = x^2yxy^2 = eyxe = yx.$$

This proves the first part of the following result.

Theorem 12.1.4 *Let G be a finite group in which every element has order two. Then G is abelian, and has order 2^m for some integer m.*

Proof We start with the finite set G. Now suppose that some element of G is a product of some of the other elements of G and their inverses; then we discard this element from G and so produce a strictly smaller set, say G_1, from which G can be obtained by taking all products of powers (including negative powers) of elements in G_1 in any order (with repetitions allowed). We now apply the same argument to G_1, and obtain (if possible) a smaller set G_2 with the property that any element of G_1, and hence also of G, can be expressed as a product of elements (and their inverses) in G_2. This process continues until we have obtained a subset G_k of G whose elements generate G in the sense above, and which is such that no proper subset of G_k has this property. We say that G_k is a *minimal generating set* for G.

Suppose that $G_k = \{a_1, \ldots, a_m\}$. So far, we have not used the fact that G is abelian. However, as G is abelian, all of the products of elements in G_k, and hence every element of G, can be expressed in the form $a_1^{u_1} \cdots a_m^{u_m}$, where each u_j is an integer. Because every element of G has order two, we may now assume that each u_j is 0 or 1. There are 2^m possibilities here for the u_j, and it remains to show that different choices of $(u_1, \ldots, u_m)$ lead to different elements of G. Suppose, then, that $a_1^{u_1} \cdots a_m^{u_m} = a_1^{v_1} \cdots a_m^{v_m}$, where each u_i, and each v_j, is 0 or 1. If $u_1 \neq v_1$ then we see that $a_1 = a_2^{t_2} \cdots a_m^{t_m}$ for some t_j, and this shows that G_k is not a minimal generating set (for we could discard a_1 from G_k and still retain a generating set). We deduce that $u_1 = v_1$. The same argument gives $u_j = v_j$ for all j, and it follows that G has order 2^m. □

Theorem 12.1.5 *Let G be a group of even order. Then G contains an element of order two.*

Proof For each g in G we form the set $\{g, g^{-1}\}$. If $\{g, g^{-1}\} \cap \{h, h^{-1}\} \neq \emptyset$, then one of the four possibilities $g = h$, $g = h^{-1}$, $g^{-1} = h$ and $g^{-1} = h^{-1}$ arises, and in all cases, $\{g, g^{-1}\} = \{h, h^{-1}\}$. It follows that G can be written as the union of a finite number of *mutually disjoint* sets $\{g, g^{-1}\}$. As $\{g, g^{-1}\}$ has two elements when $g^2 \neq e$, and one element when $g^2 = e$, there must be an even number of sets of this form with $g^2 = e$ (else the order of G would be odd). Now $e^2 = e$ so one of these sets is $\{e\}$. It follows that there must be another such set; that is, there must be some g with $g \neq e$ and $g^2 = e$. This g is an element of G of order two. □

Exercise 12.1

1. Let G be a finite group which has only one element, say x, of order two. Show that x commutes with every other element of G. [Hint: consider yxy^{-1}.]
2. Show that a group cannot have exactly two elements of order two. [Hint: Suppose that x and y have order two and consider xyx^{-1} and xy.] The group $\{I, -z, \bar{z}, -\bar{z}\}$ of isometries of $\mathbb{C}$ has exactly three elements of order two.
3. Suppose that a group G contais two elements a and b, where $a^{-1}ba = b^2$ and $b^{-1}ab = a^2$. Suppose also that every element of G can be expressed as a product of the elements a, b, a^{-1} and b^{-1}, taken as often as we wish, in any order, and with repetitions allowed. Show that $a = b = e$ and hence that $G = \{e\}$.
4. Suppose that g is in a group G. Prove that, for all integers m and n, $g^m g^n = g^{m+n}$.
5. For each positive integer n let G_n denote the group of complex n-th roots of unity under multiplication. Show that if p and q are coprime, then $G_p \cap G_q = \{1\}$. What can be said if p and q are not coprime?
6. Let S be a *finite* set of non-zero complex numbers that is closed under multiplication. Show that for some positive integer n, S is the group of n-th roots of unity. [Hint: first show that if $z \in S$ then $|z| = 1$.]

12.2 Subgroups and cosets

There are many instances in which a subset of a group is a group in its own right; for example, $\mathbb{Z}$ is a group within the group $\mathbb{R}$ with respect to addition. Similarly, a subspace of a vector space is itself an additive group, and corresponding to the idea of a subspace, we have the idea of a subgroup.

Definition 12.2.1 A non-empty subset H of G is a *subgroup* of G if it is a group with the same rule of combination as G.

This definition does not say explicitly that if H is a subgroup of G then the identity element e_H of H coincides with the identity e in G, but this must be so. Indeed, as H is a group we see that $e_H^2 = e_H$. However, by Lemma 1.2.5,

12.2 Subgroups and cosets

this means that $e_H = e$. A similar question can be asked of inverse elements, and the answer is the same: if $h \in H$ then the inverse of h as an element of H coincides with its inverse as an element of G. Indeed, if it did not there would be two elements x in G that satisfy $xh = e = hx$, and we know that there is only one.

A group G always has at least two subgroups, namely G itself and the subgroup $\{e\}$ consisting of the identity element alone. We call $\{e\}$ the *trivial subgroup* of G, and we say that H is a *non-trivial subgroup* of G if $H \neq \{e\}$. We say that H is a *proper subgroup* of G if $H \neq G$, and our interest lies in the non-trivial proper subgroups of G.

We now give a simple test for a subset to be a subgroup.

Theorem 12.2.2 *Suppose that H is a non-empty subset of a group G. Then H is a subgroup of G if and only if*

(a) *if $g \in H$ and $h \in H$, then $gh \in H$, and*
(b) *if $g \in H$ then $g^{-1} \in H$.*

Proof If H is a subgroup then (a) and (b) hold. Suppose now that H is a non-empty subset of G and that (a) and (b) hold. As the associative law holds for G it automatically holds for H. Moreover, (a) implies that H is closed under composition of its elements. As H is non-empty, there is some g in H, and so from (b) and then (a) (with $h = g^{-1}$), $e \in H$. Finally, (b) guarantees that each g in H has an inverse in H, and the proof is complete. □

Next, we describe an important property of the class of all subgroups of a given group.

Theorem 12.2.3 *Let G be any group. Then the intersection of any collection of subgroups of G is also a subgroup of G.*

Proof Let the collection of subgroups of G be H_t, where t lies in some set (of labels) T. Let H be the intersection $\cap_t H_t$ of all of the H_t; thus $g \in \cap_t H_t$ if and only if $g \in H_t$ for every t in T. Note that there is no assumption here that T is finite. As $e \in H_t$ for every t we see that $e \in H$; thus $H \neq \emptyset$. If g and h are in H, then they are in every H_t. Thus gh is in every H_t and so it is in H. Similarly, if $g \in H$, then $g \in H_t$ for every t; hence $g^{-1} \in H_t$ for every t, and so $g^{-1} \in H$. The conclusion now follows from Theorem 12.2.2. □

As a consequence of Theorem 12.2.3, we see that if G_0 is any non-empty *subset* of a group G, then we can consider the collection of all subgroups H_t of G that contain G_0 (there is at least one, namely G itself). It follows that $\cap_t H_t$ is also a subgroup that contains G_0, and it is clearly *the smallest such subgroup* because any subgroup containing G_0 is one of the H_t. This argument justifies the following definition.

Definition 12.2.4 Let G_0 be a non-empty subset of a group G. Then the subgroup of G *generated by* G_0 is the smallest subgroup of G that contains G_0 (and it is the intersection of all subgroups that contain G_0).

For example, let $G = \mathbb{R}$ with addition and let $G_0 = \{3, 5\}$. What is the subgroup H generated by G_0? First, if H_1 is a subgroup that contains G_0, then $3 + 3 - 5 \in H_1$ so that $1 \in H_1$; thus $H_1 \supset \mathbb{Z}$. As $\mathbb{Z}$ is a subgroup of $\mathbb{R}$ that contains G_0 we deduce that $\mathbb{Z}$ is the subgroup generated by G_0. In plain terms, any subgroup of $\mathbb{R}$ that contains 3 and 5 also contains $\mathbb{Z}$, and $\mathbb{Z}$ is such a subgroup.

We come now to the idea of a *coset*; this is a subset of a group G that is constructed from a given subgroup H of G. Given any two subsets, say X and Y, of a group G, we can create a new subset XY of G defined by

$$XY = \{xy : x \in X, \; y \in Y\}.$$

Of course, if we wish, we can also consider sets of the form XYZ, $X^{-1}Y$ and so on. If X has only one element, say $X = \{x\}$, we naturally write

$$xY = \{xy : y \in Y\}, \quad Yx = \{yx : y \in Y\}, \qquad (12.2.1)$$

and, similarly,

$$xYx^{-1} = \{xyx^{-1} : y \in Y\}. \qquad (12.2.2)$$

It is clear that these sets obey simple algebraic rules; for example, $u(vY) = (uv)Y$, $u^{-1}(uY) = Y$ and so on. The sets (12.2.1) and (12.2.2) take on a special significance when Y is a subgroup of G, and we begin with the following simple result about subgroups.

Theorem 12.2.5 *Let H be a subgroup of a group G, and let g be any element of G. Then $g \in H$ if and only if $gH = H$, and also if and only if $Hg = H$.*

Proof First, if $gH = H$ then $ge \in H$ so that $g \in H$. Now suppose that $g \in H$. As $gh \in H$ whenever $h \in H$ we see that $gH \subset H$. Now take any h in H and note that $h = g(g^{-1}h)$. As $g^{-1}h \in H$, we see that $h \in gH$ so that $H \subset gH$. This proves that $gH = H$. The proof that $g \in H$ if and only if $Hg = H$ is similar. □

Theorem 12.2.5 raises the question of what the sets gH and Hg look like when $g \notin H$, and the following examples suggest that these sets are indeed

12.2 Subgroups and cosets

worth studying. Of course, if we are using the additive notation here, we replace gH by $g+H$ and so on.

Example 12.2.6 Let G be the group $\mathbb{R}^3$ with respect to addition, and let

$$H = \{(x_1, x_2, x_3) \in \mathbb{R}^3 : x_3 = 0\}.$$

Then H (a horizontal plane) is a subgroup of G, and $x+H$ is the translation of the plane H by x. Theorem 12.2.5 records the geometrically obvious fact that $x+H = H$ if and only if $x \in H$. Note that any two translations of H are either equal or disjoint, and that G is the union of all of the $x+H$. □

Example 12.2.7 Let G be the group of non-zero complex numbers with respect to multiplication, and let H be the subgroup $\{z : |z| = 1\}$. For any w in G, wH is the circle with centre 0 and radius $|w|$, so that (as in Theorem 12.2.5) $wH = H$ if and only if $w \in H$. In this example the sets wH (for varying w) are circles centered at the origin; thus again, any two sets wH are either equal or disjoint, and G is the union of these sets. □

Example 12.2.8 This example is taken from modular arithmetic. Let $G = \mathbb{Z}$ (the group of integers under addition), and let H be the subgroup of multiples of 3 (we could replace 3 by any other integer here). Then

$$0+H = \{\ldots, -6, -3, 0, 3, 6, \ldots\},$$
$$1+H = \{\ldots, -5, -2, 1, 4, 7, \ldots\},$$
$$2+H = \{\ldots, -4, -1, 2, 5, 8, \ldots\}.$$

It is obvious that given any integer m, if we write $m = 3k + a$, where $a \in \{0, 1, 2\}$, then $q + H = a + H$. Thus once again we see that the two sets $u + H$ and $v + H$ are equal or disjoint, and that G is the union of the sets $u + H$ (in this case it is the union of just three of these sets). □

In each of the three previous examples the groups G were abelian, and if G is any abelian group, then $gH = Hg$ for every g in G. However, if G is not abelian, gH and Hg may be different sets. We give an example.

Example 12.2.9 Let G be the group of permutations of $\{1, 2, 3\}$; thus

$$G = \{I, (1\,2), (2\,3), (3\,1), (1\,2\,3), (1\,3\,2)\}.$$

Let $H = \{I, (1\,2)\}$ and $g = (1\,3)$; then $gH = \{(1\,3), (1\,2\,3)\}$ whereas $Hg = \{(1\,3), (1\,3\,2)\}$. □

We shall now define what we mean by a coset and, motivated by these examples, state and prove the main result on cosets.

Definition 12.2.10 A *left coset* of a subgroup H of G is a set of the form gH; a *right coset* is a set of the form Hg.

Theorem 12.2.11 *Let H be a subgroup of a group G. Then G is the union of its left cosets, and any two left cosets are either equal or disjoint. Further, the left cosets fH and gH are equal if and only if $f^{-1}g \in H$. Similar statements are true for right cosets.*

Proof Take any g in G. As $g \in gH$, g is in at least one left coset; thus G is the union of its left cosets. We will complete the proof by showing that the three statements (a) $fH \cap gH \neq \emptyset$, (b) $fH = gH$, (c) $f^{-1}g \in H$ are equivalent to each other. Clearly, (b) implies (a). Suppose now that (a) holds; then there are h_1 and h_2 in H such that $gh_1 = fh_2$; hence $f^{-1}g = h_2h_1^{-1}$, and (c) holds. Finally, if (c) holds then, from Theorem 12.2.5, $f^{-1}gH = H$ so that $gH = ff^{-1}gH = fH$ which is (b). □

We emphasize that the force of Theorem 12.2.11 is that *a group G is partitioned into disjoint subsets by its cosets with respect to any given subgroup H of G*. This is called the *coset decomposition of G*.

Exercise 12.2

1. The *centre* C of a group G is the set of elements of G which commute with all elements of G. Show that C is an abelian subgroup of G.
2. Let H be a finite non-empty subset of a group G. Show that H is a subgroup of G if and only if H is closed with respect to the group operation in G.
3. Suppose that H and K are subgroups of a group G. Show that if $Hg_1 \subset Kg_2$ for some g_1 and g_2, then $H \subset K$.
4. Let H be a subgroup of a group G. Show that if $g_1H = Hg_2$ then $g_2H = Hg_1$.
5. Let H be a subgroup of a group G. Show that if $aH = bH$ then $Ha^{-1} = Hb^{-1}$. This means that there is a (well-defined) map α from the set of left cosets to the set of right cosets given by $\alpha(gH) = Hg^{-1}$. Show that α is a bijection.
6. Let G be the group of isometries of $\mathbb{C}$. Show that the set T_0 of translations $z \mapsto z + n$, where $n \in \mathbb{Z}$, is a subgroup of G. Show that $gT_0 \neq T_0g$, where $g(z) = iz$.
7. Suppose that a and b are non-zero real numbers, and let H be the set of (x, y) in $\mathbb{R}^2$ such that $ax + by = 0$. Show that H is a subgroup of $\mathbb{R}^2$, and that the coset $(p, q) + H$ is the line $ax + by = p + q$.

8. Let G be the group of real 3×3 invertible matrices, and let D be the subgroup of diagonal matrices. Let

$$X = \begin{pmatrix} 0 & 0 & -1 \\ 0 & -1 & 0 \\ -1 & 0 & 0 \end{pmatrix}.$$

What is the coset XD?

9. Let G be $\mathbb{Z}_{16}$ (the group $\{0, 1, \ldots, 15\}$ under addition modulo 16), and let H be the subgroup of G generated by the single element 4. List all (four) cosets of G with respect to H.

12.3 Lagrange's theorem

We recall that the order $|G|$ of a group G is the number of elements in G. Lagrange's theorem is concerned with the relation between the order of a group and the order of one of its subgroups.

Theorem 12.3.1: Lagrange's theorem. *Let H be a subgroup of a finite group G. Then $|H|$ divides $|G|$, and $|G|/|H|$ is the number of distinct left (or right) cosets of H in G.*

Lagrange's theorem shows that the order of G alone influences the number of subgroups that G can have (regardless of its group theoretic structure), and the most striking application of this is to groups of prime order.

Corollary 12.3.2 *Suppose that G is a group of order p, where p is a prime. Then the only subgroups of G are G and the trivial subgroup $\{e\}$.*

This is clear because if H is a subgroup of G then, by Lagrange's Theorem, $|H|$ divides p. Thus $|H|$ is 1 or p, and H is $\{e\}$ or G.

The proof of Lagrange's theorem We consider a finite group G and a subgroup H. By Theorem 12.2.11, G is the union of a finite number of pairwise disjoint cosets, say $G = g_1 H \cup \cdots \cup g_r H$ (see Figure 12.4.1). Consequently,

$$|G| = |g_1 H| + \cdots + |g_r H|.$$

Now for any g, the map $x \mapsto gx$ is a bijection of H onto gH (with inverse $x \mapsto g^{-1}x$). It follows that H and gH have the same number of elements, namely $|H|$, so that $|G| = r|H|$. □

Another corollary of Lagrange's theorem is the following result which was mentioned in Section 12.1.

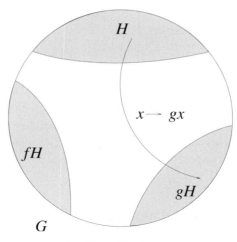

Figure 12.4.1

Theorem 12.3.3 *Let g be an element of a finite group G. Then the order of g divides the order of G.*

Proof Let g be any element of a finite group G. We know that g has finite order, say d (and $d \leq |G|$; see Section 12.1). It is easy to check that the set $\{e, g, \ldots, g^{d-1}\}$ is a subgroup of G of order d (the inverse of g^k is g^{d-k}) so, by Lagrange's Theorem, d divides $|G|$. □

We end this section with an example to show that *the converse of Lagrange's theorem is false*; it is *not* true that if an integer q divides the order of G, then G necessarily has a subgroup of order q.

Example 12.3.4 Let G be the group A_4 of all even permutations acting on $\{1, 2, 3, 4\}$. Now $|A_4| = 12$, and we are going to show that A_4 has no subgroup of order six, We suppose, then, that H is a subgroup of A_4 of order six, and we seek a contradiction. The twelve elements of A_4 are the identity I, three permutations of the form $(a\ b)(c\ d)$, where a, b, c and d are distinct, and eight three-cycles. As H has order six, it contains an element of order two (Theorem 12.1.5), so we may suppose that $\sigma \in H$, where $\sigma = (a\ b)(c\ d)$. As A_4 has only four elements that are not of order three, H must also contain a three cycle, ρ, say. Thus H contains the subgroup $H_0 = \{I, \rho, \rho^2\}$. As $\sigma \notin H_0$, the coset σH_0 is disjoint from H_0, thus

$$H = \{I, \rho, \rho^2\} \cup \sigma\{I, \rho, \rho^2\} = \{I, \rho, \rho^2, \sigma, \sigma\rho, \sigma\rho^2\}.$$

Now H contains $\rho\sigma$, so this element must be one of the six elements listed here. Clearly, $\rho\sigma$ is not I, ρ, ρ^2 or σ; thus either (i) $\rho\sigma = \sigma\rho$, or (ii) $\rho\sigma = \sigma\rho^2$. We shall now show that both (i) and (ii) lead to a contradiction. Suppose that (i) holds, and let k be the single integer in $\{1, 2, 3, 4\}$ such that $\rho(k) = k$. Then $\sigma(k) = \sigma\rho(k) = \rho\sigma(k)$, so that ρ fixes $\sigma(k)$. Thus $\sigma(k) = k$, which is false. Finally, suppose that (ii) holds. Then $(\rho\sigma)^2 = (\rho\sigma)(\sigma\rho^2) = I$, so that H contains another element of order two, namely $\rho\sigma$. Now the product of any two elements in A_4 of order two is the third element of order two; thus $(\rho\sigma)\sigma$ is also of order two, and this is false as ρ has order three. □

Exercise 12.3

1. Show that if a group G contains an element of order six, and an element of order ten, then $|G| \geq 30$.
2. The order of a group G is less than 450, and G has a subgroup of order 45, and a subgroup of order 75. What is the order of G?
3. Suppose that a group G has subgroups H and K of orders p and q, respectively, where p and q are coprime. Show that $H \cap K = \{e\}$.
4. Suppose that an abelian group G has order six. Use Theorem 12.1.4 and Lagrange's theorem to show that G has an element of order three. Now show that G has an element of order two, and deduce that G is cyclic (that is, for some g, $G = \{e, g, \ldots, g^5\}$ with $g^6 = e$).

12.4 Isomorphisms

Consider the following four groups:

$$G_1 = \{I, (1\,2)(3\,4), (1\,3)(2\,4), (1\,4)(2\,3)\};$$
$$G_2 = \{0, 1, 2, 3\};$$
$$G_3 = \{1, i, -1, -i\},$$

with respect to composition, addition modulo 4, and multiplication, respectively. Now every element of G_1 (other than the identity) has order two, and as this is not so for G_2 and G_3, there is a significant group-theoretic difference between G_1 and the other two groups. By contrast, there is no significant difference between G_2 and G_3. Indeed, if we consider the bijection θ from G_2 to G_3 given by $\theta(n) = e^{2\pi i n/4}$, then addition modulo 4 in G_2 'corresponds' to multiplication in G_3; for example, $2 + 3 = 1$ in G_2, while in G_3,

$$\theta(2)\theta(3) = e^{[2\pi i(2/4) + 2\pi i(3/4)]} = e^{2\pi i(5/4)} = e^{2\pi i/4} = \theta(1).$$

In short, while G_2 and G_3 are different sets, they are, in a group-theoretic sense, the 'same' group. Another example of two groups that are the 'same' is the group $\{1, -1\}$ under multiplication, and the group $\{I, \sigma\}$, where $\sigma(z) = -z$, of isometries of the complex plane.

These examples raise the question of deciding when we should regard two groups as being the same group. For example, later, we shall want to *list all groups of order four*, and this will be an impossible task if we insist on distinguishing between groups that are formally different from each other even though they possess the same group-theoretic structure. What we need is a formal way of identifying groups that have identical structures, and when we have this we shall see that *there are only two groups of order four* (namely, G_1 and G_3). The identification of two groups is given by a special mapping, called an isomorphism, and which we have already illustrated by the map θ above.

Definition 12.4.1 Let G and G' be groups. A map $\theta : G \to G'$ is an *isomorphism* if

(a) θ is a bijection of G onto G', and
(b) for all g and h in G, $\theta(gh) = \theta(g)\theta(h)$.

If such a θ exists, we say that G and G' are *isomorphic* groups.

A more informal view of an isomorphism is as follows. Given g and h in G we can combine these in G to obtain gh and then map this into G' to obtain $\theta(gh)$. Alternatively, we can map g and h to $\theta(g)$ and $\theta(h)$, respectively, in G' and then combine these to obtain $\theta(g)\theta(h)$. *The crucial property of the isomorphism θ (and later, of a homomorphism) is that these two operations have the same result.* Perhaps one of the most interesting and elementary examples of an isomorphism is the function e^x from $\mathbb{R}$ to itself which we now discuss.

Example 12.4.2 We assume familiarity with the function $\exp : x \mapsto e^x$ and we want to use its properties to illustrate the idea of an isomorphism. Now exp is a map from the additive group $\mathbb{R}$ to the multiplicative group $\mathbb{R}^+$ of positive real numbers. It is known (from analysis) that exp is a bijection from $\mathbb{R}$ to $\mathbb{R}^+$. The crucial property $e^{x+y} = e^x e^y$ of exp is *exactly* the condition that is needed to show that exp is an isomorphism; thus *the group $(\mathbb{R}, +)$ is isomorphic to the group $(\mathbb{R}^+, \times)$*. □

In case the reader should have the idea that it is an easy matter to decide whether or not two groups are isomorphic, we mention (but will not prove) the following amazing result (in which the operation in both groups is the usual multiplication of complex numbers).

12.4 Isomorphisms

Example 12.4.3 The group $\{z \in \mathbb{C} : |z| = 1\}$ is isomorphic to the group $\{z \in \mathbb{C} : z \neq 0\}$. Of course, the isomorphism here has nothing whatever to do with the geometric structure of these sets as subsets of $\mathbb{C}$, and this is where one's intuition may lead one astray. This example is a stark reminder that group theory exists independently of geometry (for a proof, see *Mathematics Magazine*, 72 (1999), 388–91). □

If one wants to show that two given groups G and G' are isomorphic, one has *to define a map* $\theta : G \to G'$ *and then show that θ has the required properties*. This is not always easy to do. To show that G and G' are not isomorphic it suffices *to find a property that is preserved under an isomorphism, and which is satisfied by only one of G and G'*. This is usually easier, and with this in mind we list some of the basic properties of groups that are preserved under an isomorphism.

Theorem 12.4.4 *Suppose that $\theta : G \to G'$ is an isomorphism. Then*

(a) *if G is abelian then so is G';*
(b) *$\theta(e) = e'$, where e and e' are the identities in G and G';*
(c) *for all g in G, $\theta(g)^{-1} = \theta(g^{-1})$;*
(d) *the order of g in G is equal to the order of $\theta(g)$ in G'.*

Proof Suppose that G is abelian, and take any g' and h' in G'. As θ maps G onto G' there are g and h in G with $g' = \theta(g)$ and $h' = \theta(h)$. Thus

$$g'h' = \theta(g)\theta(h) = \theta(gh) = \theta(hg) = \theta(h)\theta(g) = h'g'$$

and we have proved (a). Next, take any g' in G' and (as above) write $g' = \theta(g)$. Then $\theta(e)g' = \theta(e)\theta(g) = \theta(eg) = \theta(g) = g'$. In the same way we see that $g'\theta(e) = g'$, and as G' has a unique identity element, this element must be $\theta(e)$. This proves (b), and the proof of (c) is similar. For any g in G we have

$$\theta(g)\theta(g^{-1}) = \theta(gg^{-1}) = \theta(e) = e' = \theta(g^{-1}g) = \theta(g^{-1})\theta(g),$$

and this shows that $\theta(g^{-1}) = [\theta(g)]^{-1}$. Finally, for any g in G and any integer m, $\theta(g^m) = \theta(g)^m$. As θ is a bijection, $e, g, \ldots, g^m$ are distinct if and only if $e', \theta(g), \ldots, \theta(g)^m$ are distinct, and this proves (d). □

The idea of classes of isomorphic groups is best described in terms of equivalence relations. Briefly, an *equivalence relation* $\sim$ on a set X is a relation between the elements of x with the three properties

(a) for all x, $x \sim x$;
(b) if $x \sim y$ then $y \sim x$;
(c) if $x \sim y$ and $y \sim z$ then $x \sim z$.

These three properties say that $\sim$ is *reflexive*, *symmetric* and *transitive*, respectively. We define the *equivalence class* $E(x)$ containing x to be the set of elements of X that are related to x by $\sim$; explicitly,

$$E(x) = \{y \in X : x \sim y\}.$$

Equivalence relations are used throughout mathematics, and their main use stems from the fact that they provide a natural way of splitting a set into a union of mutually disjoint subsets.

Theorem 12.4.5 *Let X be a non-empty set, and let $\sim$ be an equivalence relation on X. Then any two equivalence classes are equal or disjoint, and X is the union of a number of mutually disjoint equivalence classes.*

The reader should notice the similarity between this result and Theorem 12.2.11 (about cosets). Indeed, the idea of an equivalence relation can be used to provide an alternative approach to cosets. Before we prove Theorem 12.4.5, we give its application to isomorphisms between groups.

Theorem 12.4.6 *Let $\mathcal{G}$ be the class of all groups, and let $\sim$ be the relation on $\mathcal{G}$ defined by $G_1 \sim G_2$ if and only if G_1 is isomorphic to G_2. Then $\sim$ is an equivalence relation on $\mathcal{G}$.*

This result is the formal statement of how we may now treat two groups as being the 'same' group. According to Theorem 12.4.6, the equivalence class containing a given group G is the class of all groups that are isomorphic to G; equally, it is a maximal collection of groups any two of which are isomorphic to each other, and each of which fails to be isomorphic to a group outside the class. From an algebraic point of view we shall no longer distinguish between isomorphic groups, and this will be reflected in our blatant and deliberate abuse of language. For example, when we say that *there is only one group of order three* we mean that any two groups of order three are isomorphic to each other. However, it should be remembered that when we are applying group theory to geometry there may be important *geometric reasons* why we should distinguish between isomorphic groups. We now prove Theorems 12.4.5 and 12.4.6.

The proof of Theorem 12.4.5 Suppose that x and y are in X and that $E(x) \cap E(y)$ contains an element z. Now take any y' in $E(y)$. Then $y' \sim y$ and $y \sim z$; thus $y' \sim z$. However, $z \sim x$; thus $y' \sim x$ and so $y' \in E(x)$. We deduce that $E(y) \subset E(x)$ and, by symmetry, $E(x) \subset E(y)$. This shows that if $E(x) \cap E(y) \neq \emptyset$ then $E(x) = E(y)$. Finally, it is obvious that X is the union of its equivalence classes because if $x \in X$ then $x \in E(x)$ so that

$$X = \bigcup_{x \in X} \{x\} \subset \bigcup_{x \in X} E(x) \subset X,$$

so that X is the union of the $E(x)$. ☐

The proof of Theorem 12.4.6 First, the identity map of a group G onto itself is an isomorphism; thus $G \sim G$. Next, it is clear that if $\theta_1 : G_1 \to G_2$ and $\theta_2 : G_2 \to G_3$ are isomorphisms then $\theta_2\theta_1$ is an isomorphism of G_1 onto G_3. Indeed,

$$\begin{aligned}(\theta_2\theta_1)(gh) &= \theta_2\bigl(\theta_1(gh)\bigr) \\ &= \theta_2\bigl(\theta_1(g)\theta_1(h)\bigr) \\ &= \theta_2\bigl(\theta_1(g)\bigr)\theta_2\bigl(\theta_1(h)\bigr) \\ &= (\theta_2\theta_1)(g)(\theta_2\theta_1)(h).\end{aligned}$$

Finally, we need to show that if $\theta : G_1 \to G_2$ is an isomorphism, then so is $\theta^{-1} : G_2 \to G_1$. First, as θ is a bijection of G_1 onto G_2, then θ^{-1} is a bijection of G_2 onto G_1. Now take any x and y in G_2. As θ maps G_1 onto G_2 there are unique g and h in G_1 such that $\theta(g) = x$ and $\theta(h) = y$. Then $\theta^{-1}(xy) = \theta^{-1}(x)\theta^{-1}(y)$ because

$$\theta^{-1}(xy) = \theta^{-1}\bigl(\theta(g)\theta(h)\bigr) = \theta^{-1}\bigl(\theta(gh)\bigr) = gh = \theta^{-1}(x)\theta^{-1}(y).$$

☐

It seems worth mentioning explicitly that Theorem 12.4.6 contains the following result.

Theorem 12.4.7 *If $\theta_1 : G_1 \to G_2$ is an isomorphism then so is $\theta_1^{-1} : G_2 \to G_1$. If, in addition, $\theta_2 : G_2 \to G_3$ is an isomorphism, then so is $\theta_2\theta_1 : G_1 \to G_3$.*

Exercise 12.4

1. Let G be the group $\{1, -1\}$ and let H be the group $\{z, -z, \bar{z}, -\bar{z}\}$ of isometries of $\mathbb{C}$. Show that H is isomorphic to $G \times G$.
2. Show that the multiplicative group of matrices of the form $\begin{pmatrix} 1 & n \\ 0 & 1 \end{pmatrix}$, where $n \in \mathbb{Z}$, is isomorphic to $(\mathbb{Z}, +)$.
3. Recall the additive groups $\mathbb{Z}, \mathbb{Q}$ and $\mathbb{R}$, and the multiplicative groups $\mathbb{Q}^*$ and $\mathbb{R}^*$ of non-zero numbers. Show that
   (a) $\mathbb{Z}$ is not isomorphic to $\mathbb{Q}$;
   (b) $\mathbb{Q}$ is not isomorphic to $\mathbb{Q}^*$;
   (c) $\mathbb{R}$ is not isomorphic to $\mathbb{R}^*$.

4. Consider the three multiplicative groups

$$\mathbb{R}^+ = \{x \in \mathbb{R} : x > 0\},$$
$$S^1 = \{z \in \mathbb{C} : |z| = 1\},$$
$$\mathbb{C}^* = \{z \in \mathbb{C} : z \neq 0\}.$$

Show that $\mathbb{R}^+ \times S^1$ is isomorphic to $\mathbb{C}^*$. As $\mathbb{R}^+$ is isomorphic to $\mathbb{R}$, this shows that $\mathbb{C}^*$ is isomorphic to $\mathbb{R} \times S^1$ (a cylinder).

5. Let G be the set of real 2×2 matrices of the form

$$M(a) = \begin{pmatrix} a & a \\ a & a \end{pmatrix},$$

where $a \neq 0$. Show that G is a group under the usual multiplication of matrices (the identity is not the usual identity matrix), and that G is isomorphic to $\mathbb{R}^*$.

12.5 Cyclic groups

Let g be an element of a group G. For each positive integer n, g^n is the composition of g with itself n times, $g^0 = e$ and $g^{-n} = (g^{-1})^n$. If we are using the additive notation for G, then g^n is replaced by $g + \cdots + g$ (with n terms) and we write this as ng. If $G = \mathbb{Z}$, then ng in this sense is indeed the product of n and g.

Definition 12.5.1 A group G is said to be *cyclic*, or a *cyclic group*, if there is some g in G such that $G = \{g^n : n \in \mathbb{Z}\}$. We then say that G is *generated* by g.

Despite its appearance, a cyclic group need not be infinite; for example, -1 generates the group $\{1, -1\}$ with respect to multiplication. Moreover, a cyclic group may be generated by many of its elements; for example, if ρ is the five-cycle (1 2 3 4 5), then (as the reader can check) each ρ^k with $k \neq 0$ generates the cyclic group $\{I, \rho, \rho^2, \rho^3, \rho^4\}$. The most familiar example of an infinite cyclic group is the additive group $\mathbb{Z}$ of integers (generated by 1). The additive group $\mathbb{R}$ of real numbers is not cyclic; this is a consequence of the next result.

Theorem 12.5.2 *Let H be a non-trivial subgroup of the additive group $\mathbb{R}$. Then H is cyclic if and only if it has a smallest positive element h, in which case it is generated by h.*

Proof Suppose that H is cyclic, say $H = \{nh : n \in \mathbb{Z}\}$. Then $h \neq 0$ (else H is trivial), and it is clear that $|h|$ is the smallest positive element of H. To establish

12.5 Cyclic groups

the other implication, suppose that H has a smallest positive element, say h. As $h \in H$, we see that $nh \in H$ for every n, so that $\{nh : n \in \mathbb{Z}\} \subset H$. Now suppose that $x \in H$ and write $x = nh + t$, where n is an integer, and t is a real number satisfying $0 \le t < h$. As H contains x and $-nh$, it also contains $x - nh$ which is t. As h is the smallest positive element of H, and $0 \le t < h$, we conclude that $t = 0$ and hence that $x = nh$. This shows that $H = \{nh : n \in \mathbb{Z}\}$. □

The next result shows that there is only one infinite cyclic group.

Theorem 12.5.3 *Any two cyclic groups of the same finite order m are isomorphic, and hence isomorphic to the additive group $\mathbb{Z}_m$. Any two infinite cyclic groups are isomorphic, and hence isomorphic to the additive group $\mathbb{Z}$.*

Proof We shall prove the second statement first. Suppose that G is an infinite cyclic group generated by, say, g. Define the map $\theta : \mathbb{Z} \to G$ by $\theta(n) = g^n$. We shall show that θ is an isomorphism. Certainly, θ maps $\mathbb{Z}$ onto G. If $\theta(m) = \theta(n)$, then $g^n = g^m$ and so $n = m$ (since otherwise, there would be some non-zero k with $g^k = e$ and then G would be finite). Thus θ is a bijection of $\mathbb{Z}$ onto G. Finally,

$$\theta(n+m) = g^{n+m} = g^n g^m = \theta(n)\theta(m)$$

so that θ is an isomorphism of G onto $\mathbb{Z}$. If G_1 and G_2 are infinite cyclic groups, they are each isomorphic to $\mathbb{Z}$, and hence also to each other. This proves the second statement.

Suppose now that G is a cyclic group, say $G = \{e, g, \ldots, g^{m-1}\}$, of finite order m, so that $g^n = e$ if and only if m divides n. Let $\theta : \mathbb{Z}_m \to G$ be the map defined by $\theta(n) = g^n$. Then θ is a surjective map between finite sets with the same number of elements and so is a bijection between these sets. Further, for all j and k, $\theta(j \oplus_m k) = g^{j+k}$ (where $\oplus_m$ is addition in $\mathbb{Z}_m$) because $g^m = e$. □

Theorem 12.5.3 implies that there is only one cyclic group of order n, and it is usual to denote this by C_n.

One of the important properties of a cyclic group is that its cyclic nature is inherited by all of its subgroups.

Theorem 12.5.4 *A subgroup of a cyclic group is cyclic.*

Proof Let H be a subgroup of the cyclic group $\{g^n : n \in \mathbb{Z}\}$. We may suppose that H is a non-trivial subgroup, so that g^n, and g^{-n}, are in H for some n. Let d be the smallest positive integer such that $g^d \in H$. Clearly H contains g^{md} for every integer m. On the other hand, if $g^k \in H$, we can write $k = ad + b$, where $0 \le b < d$, and we find that $g^b \in H$. We deduce that $b = 0$, and hence that $k = ad$. It follows that $H = \{g^{md} : m \in \mathbb{Z}\}$ and this shows that H is cyclic, and generated by g^d. □

Finally, we obtain more information about groups of prime order (see Corollary 12.3.2).

Theorem 12.5.5 *A group of prime order is cyclic, and is generated by any of its elements other than the identity e.*

Proof Let G be a group with $|G| = p$, where p is a prime, and let g be in G with $g \neq e$. As G is finite, g has finite order k, say, which divides p. As $k \neq 1$ (else $g = g^k = e$), we see that $k = p$ and hence that $G = \{e, g, \ldots, g^{p-1}\}$. Thus G is cyclic and is generated by g. □

Exercise 12.5

1. Let ρ be the cycle (1 2 3 4 5), and let G be the group generated by ρ. Show that each element of G (other than the identity) generates G. Show that this is false if (1 2 3 4 5) is replaced by the cycle (1 2 3 4 5 6).
2. Show that a finite group of rotations of $\mathbb{R}^2$ about the origin is a cyclic group. Construct a proper subgroup of the group of rotations of $\mathbb{R}^2$ about the origin that is not cyclic (in this group every finite subgroup will be cyclic).
3. Let C_k denote a cyclic group of order k. Show that if m and n are coprime, then $C_m \times C_n$ is cyclic and hence isomorphic to C_{mn}. Show, however, that $C_3 \times C_3$ is not isomorphic to C_9.
4. Suppose that G is a group and that the only subgroups of G are the trivial subgroup $\{e\}$ and G itself. Show that G is a finite cyclic group.
5. Let G be a cyclic group with exactly n elements. Show that G is generated by every x ($\neq e$) in G if and only if n is a prime.
6. Let G be a cyclic group of order n generated by, say, g. Show that if k divides n, then there is one, and only one, subgroup of G of order k, and that this is generated by $g^{n/k}$. [This is a partial converse to Lagrange's theorem.]

12.6 Applications to arithmetic

In this section we digress to show some simple applications of group theory to arithmetic. Let n be a positive integer. Then a non-zero integer d *divides* n, or is a *divisor* of n, if there is an integer k such that $n = kd$ (equivalently, if n/d is an integer). If this is so, we write $d|n$. An integer d is a *common divisor* of integers m and n if d divides both m and n. The largest of the finite number of the common divisors of m and n is the *greatest common divisor* of m and n and this is denoted by $\gcd(m, n)$. Notice that, by definition, $\gcd(m, n)$ does indeed divide both m and n. Finally, two integers m and n are *coprime* if $\gcd(m, n) = 1$; that is, if they have no common divisor except 1. As examples, we see that $\gcd(24, 15) = 3$, and $\gcd(15, 11) = 1$. In both of these cases the greatest common divisor of the two integers is an integral combination of them ($2 \times 24 - 3 \times 15$, and $3 \times 15 - 4 \times 11$, respectively), and these are special cases of the following general fact.

Theorem 12.6.1 *Given any positive integers p and q there are integers a and b with $\gcd(p, q) = ap + bq$.*

Proof Let $H = \{mp + nq : m, n \in \mathbb{Z}\}$. It is clear that H is a subgroup of $\mathbb{R}$ and so, by Theorem 12.5.2, there is a positive d such that

$$\{kd : k \in \mathbb{Z}\} = H = \{mp + nq : m, n \in \mathbb{Z}\}.$$

As $d \in H$ there are integers a and b such that $d = ap + bq$. As $p \in H$ there is an integer k such that $p = kd$; thus d divides p. Similarly, d divides q and so, as d is a common divisor of p and q, $d \leq \gcd(p, q)$. Finally, as $\gcd(p, q)$ divides p and q, it also divides d ($= ap + bq$). Thus $\gcd(p, q) \leq d$ and hence $\gcd(p, q) = d$. $\square$

The following two results are immediate corollaries of Theorem 12.6.1.

Corollary 12.6.2 *Any common divisor of p and q also divides $\gcd(p, q)$.*

Corollary 12.6.3 *The integers p and q are coprime if and only if there are integers a and b with $ap + bq = 1$.*

A similar argument enables us to discuss the *least common multiple* of integers p and q: this is the smallest integer t such that both p and q divide t, and it is denoted by $\mathrm{lcm}(p, q)$. Let

$$H = \{mp : m \in \mathbb{Z}\}, \quad K = \{nq : n \in \mathbb{Z}\}.$$

Now H and K are cyclic subgroups of $\mathbb{R}$, so their intersection $H \cap K$ is also a cyclic subgroup of $\mathbb{R}$ (see Theorems 12.2.3 and 12.5.4). Thus, by Theorem 12.5.2, there is an integer ℓ such that

$$H \cap K = \{m\ell : m \in \mathbb{Z}\}.$$

We deduce that for some integers m_1 and n_1, $m_1 p = \ell = n_1 q$, and so p and q both divide ℓ. On the other hand, if p and q divide t, say, then $t \in H \cap K$ so that $\ell \leq t$. We deduce that $\mathrm{lcm}(p, q) = \ell$, and hence that $\mathrm{lcm}(p, q)$ *is the* generator of the intersection of the two subgroups generated by p and q. We illustrate these ideas with an example.

Example 12.6.4 Suppose that p and q be coprime positive integers, and let

$$H = \{mp + nq : m, n \in \mathbb{Z}, \; m+n \text{ even}\}.$$

What is H? Clearly H is a subgroup of $\mathbb{Z}$ and so $H = \{nd : n \in \mathbb{Z}\}$ for some positive integer d. As p and q are coprime, there are integers a and b with $ap + bq = 1$. Thus $2ap + 2bq = 2$ and, as $2a + 2b$ is even, this shows that $2 \in H$, and hence that H contains all even integers. Now consider the two cases (i) $p+q$ is odd, and (ii) $p+q$ is even. If $p+q$ is odd then H contains an odd integer (namely $p+q$) so that in this case, $d = 1$ and $H = \mathbb{Z}$. Now suppose that $p+q$ is even, say $p+q = 2u$, and take any $mp + nq$ in H. As $m+n$ is even, $mp + nq$ is also even; thus in this case $d = 2$ and H is the set of all even integers. □

Exercise 12.6

1. Find the greatest common divisor, and the least common multiple, of 12, 18 and 60.
2. What is the subgroup of $\mathbb{Z}$ generated (i) by 530 and 27, and (ii) by 531 and 27 ?
3. Show that for any integers $p_1, \ldots, p_n$ there are integers $m_1, \ldots, m_n$ such that $\gcd(p_1, \ldots, p_n) = m_1 p_1 + \cdots + m_n p_n$.
4. A set with an associative binary operation with an identity is called a *monoid*. An element x in a monoid is a *unit* if its inverse x^{-1} exists. Show that the set of units of a monoid X is a group (even when X itself is not). We give applications of this idea in the next four exercises.
5. Show that the group of units of $\mathbb{Z}_8$ is $\{1, 3, 5, 7\}$. Is this cyclic? Find the group of units of $\mathbb{Z}_{10}$.
6. Show that $\mathbb{Z}_n$ is a monoid (with respect to multiplication). Show that m is a unit in $\mathbb{Z}_n$ if and only if m is coprime to n. Euler's function $\varphi(n)$ is the order of the group of units of $\mathbb{Z}_n$ (equivalently, the number of positive integers m that are less than, and coprime to, n). Deduce that if a is a unit, then $a^{\varphi(n)}$ is the identity in the group of units. Thus we have Euler's theorem: *if $(a, n) = 1$ then $a^{\varphi(n)}$ is congruent to 1 mod n.*
7. Use Euler's theorem to derive Fermat's theorem: *if p is prime then, for all integers a, a^p is congruent to a mod p.*
8. Let X be a given set and let $\mathcal{F}$ be the set of functions $f : X \to X$. Show that $\mathcal{F}$ is a monoid with respect to the composition of functions. Show that

the units in $\mathcal{F}$ are precisely the bijections of X onto itself (thus the group of units of $\mathcal{F}$ is the group of permutations of X).

12.7 Product groups

In this section we show how to combine two given groups to produce a new 'product' group. This idea can be used to produce new examples of groups, and sometimes to analyse a given group by expressing it as a product of simpler groups. Let G and G' be any two groups. Then $G \times G'$ is the set of ordered pairs (g, g') where $g \in G$ and $g' \in G'$, and there is a natural way of combining two such pairs (g, g') and (h, h'), namely

$$(g, g')(h, h') = (gh, g'h'), \qquad (12.7.1)$$

Notice that we only need to be able to combine two elements in G, and two elements in G', and that G and G' need not be related in any.

Theorem 12.7.1 *The set $G \times G'$ with this operation is a group.*

Proof The rule (12.7.1) guarantees that $(g, g')(h, h')$ is in $G \times G'$. Next, the verification of the associative law is straightforward:

$$\begin{aligned}\big((g, g')(h, h')\big)(k, k') &= (gh, g'h')(k, k') \\ &= \big((gh)k, (g'h')k'\big) \\ &= \big(g(hk), g'(h'k')\big) \\ &= (g, g')(hk, h'k') \\ &= (g, g')\big((h, h')(k, k')\big).\end{aligned}$$

The identify element of $G \times G'$ is (e, e'), where e and e' are the identities in G and G', for $(e, e')(g, g') = (eg, e'g') = (g, g')$, and similarly, $(g, g') = (g, g')(e, e')$. Finally, the inverse of (g, g') is $(g^{-1}, (g')^{-1})$. $\square$

Of course, the rule (12.7.1) extends to any number of factors, and in this way we can construct a group $G_1 \times \cdots \times G_n$ from groups $G_1, \ldots, G_n$. We now show how this construction is used to provide examples of groups.

Example 12.7.2 Consider the group $\mathbb{R} \times \mathbb{R}$. Then

$$\mathbb{R} \times \mathbb{R} = \{(x, y) : x, y \in \mathbb{R}\}, \quad (x, y) + (x', y') = (x + x', y + y'),$$

and this is the Euclidean plane $\mathbb{R}^2$ with vector addition. In a similar way, $\mathbb{R} \times \cdots \times \mathbb{R}$ (with n factors) is the additive group $\mathbb{R}^n$.

Example 12.7.3 Let $G_2 = \{1, -1\}$ and $G_3 = \{1, w, w^2\}$, where both are multiplicative groups of complex numbers, and $w = e^{2\pi i/3}$. Now both G_2 and G_3 are cyclic, and we shall show that $G_2 \times G_3$ is a cyclic group of order six that is generated by the pair $(-1, w)$. Indeed, as is easily checked, the first six powers of this element are $(-1, w)$, $(1, w^2)$, $(-1, 1)$, $(1, w)$, $(-1, w^2)$ and $(1, 1)$, and as these six elements are distinct, $G_2 \times G_3$ is cyclic. □

Example 12.7.4 Now let G_2 be as in Example 12.7.3, and let $G_4 = \{1, i, -1, -i\}$, where both are multiplicative cyclic groups of complex numbers. Now $G_1 \times G_2$ has order eight but, in contrast to Example 12.7.3, we shall see that it is *not* cyclic. A cyclic group of order eight is generated by some element g of order eight. However, every element of $G_2 \times G_4$ has order at most four, because if $(g, h) \in G_2 \times G_4$, then $(g, h)^4 = (g^4, h^4) = (1, 1)$. □

Example 12.7.5 Let G_1 be the multiplicative group $\{z \in \mathbb{C} : |z| = 1\}$, and let $G_2 = \mathbb{R}$. We shall identify $\mathbb{C}$ with the horizontal coordinate plane in $\mathbb{R}^3$; then $G_1 \times G_2$ is identified with the vertical cylinder

$$C = \{(x, y, t) \in \mathbb{R}^3 : x^2 + y^2 = 1\} = \{(z, t) : |z| = 1, t \in \mathbb{R}\}$$

in $\mathbb{R}^3$. According to Theorem 12.7.1, this cylinder can be given the structure of a group whose rule of combination is $(z, t)(w, s) = (zw, t + s)$. Note that the map $(z, t) \mapsto (e^{i\theta}, \ell)(z, t)$ corresponds to a rotation of C by an angle θ about its vertical axis followed by a vertical (upwards) translation of ℓ (this is a screw-motion). □

Exercise 12.7

1. Show that if G has p elements and G' has q elements, then $G \times G'$ has pq elements.
2. Show that $G \times G'$ is abelian if and only if both G and G' are abelian.
3. Let G be the group of all isometries of $\mathbb{C}$ consisting of all real translations, and all glide reflections with axis $\mathbb{R}$. Show that G is isomorphic to $\{1, -1\} \times \mathbb{R}$.
4. The surface of a torus T (a doughnut ring) can be obtained by rotating a circle in $\mathbb{R}^3$ about a line lying in the plane of the circle (and not meeting the circle). It follows that the points on the torus can be parametrised by two coordinates $(e^{i\theta}, e^{i\phi})$, where θ and ϕ are real. Show how to make the torus T into a group.

12.8 Dihedral groups

The *dihedral group* D_{2n}, where $n \geq 2$, is any group that is isomorphic to the group of symmetries of a regular polygon with n sides (note: some authors use D_n for this group). This section is devoted to a geometric, and an algebraic, description of D_{2n}. Consider a regular polygon $\mathcal{P}$ with n sides. By translating, rotating, and scaling $\mathcal{P}$ we may assume that the set V of vertices of $\mathcal{P}$ is the set of n-th roots of unity; thus $V = \{1, \omega, \ldots, \omega^{n-1}\}$, where $\omega = \exp(2\pi i/n)$. If $n = 2$, then $\mathcal{P}$ is the segment from -1 to 1.

A *symmetry* of $\mathcal{P}$ is an isometry of $\mathbb{C}$ that maps the set V (or, equivalently, $\mathcal{P}$) onto itself, and D_{2n} is the group of symmetries of $\mathcal{P}$ (we leave the reader to prove that this is a group with the usual composition of functions). Now if an isometry of $\mathbb{C}$ leaves the set V unchanged then it necessarily fixes 0 (for 0 is the only point of $\mathcal{P}$ that is a distance one from each of the vertices). As any isometry of $\mathbb{C}$ is of the form $z \mapsto az + b$, or $z \mapsto a\bar{z} + b$, it follows that each element of D_{2n} is either a rotation $z \mapsto az$, or a reflection $z \mapsto a\bar{z}$ in some line through the origin, where, in each case, $|a| = 1$. It is clear that there are exactly two maps in D_{2n} that map 1 (in V) to a chosen point of V, say ω^k, namely $z \mapsto \omega^k z$ and $z \mapsto \omega^k \bar{z}$, and we deduce from this that D_{2n} *has order* $2n$.

Now let R_n be the subgroup of symmetries of $\mathcal{P}$ that are rotations; then R_n is the cyclic group of order n and generated by r, where $r(z) = \omega z$ and $r^n = I$. Next, let $\sigma(z) = \bar{z}$. Then σ is in D_{2n} but not R_n, so that D_{2n} contains the two disjoint cosets R_n and σR_n. As these cosets have n elements each, we conclude that D_{2n} is the union of these two (disjoint) cosets. A similar argument shows that D_{2n} is also the union of the two disjoint cosets R_n and $R_n \sigma$; thus

$$\sigma R_n \cup R_n = D_{2n} = R_n \cup R_n \sigma.$$

As R_n is disjoint from each of the cosets σR_n and $R_n \sigma$, we see that $\sigma R_n = R_n \sigma$; thus $\sigma r = r^a \sigma$ for some a. In fact, the elements r and σ (which generate D_{2n}) are related by the special relation

$$\sigma r = r^{-1} \sigma = r^{n-1} \sigma$$

which can be verified directly from $r(z) = \omega z$ and $\sigma(z) = \bar{z}$. To summarize, D_{2n} *is a group of order* $2n$ *that is generated by two elements* r *and* σ *which are subject to the relations*

$$r^n = I, \quad \sigma^2 = I, \quad \sigma r = r^{-1} \sigma. \tag{12.8.1}$$

These are not the only relations satisfied by r and σ; in fact, $\sigma R = R^{-1} \sigma$ for any rotation R about the origin.

Example 12.8.1 Let us briefly consider the symmetries of a square. We may locate the square so that its vertices are $1, i, -1$ and $-i$, and we now relabel these vertices by the integers 1, 2, 3 and 4, respectively. Now any symmetry of the square can be written as a permutation of $\{1, 2, 3, 4\}$ and this is an efficient way to describe the geometry. We have seen that D_8 is generated by r and σ, where $r(z) = iz$ and $\sigma(z) = \bar{z}$. In terms of permutations, $r = (1\,2\,3\,4)$, and $\sigma = (2\,4)$. We know from the discussion above that there are exactly four reflective symmetries of the square, namely $\sigma, \sigma r, \sigma r^2$ and σr^3; in terms of permutations, these are $(2\,4), (1\,4)(2\,3), (1\,3)$ and $(1\,2)(3\,4)$, respectively. Of course, $r^2 = (1\,3)(2\,4)$ and $r^3 = (1\,4\,3\,2)$. As $\sigma^2 = I$ and $r^4 = I$, it may be tempting to think that D_8 is isomorphic to $C_2 \times C_4$; however, this is not so as $C_2 \times C_4$ is abelian whereas D_8 is not. □

We have said that D_{2n} is a group of order $2n$ that is generated by two elements r and σ which are subject to the relations (12.8.1). In fact, this definition of D_{2n} is incomplete until we have established the next result.

Theorem 12.8.2 *There is only one group G of order $2n$ that is generated by two elements a and b, of orders 2 and n, respectively, with $ab = b^{-1}a$.*

Proof Each element of G is a finite word in the elements a, a^{-1}, b and b^{-1}; for example, $aab^{-1}a^{-1}bbba$. In fact, as $a^{-1} = a$ and $b^{-1} = b^{n-1}$, every element of G is a word in a and b. Suppose such a word w contains within it the consecutive terms ab, say $w = xaby$ for some x and y. Then $w = xb^{n-1}ay$, and this particular occurrence of a has been moved to the right (passing b in the process). By repeating this operation as many times as is necessary, we can move all the occurrences of a to the extreme right of the word w (the powers of b will change, but this does not matter), and so we see that any element of G can be written in the form $b^p a^q$ for some non-negative integers p and q. As $b^n = a^2 = e$, and G has $2n$ elements, it is clear that $G = \{b^u a^v : 0 \leq u \leq n,\ v = 0, 1\}$. It is also clear that any two groups of this form, with the same relations, are isomorphic to each other. □

Definition 12.8.3 The *dihedral group* D_{2n} is the group of order $2n$ that is generated by an element r of order n, and an element σ of order two, where $\sigma r = r^{-1}\sigma$.

It is clear that D_{2n} is generated by the two elements $r\sigma$ and σ, both of order two. There is a converse to this.

Theorem 12.8.4 *Suppose that G is a finite group that is generated by two (distinct) elements of order two. Then G is a dihedral group.*

12.8 Dihedral groups

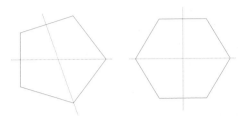

Figure 12.8.1

Proof We suppose that G is generated by x and y, where $x^2 = e = y^2$. In order to match our earlier notation, we let $r = xy$ and $\sigma = x$. As G is finite, r has some finite order n, say. Then G is generated by r and σ, where $r^n = e = \sigma^2$. Moreover, $\sigma r = y = y^{-1} = r^{-1}\sigma$, which is the other relation that needs to be satisfied in a dihedral group. According to Theorem 12.8.2, it only remains to show that G has order $2n$. Now we can argue exactly as in the proof of Theorem 12.8.2, and deduce that

$$G = \{r^u \sigma^v : 0 \leq u < n, \ v = 0, 1\}.$$

If these $2n$ elements are distinct, then G is of order $2n$ and we are finished. If not, then two are equal to each other and we must then have $r^p = \sigma^q \neq e$ for some p and q with $0 < p < n$. Now $q = 1$ (else $r^p = e$), and this means that $(xy)^p = x$. We shall complete the proof in the case when $p = 3$, but the same method will obviously work for any p. If $p = 3$ we have $xyxyxyx = (xy)^3 x = e$. Now $xyxyxyx = e$ implies that $yxyxy = e$, then $xyx = e$ and, finally, that $y = e$ which is false. □

There is a significant difference between the groups D_{2n} for even n and the groups D_{2n} for odd n, and this is most easily explained in terms of the geometry of the regular n-gon. After sketching a few diagrams it should be clear that when n is odd each of the n reflections in D_{2n} fixes exactly one vertex of the polygon. By contrast, when n is even, exactly half of the reflections fix two vertices, and the remaining reflections fix none (see Figure 12.8.1). We shall examine this difference from a group-theoretic point of view in Theorem 12.10.7.

Exercise 12.8

1. Show that D_{2n} is abelian if and only if σ and r commute. Deduce that D_{2n} is abelian if and only if $n = 2$, and hence that D_{2n} is isomorphic to $C_2 \times C_n$ (where C_k denotes a cyclic group of order k) if and only if $n = 2$.
2. As in the text, let $\sigma(z) = \bar{z}$ and $r(z) = \omega z$, where $\omega = \exp(2\pi i/n)$. Show *analytically* that if n is odd then every reflection $r^\ell \sigma$ in the group fixes

exactly one vertex of $\mathcal{P}$, while if n is even, then exactly half of the reflections $r^\ell \sigma$ fix two vertices of $\mathcal{P}$ whereas the other half do not fix any vertex of $\mathcal{P}$.
3. Show that D_4 is isomorphic to the group $\{I, \bar{z}, -z, -\bar{z}\}$ of isometries of $\mathbb{C}$. This group is known as the Klein 4-group.
4. For which values of n does the dihedral group D_n contain an element (not the identity) that commutes with every element of D_n?
5. Show that the only finite abelian groups of isometries of $\mathbb{C}$ are the cyclic groups of rotations and groups that are isomorphic to the Klein 4-group.

12.9 Groups of small order

In this section we shall use the theory we have developed so far to find all groups of order at most six; by this we mean that we want to produce a list of groups such that any group of order at most six is isomorphic to one of the groups in our list. By Theorem 12.5.5, every group of prime order is cyclic, and as any two cyclic groups of the same finite order are isomorphic to each other, there is only one group of order p, where $p = 2, 3, 5, 7, 11, 13, \ldots$. Nevertheless, it is still of interest to see how one can analyze groups of small order from 'first principles', and we shall illustrate this by considering groups of order three.

Example 12.9.1 Let G be a group of order three, say $G = \{e, a, b\}$. Now $ab \neq a$ (else $b = e$) and $ab \neq b$ (else $a = e$); thus $ab = e$ and $b = a^{-1}$. It follows that $b = ba^3 = a^2$, so that $G = \{e, a, a^2\}$, a cyclic group of order three. □

Next, we show that to within an isomorphism there are exactly two groups of order four.

Theorem 12.9.2 *Every group of order four is either cyclic or is isomorphic to the dihedral group D_4. In particular, every group of order four is abelian.*

Proof Let G be a group of order four. By Theorem 12.3.3, every element of G (other than e) has order two or four. If there is an element, say g, of order 4, then G is the cyclic group $\{e, g, g^2, g^3\}$ generated by g.

Suppose now that G is not cyclic. Then every element of G has order two and, by Theorem 12.1.4, G is abelian. Thus $a^2 = b^2 = c^2 = e$, and $ab = ba = c$, and so on, and it is clear that G is isomorphic to D_4. □

Finally, we consider groups of order six; in fact, we shall prove the following result which includes these groups as a special case.

Theorem 12.9.3 *Let G be a group of order $2p$, where p is a prime. Then G is either cyclic, or isomorphic to the dihedral group D_{2p}.*

Proof By Theorem 12.9.2, we may assume that $p \geq 3$, and hence that p is odd. As p is prime, every element of G (other than e) has order 2, p, or $2p$ (Theorem 12.3.3). Theorem 12.1.5 implies that G contains an element a of order two. However, not every element of G has order 2 for if it does then, by Theorem 12.1.4, $2p = |G| = 2^n$ for some n, which is false as p is odd. Thus G must contain an element of order p, or an element of order $2p$, and in both cases, G contains an element, say b, of order p.

Let $H = \{e, b, b^2, \ldots, b^{p-1}\}$, a cyclic subgroup of G of order p. Next, we show that $a \notin H$. If $a \in H$ then $a = b^t$, say, so that $b^{2t} = e$. This implies that p divides t, and hence that $a = b^t = e$ which is false. It follows that $a \notin H$ and so if we write G in terms of its coset decomposition, we have

$$G = H \cup Ha = H \cup aH,$$

so that $aH = Ha$. We deduce that $aHa^{-1} = H$, so there is some integer k in $\{0, 1, \ldots, p-1\}$ such that $aba^{-1} = b^k$. As $a^2 = e$, this implies that $aba = b^k$, and hence that

$$b = ab^k a = (aba)^k = (b^k)^k = b^{k^2},$$

so that $b^{k^2-1} = e$. As b has order p we see that p divides $(k-1)(k+1)$ and, as $0 \leq k < p$, we conclude that $k = 1$ or $k = p - 1$. These two cases correspond to $ba = ab$, and $ab = b^{-1}a$, respectively. In the first case, $(ba)^m = b^m a^m$ so that ba has order $2p$ (the lowest common multiple of 2 and p) and G is cyclic. In the second case, G is of order $2p$ and is generated by b and a, where b has order p, a has order two, and $ab = b^{-1}a$; thus $G = D_{2p}$. □

Exercise 12.9

1. Show that the set $\{2, 4, 8\}$ with the operation of multiplication modulo 14 is a group of order three. What is the identity element?
2. Show that the set $\{1, 3, 5, 7\}$, with multiplication modulo 8 is group of order four. Is this group isomorphic to C_4 or to D_4?
3. Let G be the product group $C_2 \times C_3$ (of order six), where C_n denotes the cyclic group of order n. Is G isomorphic to C_6 or to D_6?
4. Let G be the multiplicative group of non-zero elements in $\mathbb{Z}_7$. Is G isomorphic to C_6 or to D_6?
5. What is the smallest order a non-abelian group can have?

12.10 Conjugation

We begin with the following geometric problem. Suppose that R_L is the reflection across a line L in the complex plane $\mathbb{C}$, and that f is an isometry of $\mathbb{C}$. What is the formula for the reflection $R_{f(L)}$ across the line $f(L)$? To answer this, consider the mapping $f R_L f^{-1}$. This mapping is an isometry, and it fixes every point of the form $f(z)$, where $z \in L$. Thus $f R_L f^{-1}$ is an isometry that fixes every point of $f(L)$ and so it is either the identity I or $R_{f(L)}$. Now $f R_L f^{-1} \neq I$ (else $R_L = I$), so $f R_L f^{-1} = R_{f(L)}$. This is illustrated in Figure 12.10.1.

A similar argument shows that if $f(z) = e^{i\theta} z$, a rotation about the origin, and if $g(z) = z - a$, then gfg^{-1} is a rotation about the point a in $\mathbb{C}$. Finally, we have seen earlier that if a linear transformation is represented by a matrix A with respect to one basis in a vector space, then it is represented by a matrix of the form BAB^{-1} with respect to another basis. These are all examples of *conjugacy* in a group, and this is a very fruitful idea even in an abstract group.

Definition 12.10.1 Let G be a group and suppose that f and g are in G. We say that f and g are *conjugate in* G, or that f *is a conjugate of* g, if there is some h in G such that $f = hgh^{-1}$. The *conjugacy class* of g is the set $\{hgh^{-1} : h \in G\}$ of all conjugates of g, and we shall denote this by $[g]$. Finally, subgroups H_1 and H_2 are *conjugate subgroups* of G if, for some h in G, $H_1 = hH_2h^{-1}$.

Some elementary remarks will help the idea to settle, and we leave the reader to provide the proofs. First, if f and g are conjugate, say $f = hgh^{-1}$, then $f^n = (hgh^{-1})^n = hg^n h^{-1}$, so that $f^n = e$ if and only if $g^n = e$; thus *conjugate elements have the same order*. Next, as $heh^{-1} = e$, we see that $[e] = \{e\}$, and as as $g = ege^{-1}$, we see that $g \in [g]$. If G is abelian, then $[g] = \{g\}$. More

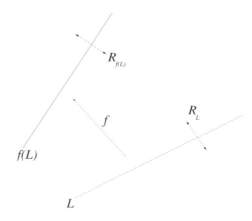

Figure 12.10.1

12.10 Conjugation

generally, $[g] = \{g\}$ *if and only if g commutes with every element of G.* This last observation tells us that the conjugates of g are giving us important information about the element g (for example, the more elements g commutes with, the smaller the conjugacy class $[g]$ is). It seems natural, then, to extend this idea to subgroups, and this turns out to be one of the most important ideas in group theory.

Definition 12.10.2 A subgroup H of G is *normal*, or *self-conjugate*, if every conjugate subgroup of H is equal to H; that is, if $gHg^{-1} = H$ for every g in G.

We shall develop the theory of normal subgroups in the rest of this chapter. There is an immediate consequence of this definition which relates normal subgroups and cosets. We recall that if H is a subgroup of G, then the left coset gH need not be the same as the right coset Hg.

Theorem 12.10.3 *let H be a subgroup H of a group G; then the following are equivalent:*

(a) *H is normal;*
(b) $gH = Hg$ *for every g in G;*
(c) *every left coset is a right coset;*
(d) *every right coset is a left coset.*

Proof It follows immediately from Definition 12.10.2 that H is normal if and only if $gH = Hg$ for every g, and this implies that every left coset is a right coset. Now suppose that every left coset is a right coset, and take any left coset gH. Then there is some f in G such that $gH = Hf$. As $g \in gH$ we see that $g \in Hf$ so that $Hf \cap Hg \neq \emptyset$. As any two right cosets are disjoint or equal, we have $Hf = Hg$ so that $gH = Hg$. This proves that (a), (b) and (c) are equivalent, and a similar argument can be used for (d). □

We recall that given a subgroup H of G, the group G is partitioned by its cosets into a disjoint union of sets. The conjugacy classes provide a similar splitting of G.

Theorem 12.10.4 *The relation of conjugacy in a group G is an equivalence relation; thus G is the disjoint union of mutually disjoint conjugacy classes.*

Proof Let us write $f \sim g$ to mean that f is conjugate to g; that is, $f = hgh^{-1}$ for some h or, equivalently, $[f] = [g]$. If $f = hgh^{-1}$ then $g = h^{-1}fh$ so that $f \sim g$ implies that $g \sim f$. Next, as $g = ege^{-1}$ we see that $g \sim g$. Finally, suppose that $f = ugu^{-1}$ and $g = vhv^{-1}$; then $f = (uv)h(uv)^{-1}$, so that $f \sim g$ and $g \sim h$ implies $f \sim h$. These facts show that $\sim$ is an equivalence relation on G, and hence the equivalence classes partition G as described in Theorem 12.4.5. □

The conjugacy classes of elements in G can be used to determine whether a given subgroup of G is normal or not.

Theorem 12.10.5 *A subgroup H of G is a normal subgroup if and only if it is a union of conjugacy classes.*

Proof Suppose that H is a normal subgroup of G, and take any element h of H. Then, for every g in G, $gHg^{-1} = H$ so that $ghg^{-1} \in H$. This shows that $[h] = \{ghg^{-1} : g \in G\} \subset H$, so that H is the union of the conjugacy classes $[h]$ taken over all h in H. Conversely, suppose now that H is a union of conjugacy classes. Then, for any h in H, $[h] \subset H$. Now for all g in G, $ghg^{-1} \in [h]$ so that if $h \in H$ then $ghg^{-1} \in H$. Thus for all g in G, $gHg^{-1} \subset H$. If we now replace g by g^{-1}, we see that $g^{-1}Hg \subset H$ and hence that $H \subset gHg^{-1}$. We deduce that $gHg^{-1} = H$ for all g in G, so that H is a normal subgroup of G. □

We end this section with two applications of the idea of conjugacy. The first of these is to permutation groups; the second is to dihedral groups, and this provides the algebraic account of the distinct between D_{2n} with n even and n odd that was discussed in the last paragraph in Section 12.8.

Theorem 12.10.6 *Two permutations of $\{1, \ldots, n\}$ are conjugate in S_n if and only if they have the same cycle type.*

Proof We need to define the term *cycle type*, and we shall be content here with a definition and a sketch of the proof. The idea of the proof is competely straightforward, and a formal proof is unlikely to help. Let σ be a permutation of $\{1, \ldots, n\}$; then σ is a product, say $\rho_1 \cdots \rho_r$, of disjoint cycles. Suppose that the cycle ρ_j has length ℓ_j; the *cycle type* of σ is the vector $(\ell_1, \ldots, \ell_r)$, where (because the ρ_j commute) this is only determined up to a permutation of its entries. For example, the two permutations $(1\ 2)(3\ 4\ 5)$ and $(1\ 2\ 3)(4\ 5)$ have the same cycle type.

Let us consider a typical cycle, say $\rho = (a\ b\ c)(d\ e)$. Then (by inspection) for any permutation μ,

$$\mu\rho\mu^{-1} = \big(\mu(a)\,\mu(b)\,\mu(c)\big)\big(\mu(d)\,\mu(e)\big).$$

Thus the conjugate cycle $\mu\rho\mu^1$ has the same cycle type as ρ. To prove the converse, we consider two permutations of the same cycle type and create a bijection to show that they are conjugate; we omit the details. □

12.10 Conjugation

Theorem 12.10.7 *The dihedral group D_{2n} has one conjugacy class of reflections if n is odd, and two conjugacy classes if n is even.*

Proof The group D_{2n} is generated by the rotation $r(z) = e^{2\pi i/n} z$ and the reflection $\sigma(z) = \bar{z}$ in the real axis, and these satisfy $r^m \sigma = \sigma r^{-m}$ for every integer m. Now D_{2n} contains exactly n reflections, namely $\sigma, r\sigma, \ldots, r^{n-1}\sigma$, and the result follows from the following three facts:

(a) if k is even then $r^k \sigma$ is conjugate to σ;
(b) if k is odd then $r^k \sigma$ is conjugate to $r\sigma$;
(c) $r\sigma$ is conjugate to σ if and only if n is odd.

First, if $k = 2q$, say, then $r^k \sigma = r^q (r^q \sigma) = r^q (\sigma r^{-q}) = r^q \sigma r^{-q}$, which proves (a). Next, if $k = 2q + 1$, then $r^k \sigma = (r^{q+1})(r^q \sigma) = r^q (r\sigma) r^{-q}$, which proves (b). Finally, if $n = 2m - 1$, then $r^m \sigma r^{-m} = r^{2m} \sigma = r^{n+1} \sigma = r\sigma$. Conversely, if $r\sigma = r^p \sigma r^{-p}$ for some p, then $r\sigma = r^{2p} \sigma$ so that $r^{2p-1} = e$. Thus n divides $2p - 1$ so that n must be odd. □

Exercise 12.10

1. Suppose that H is a subgroup of order m in a group G of order $2m$. Show that if $g \in G$ but $g \notin H$, then $G = H \cup gH = H \cup Hg$. Use this to show that H is a normal subgroup of G. Deduce that the alternating group A_n is a normal subgroup of the permutation group S_n. Now verify this by using Theorem 12.10.6.
2. Let $\alpha = (1\,2\,3\,4)(5\,6)$ and $\beta = (2\,4\,6\,3)(1\,5)$ be elements of S_6. Construct an explicit element σ of S_6 such that $\beta = \sigma \alpha \sigma^{-1}$.
3. Suppose that $m < n$ and regard the permutation group S_m of $\{1, \ldots, m\}$ as a subgroup of S_n. How many transpositions are in S_m, and how many in S_n? Show that S_m is not a normal subgroup of S_n.
4. Suppose that G is a multiplicative group of $n \times n$ real matrices, and that for every A in G, $\det(A) = \pm 1$. Show that the subgroup H of matrices A for which $\det(A) = 1$ is a normal subgroup of G.
5. Let 1, 2, 3 and 4 be a labelling of the vertices of a square S in $\mathbb{C}$ so that any symmetry of S can be regarded as a permutation of $\{1, 2, 3, 4\}$. In this sense, D_8 is a subgroup of S_4. Show that D_8 is not a normal subgroup of S_4.
6. Let G be a group, and let D (the 'diagonal' group) be the subgroup $\{(g, g) : g \in G\}$ of $G \times G$. Show that D is a normal subgroup of $G \times G$ if and only if G is abelian.
7. Show that every subgroup of rotations in a dihedral group D_n is normal in D_n.

12.11 Homomorphisms

We begin by recalling that a map $\theta : G \to G'$ is an isomorphism between the groups G and G' if

1. θ is a bijection of G onto G' and
2. for all g and h in G, $\theta(gh) = \theta(g)\theta(h)$.

A homomorphism between groups is a map that satisfies (2) but not necessarily (1). In particular, while the inverse of an isomorphism is an isomorphism, the inverse of a homomorphism may not exist.

Definition 12.11.1 A map $\theta : G \to G'$ between groups G and G' is a *homomorphism* if $\theta(gh) = \theta(g)\theta(h)$ for all g and h in G.

Although the inverse of a homomorphism may not exist, a homomorphism does share some properties in common with isomorphisms. For example, if $\theta : G \to G'$ is a homomorphism, and if e and e' are the identities in G and G', respectively, then $\theta(e) = \theta(e^2) = \theta(e)\theta(e)$, so that $\theta(e) = e'$. Thus *a homomorphism maps the identity in G to the identity in G'*. Similarly, as $e' = \theta(e) = \theta(gg^{-1}) = \theta(g)\theta(g^{-1})$, we see that for all g in G,

$$\theta(g)^{-1} = \theta(g^{-1}).$$

The most obvious example of a homomorphism is *a linear map between vector spaces*. Indeed, a vector space is a group with respect to addition, and as any linear map $\alpha : V \to W$ satisfies

$$\alpha(u + v) = \alpha(u) + \alpha(v)$$

for all u and v in V, it is a homomorphism from V to W. The fact that the kernel is an important concept for vector spaces suggests that we should seek a kernel of a homomorphism between groups. The kernel of a linear map $\alpha : V \to W$ is the set of vectors v which map to the identity element (the zero vector) in the additive group W, so the following definition is quite natural.

Definition 12.11.2 The kernel $\ker(\theta)$ of a homomorphism $\theta : G \to G'$ is the set $\{g \in G : \theta(g) = e'\}$, where e' is the identity element of G'.

The next result is predictable (recall that the kernel of a linear map $\alpha : V \to W$ is a subspace of V).

Theorem 12.11.3 *Let $\theta : G \to G'$ be a homomorphism. Then $\ker(\theta)$ is a subgroup of G.*

12.11 Homomorphisms

Proof Let $K = \ker(\theta) = \{g \in G : \theta(g) = e'\}$. We have seen above that $e \in K$, so that K is non-empty. If g and h are in K then $\theta(gh) = \theta(g)\theta(h) = e'e' = e'$ so that $gh \in K$. Finally, if $g \in K$ then

$$\theta(g^{-1}) = (\theta(g))^{-1} = (e')^{-1} = e'$$

so that $g^{-1} \in K$. The result now follows from Theorem 12.2.2. □

Let us illustrate these ideas in an example.

Example 12.11.4 Consider the group $\mathbb{C}^*$ of nonzero complex numbers under multiplication, and let $\theta(z) = z/|z|$. As $\theta(zw) = \theta(z)\theta(w)$, we see that θ is a homomorphism of $\mathbb{C}^*$ onto the group $\{z : |z| = 1\}$, and its kernel consists of all non-zero z such that $z = |z|$. Thus the $\ker(\theta)$ is the subgroup $\mathbb{R}^+$ of positive real numbers. Consider also the map $\varphi : \mathbb{C}^* \to \mathbb{R}^+$ given by $\varphi(z) = |z|$. Then φ is a homomorphism whose kernel is the group $\{z : |z| = 1\}$, □

We recall that a subgroup H of a group G is a normal subgroup if and only if $gH = Hg$ for every g. We know that not every subgroup is normal, and here is a simple geometric example of a non-normal subgroup.

Example 12.11.5 Let G be the group $\{x \mapsto \varepsilon x + n : n \in \mathbb{Z}, \ \varepsilon = \pm 1\}$ of isometries of $\mathbb{R}$, and let H be the subgroup $\{g \in G : g(0) = 0\}$. Then $H = \{I, h\}$, where I is the identity and $h(x) = -x$. Now let $f(x) = x + 1$; then $fH = \{f, fh\} \neq \{f, hf\} = Hf$. □

The following result is extremely important.

Theorem 12.11.6 *Let $\theta : G \to G'$ be a homomorphism with kernel K. Then K is a normal subgroup of G.*

Proof Consider any k in K. Then, for any g in G,

$$\theta(gkg^{-1}) = \theta(g)\theta(k)\theta(g^{-1}) = \theta(g)e'\theta(g^{-1}) = \theta(g)\theta(g^{-1}) = \theta(gg^{-1}) = e',$$

so that $gkg^1 \in K$. As k was any element in K, this shows that $gKg^{-1} \subset K$, and hence that $gK \subset Kg$. As this holds for all g, we also have $g^{-1}K \subset Kg^{-1}$, equivalently $Kg \subset gK$, so we deduce that $gK = Kg$. □

The fact that the kernel of a homomorphism is a normal subgroup, enables us to take another result for vector spaces and provide an analogous result for groups. The reader will recall that if $\alpha : V \to W$ is a linear map then, for a given w in W, the set of solutions v of $\alpha(v) = w$ (if any) are of the form $v_0 + K$, where K is the kernel of α. Exactly the same result is true for homomorphisms between groups (see Figure 12.11.1).

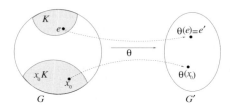

Figure 12.11.1

Theorem 12.11.7 *Suppose that $\theta : G \to G'$ is a homomorphism with kernel K, and that $y \in G'$. Suppose also that $x_0 \in G$ and that $\theta(x_0) = y$. Then the set of solutions x of the equation $\theta(x) = y$ is the coset $x_0 K$.*

Proof First, every element of $x_0 K$ is a solution to the given equation, for the general element of $x_0 K$ is $x_0 k$, where $k \in K$, and $\theta(x_0 k) = \theta(x_0)\theta(k) = ye' = y$. Next suppose that $\theta(x) = y$, and consider the element $x_0^{-1} x$ in G. Then

$$\theta(x_0^{-1} x) = \theta(x_0^{-1})\theta(x) = \big(\theta(x_0)\big)^{-1} y = y^{-1} y = e',$$

so that $x_0^{-1} x \in K$. This means that $x \in x_0 K$ and so the set of solutions of $\theta(x) = y$ is the coset $x_0 K$. □

Let us apply Theorem 12.11.7 to a situation that the reader is already familiar with.

Example 12.11.8 Let $\mathbb{C}^*$ be the multiplicative group of non-zero complex numbers, and let $\theta(z) = z^n$. As $(zw)^n = z^n w^n$, θ is a homomorphism of G into itself, and the kernel of θ is the group K of n-th roots of unity. According to Theorem 12.11.7, the solutions of the equation $\theta(z) = w$, that is, the equation $z^n = w$, are simply the points of the form $z_1 \zeta$, where z_1 is some n-th root of w, and ζ is any n-th root of unity. Of course, we know this but the point is to illustrate Theorem 12.11.7. □

It is a remarkable fact that the converse of Theorem 12.11.6 holds; that is, not only is the kernel of a homomorphism a normal subgroup, but *every normal subgroup is the kernel of a homomorphism*. We shall state this as a theorem now; it will be proved in the next section.

Theorem 12.11.9 *Let H be a subgroup of G. Then H is a normal subgroup of G if and only if there is some group G', and some homomorphism $\theta : G \to G'$, whose kernel is H.*

We end with another example.

Example 12.11.10 Consider the permutation group S_n of $\{1, 2, \ldots, n\}$, and the alternating group A_n (of all even permutations in S_n). If σ is an even permutation, then $\sigma A_n = A_n = A_n\sigma$; if σ is an odd permutation, then $A_n \cup \sigma A_n = S_n = A_n \cup A_n\sigma$, so that again, $\sigma A_n = A_n\sigma$. We deduce that $\sigma A_n = A_n\sigma$ for every σ; thus A_n is a normal subgroup of S_n. Here is an alternative proof. The map $\varepsilon : S_n \to \{1, -1\}$, where $\varepsilon(\rho)$ is the sign of ρ, is a homomorphism from S_n to $\{1, -1\}$, for $\varepsilon(\alpha\beta) = \varepsilon(\alpha)\varepsilon(\beta)$. Its kernel is the set of ρ such that $\varepsilon(\rho) = 1$ and this is A_n; thus A_n is a normal subgroup of S_n. □

Exercise 12.11

1. Let G and H be groups, and let $\theta : G \times H \to G$ be the map $\theta(g, h) = g$. Show that θ is a homomorphism. What is its kernel?
2. Show that if H is a subgroup of an abelian group G, then H is a normal subgroup of G. In particular, every subspace U of a vector space V is a normal subgroup of V. Find a homomorphism (a linear map) of V into itself with kernel U.
3. Show that there are exactly two homomorphisms from C_6 to C_4.
4. Show that the alternating group A_n is a normal subgroup of S_n. More generally, prove that if a group G of order $2n$, and if H is a subgroup of order n, then H is a normal subgroup of G.
5. Let G be the group of similarities of the complex plane of the form $f(z) = az + b$, where $a \neq 0$, and let $\mathbb{C}^*$ be the multiplicative group of non-zero complex numbers. Show that the map $\phi(f) = a$ is a homomorphism of G onto $\mathbb{C}^*$. What (in geometric terms) is its kernel K?
6. Suppose that $\theta : G \to H$ is a homomorphism. Show that, for any g in G, the order of $\theta(g)$ divides the order of g.
7. Suppose that a subgroup H of G has the property that there is no other subgroup of G that is isomorphic to H. Show that H is a normal subgroup of G.
8. Let m be a positive integer, and let G be an abelian group. Show that the map $\theta : G \to G$ defined by $\theta(g) = g^m$ is a homomorphism. Deduce that the set of elements of order m in G is a normal subgroup of G.

12.12 Quotient groups

This section contains the most important idea about groups that is in this text. We begin with two examples; then we develop the theory.

Example 12.12.1 Let G be the additive group of vectors $\mathbb{R}^3$, and let H be the subgroup given by $x_3 = 0$ (the horizontal coordinate plane). We consider two planes that are parallel to H, say $x + H$ and $y + H$, and attempt to 'add' these by the rule

$$(x + H) = (y + H) = (x + y) + H.$$

Does this make any sense, and if it does, is it useful? First, if we choose planes H' and H'' parallel to H we can certainly write them in the form $H' = x + H$ and $H'' = y + H$, but there is a problem because the x and y here are not uniquely determined by the planes. Suppose we write them in another way, say $H' = x_1 + H$ and $H'' = y_1 + H$. Then, in order that these two representations of H' are the same, we must have $x_1 - x \in H$ and, similarly, $y_1 - y \in H''$. Suppose that $x_1 - x = h'$ and $y_1 - y = h''$; then

$$(x_1 + H) + (y_1 + H) = (x + h' + H) + (y + h'' + H)$$
$$= (x + y) + (h + h' + H)$$
$$= (x + y) + H,$$

so that this 'addition' of planes (parallel to H) is properly defined, and the 'sum' is again a plane parallel to H. Note that $H' + H = H' = H + H'$, so that H acts as the identity for this addition. It is also worth noting that among the class of all planes that are parallel to H, the plane H itself is the only one of these planes that is a subgroup of $\mathbb{R}^3$. It is not difficult to see that the set of these planes, with the addition defined above, is a group, with identity H, and the inverse of $x + H$ is $(-x) + H$.

Continuing with this example, consider now the map $\theta : \mathbb{R}^3 \to \mathbb{R}$ given by $\theta(x_1, x_2, x_3) = x_3$. It is easy to check that this map is a homomorphism of $\mathbb{R}^3$ onto the 'vertical' axis, say Z and, of course, each plane H' is mapped by θ onto the point $H' \cap Z$. The addition of planes corresponds precisely to the addition along Z, and H is the kernel of the homomorphism θ.

Let us summarize what we have found in this example:

(a) the plane H is a subgroup of the group $\mathbb{R}^3$;
(b) The cosets $x + H$ of H can be added together to form a group;
(c) there is a homomorphism of $\mathbb{R}^3$ onto another group Z (here, the third axis) whose kernel is H;
(d) Z is isomorphic to the group of cosets of H. □

The next example is from number theory, but the conclusions are the same.

Example 12.12.2 Consider the additive group $\mathbb{Z}$ of integers, and the subgroup H that is the set of multiples of a given integer k. Now let $\theta : \mathbb{Z} \to \mathbb{Z}_k$ be the map taking an integer n to its remainder modulo k (if $n = ak + b$, where $0 \le b < k$, then $\theta(n) = b$). Clearly, the kernel of θ is H, and the cosets of H

12.12 Quotient groups

may be regarded as the elements of $\mathbb{Z}_k$; for example,

$$a + H = \{\ldots, -k+a, a, k+a, \ldots\}.$$

The addition of cosets obviously 'mirrors' the addition in $\mathbb{Z}_k$, and the four conclusions drawn at the end of Example 12.12.1 are valid here. □

We are now going to show that the examples we have just examined are typical of the situation for a general group *with just one proviso*: in the examples above, the groups are abelian, but in the general case we have to compensate for G failing to be abelian by insisting that *the subgroup H shall be a normal subgroup*. If H is a normal subgroup of G we can obtain all of these four conclusions, but some considerable effort is needed, and this effort is the culmination of our work on groups.

Consider a homomorphism θ from a group G into a group G', and let $\theta(G)$ be the image of G; that is $\theta(G) = \{\theta(g) : g \in G\}$, and let K be the kernel of θ. The first step is to prove the following result (which is suggested from our results on linear maps between vector spaces).

Theorem 12.12.3 *Suppose that $\theta : G \to G'$ is a homomorphism. Then the image $\theta(G)$ is a subgroup of H.*

The proof follows from Theorem 12.2.2. We know that $\theta(G)$ contains the identity e' of G'. If g' and h' are in $\theta(G)$, then we may write $g' = \theta(g)$ and $h' = \theta(h)$; then $g'h' = \theta(g)\theta(h) = \theta(gh)$, so that $g'h' \in \theta(G)$. A similar argument shows that $(g')^{-1} \in \theta(G)$. □

Next, we make an elementary, but useful, observation. Suppose that G is a group and that φ is a bijection of a set X onto G. Then X inherits the structure of a group from G by the definition

$$x * y = \varphi^{-1}\big(\varphi(x)\varphi(y)\big)$$

(that is to combine x and y, we map them to G, combine their images in G and then map the result back into X). The proof of this is entirely straightforward and is left to the reader. Notice that if we make X into a group in this way, then φ is an isomorphism from X to G.

Let us now apply this to the situation considered above in which θ is a homomorphism from G to G', $\theta(G)$ is the image of G, and K is the kernel of θ. Theorem 12.11.7 says that if $y \in \theta(G)$ then the set of g in G such that $\theta(g) = y$ is some coset $x_0 K$. It follows that θ acts as a bijection, namely $x_0 K \mapsto y = \theta(x_0)$ from the set $\mathcal{C}$ of all these cosets onto the group $\theta(G)$. Note carefully that $\mathcal{C}$ is the set whose elements are the cosets gK; thus $\mathcal{C} = \{gK; g \in G\}$ (and this is not the same as the union $\cup_g gK$ of the cosets, for this union is G; see Figure 12.11.1).

Because θ acts as a bijection between $\mathcal{C}$ and $\theta(G)$ it follows (from the observation above) that the set $\mathcal{C}$ of cosets inherits a group structure from the group $\theta(G)$. What, then, is the rule of composition in this group $\mathcal{C}$ of of cosets? According to the remarks made at the start of this section, this rule is given by

$$(gK)*(hK) = \theta^{-1}\big(\theta(g)\theta(h)\big). \tag{12.12.1}$$

As $\theta(g)\theta(h) = \theta(gh)$ the set on the right in (12.12.1) contains the element gh, and it a coset (as we know that $*$ is a group operation that combines left cosets to give a left coset). It must therefore be the coset ghK, so the rule of composition of left cosets is now seen to be

$$(gK)*(hK) = (gh)K. \tag{12.12.2}$$

We state this as a theorem (and the last statement in this should be clear).

Theorem 12.12.4 *Suppose that $\theta : G \to G'$ is a homomorphism with kernel K. Then the set G/H of left cosets $\{gK; g \in G\}$ is a group with respect to the operation $(gK)*(hK) = ghK$. The identity in G/H is eK (which is K), and G/H is isomorphic to $\theta(G)$.*

As the kernel K of the homomorphism is a normal subgroup, we see that $gK = Kg$ for every g so that it does not matter whether we use left cosets or right cosets in this argument. The group of cosets is important and is given a name.

Definition 12.12.5 *The group of cosets given in Theorem 12.12.4 is the quotient group G/K.*

Theorem 12.11.9 (which we have yet to prove) asserts that a subgroup H of a group G is normal if and only if it is the kernel of a homomorphism. Theorem 12.11.6 shows that if a subgroup H is the kernel of a homomorphism then it is normal, so we only have to prove the reverse implication, namely that *if a subgroup H is normal, then it is the kernel of a homomorphism*. We can now complete this proof.

The Proof of Theorem 12.11.9 Suppose that a subgroup H of G is normal. Then, by Theorem 12.12.4, the quotient group G/H exists. Now define the map $\theta : G \to G/H$ to be the map $\theta(g) = gH$. Now θ is a homomorphism, for

$$\theta(g_1 g_2) = (g_1 g_2 H) = (g_1 H)*(g_2 H) = \theta(g_1)*\theta(g_2),$$

as required. Finally, what is the kernel K of θ? Now $g \in K$ if and only if $\theta(g)$ is the identity in G/H, and by Theorem 12.12.4, this identity is H. Thus $g \in K$

if and only if $gH = H$, and this is so if and only if $g \in H$. We have thus shown that $H = K$, and hence that H is the kernel of the homomorphism θ. □

According to Theorem 12.11.9, we can now rewrite Theorem 12.12.4 in the following equivalent form.

Theorem 12.12.4A *Suppose that K is a normal subgroup of G. Then the set of left cosets $\{gK; g \in G\}$ is a group with respect to the operation $(gK)*(hK) = ghK$.*

Exercise 12.12

1. Suppose that H is a (cyclic) subgroup of order m of a cyclic (abelian) group G of order n. What is G/H?
2. Let $H = \{I, (1\,2)(3\,4), (1\,3)(2\,4), (1\,4)(2\,3)\}$. Show that H is a normal subgroup of S_4, so that S_4/H has order six. Write down the six cosets with respect to H. Is S_4/H is isomorphic to C_6 or to D_6?
3. Let K be a normal subgroup of a group G, and let A and B be left cosets of K (we deliberately write these in this way as there is no unique way to write a coset in the form gK). We want to define a product AB, and we can try to do this by writing $A = gK$, $B = hK$ and $AB = ghK$. Hoewever, in order to make this 'definition' legitimate, we must show that if $gK = g'K$ and $hK = h'K$ then $g'h'K = ghK$. Show that this is so.
4. Show that $\mathbb{Q}/\mathbb{Z}$ is an infinite group in which every element has finite order.
5. Suppose that H is a normal subgroup of a group G, and that the quotient group G/H has order n. Prove that for every g in G, $g^n \in H$.
6. Let G be an abelian group and let K be the set of elements of G that have finite order. Show that no elements of G/T (except its identity) have finite order.
7. Show that the quotient group $\mathbb{R}/\mathbb{Z}$ is isomorphic to the 'circle group' $\{z \in \mathbb{C} : |z| = 1\}$. [Hint: the homomorphism here is $x \mapsto e^{2\pi i x}$, and you may assume standard properties of the exponential function].
8. Consider the additive group $\mathbb{C}$ and the subgroup Γ consisting of all Gaussian integers $m + in$, where $m, n \in \mathbb{Z}$. By considering the map
$$x + iy \mapsto (e^{2\pi i x}, e^{2\pi i y}),$$
show that the quotient group $\mathbb{C}/\Gamma$ is isomorphic to $S \times S$, where S is the circle group $\{z : |z| = 1\}$.

13
Möbius transformations

13.1 Möbius transformations

A *Möbius transformation* or *map* is a function f of a complex variable z that can be written in the form

$$f(z) = \frac{az+b}{cz+d}, \qquad (13.1.1)$$

for some complex numbers a, b, c and d with $ad - bc \neq 0$. It is easy to see why we require that $ad - bc \neq 0$, for

$$f(z) - f(w) = \frac{(ad-bc)(z-w)}{(cz+d)(cw+d)}, \qquad (13.1.2)$$

so that f is constant when $ad - bc = 0$. Notice that this also shows that f is injective.

The deceptively simple form of (13.1.1) conceals two problems. First, a Möbius transformation f can be written in the form (13.1.1) in many ways (just as a rational number can be written as p/q in many ways); thus given f, we *cannot* say what its coefficients a, b, c and d are. For example, if f maps z to $2z$, its coefficients might be 2π, 0, 0 and π, respectively. Of course, given numbers a, b, c and d we can construct a Möbius map f from these, but that is another matter.

The second problem stems from the fact that, for example, $1/(z - z_0)$ is not defined at the point z_0. This means that *there is no subset of $\mathbb{C}$ on which all Möbius maps are defined*, and this presents difficulties when we try to form the composition of Möbius maps.

Example 13.1.1 Let $f(z) = (z+2)/z$ and $g(z) = (z+1)/(z-1)$. Then, apparently,

$$f(g(z)) = \frac{g(z)+2}{g(z)} = \frac{(z+1)+2(z-1)}{z+1} = \frac{3z-1}{z+1},$$

so that fg fixes the point 1. However, how can this be so as g *is not defined when* $z = 1$? Worse still, if $h(z) = 1/z$ then apparently $hfg(z) = (z+1)/(3z-1)$, although g is not defined when $z = 1$, $fg(z)$ is not defined when $z = -1$, and $hfg(z)$ is not defined when $z = 1/3$. More generally, a composition $f_1 \cdots f_n$ of Möbius maps will (in general) not be defined at n distinct points in the complex plane. □

Before we can develop the rich theory of Möbius maps we must resolve these two issues. Neither are difficult to deal with, but we must attend to the details. Our first result shows that although f in (13.1.1) does not determine (a, b, c, d), it does so to within a non-zero scalar multiple.

Theorem 13.1.2 *Suppose that a, b, c, d, α, β, γ and δ are complex numbers with $(ad - bc)(\alpha\delta - \beta\gamma) \neq 0$, and such that for at least three values of z in $\mathbb{C}$, $cz + d \neq 0$, $\gamma z + \delta \neq 0$, and*

$$\frac{az+b}{cz+d} = \frac{\alpha z + \beta}{\gamma z + \delta}. \tag{13.1.3}$$

Then there is some non-zero complex number λ such that

$$\begin{pmatrix} \alpha & \beta \\ \gamma & \delta \end{pmatrix} = \lambda \begin{pmatrix} a & b \\ c & d \end{pmatrix}. \tag{13.1.4}$$

Proof Let z_1, z_2 and z_3 be the values of z for which (13.1.3) holds with non-zero denominators. Then the quadratic equation

$$(az+b)(\gamma z + \delta) = (cz+d)(\alpha z + \beta),$$

has three distinct solutions z_j, so we can equate coefficients and deduce that $a\gamma = c\alpha$, $b\gamma + a\delta = c\beta + d\alpha$ and $b\delta = d\beta$. These conditions are equivalent to the existence of a complex number μ such that

$$\begin{pmatrix} d & -b \\ -c & a \end{pmatrix} \begin{pmatrix} \alpha & \beta \\ \gamma & \delta \end{pmatrix} = \begin{pmatrix} \mu & 0 \\ 0 & \mu \end{pmatrix},$$

where (by considering determinants) $\mu^2 = (ad-bc)(\alpha\delta - \beta\gamma) \neq 0$. However, this matrix identity is equivalent to (13.1.4) in the form

$$\begin{pmatrix} \alpha & \beta \\ \gamma & \delta \end{pmatrix} = \frac{\mu}{ad-bc} \begin{pmatrix} a & b \\ c & d \end{pmatrix}.$$

□

Informally, then, the first difficulty is resolved by saying the the vector (a, b, c, d) is determined to within a (complex) scalar multiple. The second difficulty (illustrated in Example 13.1.1) is resolved by joining an extra point, which is called *the point at infinity*, to $\mathbb{C}$. This new point is denoted by ∞. We have an intuitive notion of a complex number tending to ∞ (that is, $1/z$ tending to 0) and if $c \neq 0$ then

$$\lim_{z \to \infty} \frac{az+b}{cz+d} = \frac{a}{c}, \quad \lim_{z \to -d/c} \frac{az+b}{cz+d} = \infty.$$

These limits motivate the following definition (and as this is their only role, we need not enter into a formal discussion of limits).

Definition 13.1.3 Let f be the Möbius map given (13.1.1). If $c \neq 0$ we define $f(\infty) = a/c$ and $f(-d/c) = \infty$; if $c = 0$ we define $f(\infty) = \infty$.

There is a subtle point here. Given a Möbius map f, we can write it in the form (13.1.1) in many ways. However, Theorem 13.1.2 guarantees that the condition '$c \neq 0$' holds for all or none of these ways, and that when it holds, the values a/c and $-d/c$ are independent of the coefficients we choose. Without Theorem 13.1.2, Definition 13.1.3 would not be a legitimate definition. It is important to understand that Definition 13.1.3 is based on the idea of a function as a 'rule', and that we are *not* introducing algebraic rules for handling ∞. We shall give a geometric argument which supports Definition 13.1.3 in Section 13.8, and an algebraic argument (involving the vector space $\mathbb{C}^2$) in Section 13.6. In any event, from now on $\mathbb{C} \cup \{\infty\}$ plays the central role in the theory, and we shall consider every Möbius map to be a map of $\mathbb{C} \cup \{\infty\}$ into itself.

Definition 13.1.4 The set $\mathbb{C} \cup \{\infty\}$ is called the *extended complex plane* and is denoted by $\mathbb{C}_\infty$.

The major benefit of Definition 13.1.3 is that every Möbius tranformation is now defined on the same set, namely $\mathbb{C}_\infty$, so that the composition of any two Möbius maps is properly defined. The previous examples suggest that the composition of two Möbius maps is again a Möbius map; in fact, we have the following stronger result.

Theorem 13.1.5 *Each Möbius map is a bijection of $\mathbb{C}_\infty$ onto itself, and the Möbius maps form the Möbius group $\mathcal{M}$ with respect to composition.*

Proof We know that the composition of functions is associative, and the identity map I is a Möbius map because $I(z) = (z+0)/(0z+1)$ and $I(f(z)) = f(z) = f(I(z))$. Next, let

$$f(z) = \frac{az+b}{cz+d}, \quad f^*(z) = \frac{dz-b}{-cz+a}.$$

We shall assume that $c \neq 0$ here (and leave the easier case $c = 0$ to the reader). If $z \neq -d/c, \infty$ then $f(z) \in \mathbb{C}$, and we see (from elementary algebra, and the fact that $ad - bc \neq 0$) that $f^*(f(z)) = z$. Similarly, if $z \neq a/c, \infty$ then $f(f^*(z)) = z$. It follows that

$$f : \{z \in \mathbb{C}_\infty : z \neq -d/c, \infty\} \to \{z \in \mathbb{C}_\infty : z \neq a/c, \infty\}$$

is a bijection. As $f^*(f(-d/c)) = f^*(\infty) = -d/c$ and $f^*(f(\infty)) = f^*(a/c) = \infty$, we can now assert that f is a bijection of $\mathbb{C}_\infty$ onto itself with inverse f^*. Thus *each Möbius map f has a Möbius inverse*

$$f^{-1}(z) = \frac{dz - b}{-cz + a}. \tag{13.1.5}$$

We still need to show that the composition of two Möbius maps is again a Möbius map and, although this is not difficult, at this stage there is no elegant way to do this. Briefly, we consider the Möbius maps f, g and h, where f is given by (13.1.1), and

$$g(z) = \frac{\alpha z + \beta}{\gamma z + \delta}, \quad h(z) = \frac{(a\alpha + b\gamma)z + (a\beta + b\delta)}{(c\alpha + d\gamma)z + (c\beta + d\delta)}.$$

It is easy to see that $f(g(z)) = h(z)$ for all those z for which the obvious algebraic manipulation is valid. This leaves only a finite number of exceptional values of z to check (just as we did when considering the inverse of f). This is tedious, and elementary, and we omit the details. □

It is useful to know that any Möbius map can be written as the composition of simple transformations.

Theorem 13.1.6 *Every Möbius transformation can be expressed as the composition of at most four maps, each of which is of one of the forms $z \mapsto az$, $z \mapsto z + b$ and $z \mapsto 1/z$.*

Proof Notice that these simple maps are rotations, dilations (expansion from the origin), translations, and the *complex inversion $z \mapsto 1/z$*. Suppose that $f(z) = (az + b)/(cz + d)$. If $c = 0$ then $d \neq 0$ and $f = f_2 f_1$, where $f_1(z) = (a/d)z$ and $f_2(z) = z + b/d$. If $c \neq 0$ then $f = f_4 f_3 f_2 f_1$, where $f_1(z) = z + d/c$, $f_2(z) = 1/z$, $f_3(z) = kz$, $k = -(ad - bc)/c^2$, and $f_4(z) = z + a/c$. □

We now return to the connection between Möbius maps and 2×2 complex matrices. If A is a 2×2 non-singular complex matrix, we can use the coefficients a, b, c and d of A to construct f given in (13.1.1). The group of 2×2 non-singular complex matrices with respect to matrix multiplication is the *General Linear group* $\mathrm{GL}(2, \mathbb{C})$, and this construction defines a map

$\Phi : \mathrm{GL}(2, \mathbb{C}) \to \mathcal{M}$ which is given explicitly by

$$\Phi : \begin{pmatrix} a & b \\ c & d \end{pmatrix} \mapsto f, \quad f(z) = \frac{az+b}{cz+d}. \qquad (13.1.6)$$

Theorem 13.1.7 *The mapping Φ is a homomorphism from the group $\mathrm{GL}(2, \mathbb{C})$ onto the group $\mathcal{M}$ of Möbius maps.*

Proof Given matrices

$$A = \begin{pmatrix} a & b \\ c & d \end{pmatrix}, \quad B = \begin{pmatrix} \alpha & \beta \\ \gamma & \delta \end{pmatrix},$$

we let $f = \Phi(A)$ and $g = \Phi(B)$. Let $h = \Phi(AB)$; this is the Möbius map constructed from the matrix product AB. We have to show that $h = \Phi(AB) = \Phi(A)\Phi(B) = fg$, and this was proved as part of the proof of Theorem 13.1.5. □

Informally, this result says:

$$(\text{matrix of } f) \times (\text{matrix of } g) = (\text{matrix of } fg)$$

What is the kernel of the homomorphism Φ in (13.1.6)? By definition, it is the set of matrices A such that the Möbius map $\Phi(A)$ is the identity map I. Now, from Theorem 13.1.2, I is represented by (and only by) scalar multiples of the identity matrix, and this proves the next result.

Theorem 13.1.8 *The kernel of the homomorphism Φ is $\{\lambda I : \lambda \in \mathbb{C}\}$, where I is the 2×2 identity matrix.*

Given the Möbius transformation f in (13.1.1), we can change each of the coefficients by the same non-zero scalar multiple without changing the map; thus we may assume that $ad - bc = 1$. The space of 2×2 complex matrices with determinant one is the *Special Linear group* $\mathrm{SL}(2, \mathbb{C})$ and, as this is a subgroup of $\mathrm{GL}(2, \mathbb{C})$, Φ also acts as a homomorphism from $\mathrm{SL}(2, \mathbb{C})$ onto $\mathcal{M}$. The kernel of this homomorphism consists of those scalar multiples of the identity matrix that have determinant one; thus we also have the following result (see Theorem 12.12.4).

Theorem 13.1.9 *The homomorphism $\Phi : \mathrm{SL}(2, \mathbb{C}) \to \mathcal{M}$ has kernel $\{\pm I\}$, and $\mathcal{M}$ is isomorphic to the quotient group $\mathrm{SL}(2, \mathbb{C})/\{\pm I\}$.*

Exercise 13.1

1. Show that the set of Möbius transformations of the form (13.1.1), where a, b, c and d are integers with $ad - bc = 1$, is a subgroup of $\mathcal{M}$. This is the *Modular group* $\mathrm{SL}(2, \mathbb{Z})$.

2. Let $f(z) = (3z + 2)/(z + 1)$ and $g(z) = (z + 4)/(z - 1)$. Find matrices A and B *with determinant one* that represent f and g, respectively, and verify that $f(g(z))$ is derived from the matrix AB. What is $\det(AB)$?
3. Suppose that f is given by (13.1.1). Show that $f^2 = I$ but $f \neq I$ if and only if $a + d = 0$.
4. Let $f(z) = (2z + 1)/(3z + 4)$. Express f as a composition of rotations, dilations, translations and a complex inversion.
5. Show that a Möbius map f can be written in the form (13.1.1) with a, b, c and d real if and only if f maps $\mathbb{R} \cup \{\infty\}$ into itself.
6. Let $f(z) = e^{2\pi i/n} z$ and $g(z) = 1/z$. Show that the subgroup G of $\mathcal{M}$ generated by f and g is a dihedral group.

13.2 Fixed points and uniqueness

Roughly speaking, Theorem 13.1.2 implies that the general Möbius map has three degrees of freedom. This suggests that there should be a unique Möbius map that takes three given distinct points z_1, z_2, z_3 to three given distinct points w_1, w_2, w_3. This is indeed the case, and we have the following *existence and uniqueness* theorem.

Theorem 13.2.1 *Let $\{z_1, z_2, z_3\}$ and $\{w_1, w_2, w_3\}$ be triples of distinct points in $\mathbb{C}_\infty$. Then there is a unique Möbius map f with $f(z_j) = w_j$ for $j = 1, 2, 3$.*

Proof Suppose first that none of the z_j are ∞, and let

$$g(z) = \left(\frac{z_3 - z_2}{z_3 - z_1}\right) \frac{z - z_1}{z - z_2}.$$

Then $g(z_1) = 0$, $g(z_2) = \infty$ and $g(z_3) = 1$. Now suppose that one of the z_j is ∞. Choose a point z_4 distinct from z_1, z_2, z_3 and let $s(z) = 1/(z - z_4)$. Then $s(z) = \infty$ if and only if $z = z_4$, so that none of $s(z_1), s(z_2), s(z_3)$ are ∞. Then (by the previous paragraph) there is a Möbius map g_1 which maps $s(z_1), s(z_2)$ and $s(z_3)$ to 0, 1 and ∞. Thus in all cases, there is a Möbius map (either g or $g_1 s$) that maps z_1, z_2 and z_3 to 0, 1 and ∞, respectively.

In the same way, there is a Möbius map h such that $h(w_1) = 0$, $h(w_2) = \infty$ and $h(w_3) = 1$. Now let $f = h^{-1} g$. Then f is the required map because, for each j, $f(z_j) = h^{-1} g(z_j) = w_j$; for example, $f(z_1) = h^{-1} g(z_1) = h^{-1}(0) = w_1$.

We have to prove uniqueness, so suppose that f and F are Möbius maps such that $f(z_j) = w_j = F(z_j)$ for $j = 1, 2, 3$. Then $F^{-1} f$ fixes each z_j. Now let v be the Möbius map that maps z_1, z_2 and z_3 to 0, 1 and ∞, respectively. Then $v^{-1}(F^{-1} f)v$ is a Möbius map that fixes 0, 1 and ∞. Now it is self-evident that any such map is the identity map I; thus $v^{-1}(F^{-1}f)v = I$, and this implies (in a group) that $f = F$. □

Note that the proof of Theorem 13.2.1 avoids computation. In general, one should resist the temptation to solve a problem on Möbius maps by long computations (apart from finding composite maps), for almost always there is a short, elegant and easy proof available. There is a useful corollary of Theorem 13.2.1.

Corollary 13.2.2 *If a Möbius map has three fixed points then it is the identity map.*

This can also be proved by noting that the quadratic equation $az + b = z(cz + d) = 0$, where f is given by (13.1.1), has three distinct solutions when f has three distinct fixed points, so in this case $c = 0$, $a = d$ and $b = 0$. We end this section with some examples.

Example 13.2.3 We construct the Möbius map f such that $f(i) = 0$, $f(-i) = \infty$ and $f(1) = 1 + i$. The numerator of f must be zero at i, and its denominator must be zero at $-i$; thus $f(z) = k(z - i)/(z + i)$ for some constant k. The condition $f(1) = 1 + i$ determines k and we find that $k = 2i/(1 - i)$. □

Example 13.2.4 We construct the Möbius map f such that $f(0) = 3$, $f(1) = 1 + i$ and $f(2) = 1 - i$. As in our proof of Theorem 13.2.1, $f = h^{-1}g$, where g maps $0, 1, 2$ to $0, 1, \infty$, and h maps $0, 1, 2$ to $0, 1, \infty$. We can find g and h by the technique used in Example 13.2.3 and, after some computation, we see that $g(z) = -z/(z - 2)$, $h(z) = k(z - 3)/(z - [1 - i])$, where $k = (2 - 4i)/5$. A further computation gives

$$f(z) = \frac{(19 - 7i)z + (-24 + 12i)}{(8 + i)z + (-8 + 4i)}.$$

□

Example 13.2.5 We construct the Möbius map f such that $f(1) = 1$, $f(\infty) = i$ and $f(0) = -i$. In this case we first construct f^{-1} by the method in Example 13.2.3, and we find that $f^{-1}(z) = (-iz + 1)/(z - i)$. Using (13.1.5), we see that $f(z) = (iz + 1)/(z + i)$. Notice that f cyclically permutes the four points $0, -i, \infty, i$. This shows that f^4 (the fourth iterate of f) fixes these four points; thus, by Corollary 13.2.2, $f^4(z) = z$ for all z. One can confirm this, and at the same time illustrate Theorem 13.1.6, by showing that

$$\begin{pmatrix} i & 1 \\ 1 & i \end{pmatrix}^4 = \lambda \begin{pmatrix} 1 & 0 \\ 0 & 1 \end{pmatrix}$$

for some non-zero λ. □

Exercise 13.2

1. Find the Möbius map that fixes 1 and -1 and maps 0 to ∞.
2. Find the Möbius map f that cyclically permutes the points $-1, 0$ and 1. Verify directly that $f^3(z) = z$ for all z.
3. The Möbius map $f(z) = iz$ generates a cyclic group of order four. Now let h be any Möbius map. Show that hfh^{-1} generates a cyclic group of order four, and also fixes $h(0)$ and $h(\infty)$. Use this to construct a Möbius map of order four that fixes 1 and -1.
4. Let $f(z) = (az + b)/(cz + d)$. If we are to have $f(z_j) = w_j$ (see Theorem 13.2.1) then the unknown coeficients a, b, c and d of f must satisfy the homogeneous linear equation

$$\begin{pmatrix} z_1 & 1 & -z_1 w_1 & -w_1 \\ z_2 & 1 & -z_2 w_2 & -w_2 \\ z_3 & 1 & -z_3 w_3 & -w_3 \end{pmatrix} \begin{pmatrix} a \\ b \\ c \\ d \end{pmatrix} = \begin{pmatrix} 0 \\ 0 \\ 0 \end{pmatrix}.$$

Suppose that none of the z_j, or the w_j, are ∞. Show that this matrix has rank three, and deduce (algebraically) that the solution (a, b, c, d) is determined to within a scalar multiple.

13.3 Circles and lines

The main result in this section is that a Möbius map takes a circle into a circle or straight line, and a straight line into a circle or straight line. First, however, we need to amend our definitions of lines and circles for, at the moment, the image of a circle can never be a line (because, roughly speaking, a line will fall into two pieces if a point is removed, but a circle will not). We shall refer to the usual circles as *Euclidean circles*, and the usual lines as *Euclidean lines*. We are going to add the point ∞ to each Euclidean line and, from now on, a *circle* is either a Euclidean circle, or a Euclidean line with ∞ attached. Now we can assert that *the image of a circle is a circle*. Euclidean circles and Euclidean lines are objects in the Euclidean geometry of $\mathbb{C}$, but we must now consider circles (which may or may not contain the point ∞) in the geometry of $\mathbb{C}_\infty$. We often call a Euclidean circle 'a circle in $\mathbb{C}$', and similarly for lines.

Definition 13.3.1 A *Euclidean circle* is the set of points in $\mathbb{C}$ given by some equation $|z - z_0| = r$, where $r > 0$. A *Euclidean line* is the set of points in $\mathbb{C}$

given by some equation $|z - a| = |z - b|$, where $a \neq b$. A *circle* (in $\mathbb{C}_\infty$) is either a Euclidean circle, or a set $L \cup \{\infty\}$, where L is a Euclidean line.

It is immediate that *there is a unique circle that passes through any three given distinct points in* $\mathbb{C}_\infty$. This fact alone is enough to justify Definition 13.3.2, but it is the next result that is of real interest.

Theorem 13.3.2 *Let f be a Möbius map and C a circle. Then $f(C)$ is a circle.*

Proof First, this result asserts not just that $f(C)$ is a subset of some circle C_1, but that $f(C) = C_1$. However, it is enough to prove that for any circle C there is a circle C_1 such that $f(C) \subset C_1$. Indeed, if this is so, then $C \subset f^{-1}(C_1)$. As $f^{-1}(C_1)$ lies in some circle, this circle must be C; thus $f^{-1}(C_1) = C$ and hence $f(C) = C_1$. $\square$

We now need to show that $f(C)$ lies in some circle C_1. In view of Theorem 13.1.6, we need only do this in the cases when f is one of the maps $z \mapsto az$, $z \mapsto z + a$ and $z \mapsto 1/z$. In the first two cases f maps each Euclidean circle onto a Euclidean circle, and each Euclidean line onto a Euclidean line, and as $f(\infty) = \infty$, in each case f maps each circle onto a circle. For the rest of this proof, then, we take $f(z) = 1/z$.

Suppose first that C is a circle that does not pass through the origin; then

$$C = \{z \in \mathbb{C} : z \neq 0, \ az\bar{z} + bz + \bar{b}\bar{z} + c = 0\}. \quad (13.3.1)$$

In this case, we write $w = f(z) = 1/z$, and then, from algebra, w satisfies $a + b\bar{w} + \bar{b}w + cw\bar{w} = 0$. Thus $f(C) \subset C_1$, where

$$C_1 = \{w \in \mathbb{C} : a + b\bar{w} + \bar{b}w + cw\bar{w} = 0\}.$$

Now suppose that C is a circle that does pass though the origin. In this case we let the set in (13.3.1) be C^*, and then $C = C^* \cup \{0\}$. If $z \in C^*$ the algebra can be carried out as above, and again $f(C^*) \subset C_1$. However, in this case $c = 0$ so that C_1 is a Euclidean line. The result now follows as

$$f(C) = f(C^* \cup \{0\}) = f(C^*) \cup \{f(0)\} \subset C_1 \cup \{\infty\}.$$

The cases in which C is a Euclidean line are similar. If $C = L \cup \{\infty\}$ for some Euclidean line L that does not pass through the origin, we write

$$L = \{z \in \mathbb{C} : z \neq 0, \quad bz + \bar{b}\bar{z} + c = 0\}, \quad (13.3.2)$$

where $c \neq 0$. Then $f(C) \subset f(L) \cup \{f(\infty)\} \subset C_1 \cup \{0\} = C_1$, where the Euclidean circle C_1 is given by

$$C_1 = \{w \in \mathbb{C} : b\bar{w} + \bar{b}w + cw\bar{w} = 0\}.$$

Finally, if $C = L \cup \{\infty\}$, where L is a Euclidean line that does pass through the origin, we let L^* be the set in (13.3.2), with $c = 0$. Then C_1 is a Euclidean line, and $f(L) \subset C_1 \cup \{\infty\}$. □

We have carried out the proof of Theorem 13.3.2 in detail, to show how extra steps are necessary to cope with the fact that the algebra of $\mathbb{C}$ cannot cope with ∞, nor with 'division' by zero. Actually, what is missing here is *continuity* for if we verify the result for all z other than 0 or ∞, the required steps for these points would follow from the fact that f is continuous. We discuss this in Section 13.8, and in Section 13.6 we shall see that there is also an algebraic approach which avoids these tiresome extra steps. Having said this, we admit that, in practice, hardly anyone ever carries out these extra steps.

We now consider the following two problems.

(1) Given a Möbius f and circle C, how do we find $f(C)$?
(2) Given circles C and C' how do we find an f with $f(C) = C'$?

Neither of these are difficult to do, and *it is not necessary to perform elaborate calculations*. In the first case it suffices to select three points on C and find their images under f; then (by Theorem 13.3.2) $f(C)$ will be the unique circle through these three image points. In the second case, we select three points z_j on C, and three points w_j on C'. Then (by Theorem 13.2.1) there is a Möbius map f with $f(z_j) = w_j$, and then $f(C) = C'$. We give one example of each of these problems.

Example 13.3.3 Let $f(z) = (z - i)/(z + i)$. What is $f(C)$, where $C = \mathbb{R} \cup \{\infty\}$? As C contains 0, 1 and ∞, $f(C)$ contains -1, $-i$ and 1; thus $f(C) = \{z : |z| = 1\}$. What is $f(C')$, where C' is the imaginary axis with ∞ attached? As C' contains 0, i and ∞, $f(C')$ contains -1, 0 and 1 so that $f(C') = \mathbb{R} \cup \{\infty\}$. □

Example 13.3.4 We construct a Möbius map f which maps $|z| = 1$ onto $L \cup \{\infty\}$, where L is given by $y = x$. Now f will have this property if $f(i) = 0$, $f(-i) = \infty$ and $f(1) = 1 + i$, and there is only one such Möbius map. By inspection, it is $f(z) = (i - 1)(z - i)/(z + i)$. Of course, other choices of points would be equally valid and this is not the only Möbius map with the given property. □

Let us now discuss the mapping of *complements* of circles. If C is the Euclidean circle $|z - z_0| = r$, where $r > 0$, then the complement of C in $\mathbb{C}_\infty$ consists of two 'connected' pieces, namely the Euclidean disc given by $|z - z_0| < r$, and the set $\{z : |z - z_0| > r\} \cup \{\infty\}$. If $C = L \cup \{\infty\}$, where L is a Euclidean line, the complement of C consists of the two 'connected' Euclidean half-planes bounded by L. In each case, the complement of C consists of two disjoint 'connected' sets; these are called the *complementary components* of C. We now have the following extension of Theorem 13.3.3.

Theorem 13.3.5 *Let f be a Möbius map. Let C be a circle, and let C' be the circle $f(C)$. Then f maps each complementary component of C onto a complementary component of C'.*

It is possible to give a completely elementary *ad hoc* proof of this, but it is hardly worthwhile to do so for the proof is easy after we have studied a little about 'connected sets', and continuity, in topology. We shall content ourselves here with a continuation of Example 13.3.3.

Example 13.3.6 (continued) We have seen that $f(z) = (z - i)/(z + i)$ maps $\mathbb{R} \cup \{\infty\}$ onto $\{z : |z| = 1\}$. As $f(i) = 0$ we see (from Theorem 13.3.6) that f maps $\{x + iy : y > 0\}$ onto $\{z : |z| < 1\}$. In fact, this is clear if we note that $|f(z)| < 1$ if and only if z is closer to i than it is to $-i$. Finally, as $f(-1) = i$ we see that f maps $\{x + iy : x < 0\}$ onto $\{x + iy : y > 0\}$. $\square$

We end with a result about the set of Möbius maps that map a disc onto itself. It is clear that if D is a Euclidean disc, or a Euclidean half-plane, then the set of Möbius maps f that satisfy $f(D) = D$ is a group.

Theorem 13.3.7 *Let a and b be points in a disc D. Then there is a Möbius map f such that $f(D) = D$ and $f(a) = b$.*

Proof We shall prove this first in the case when D is the *unit disc* $\mathbb{D}$ given by where $\mathbb{D} = \{z : |z| < 1\}$. Let $g(z) = (z - a)/(1 - \bar{a}z)$. Then $g(C) = C$, where C is the *unit circle* $\{z : |z| = 1\}$, because the general point on C is $e^{i\theta}$, where θ is real, and

$$|g(e^{i\theta})| = \frac{|e^{i\theta} - a|}{|1 - \bar{a}e^{i\theta}|} = \frac{|e^{i\theta} - a|}{|e^{i\theta}||e^{-i\theta} - \bar{a}|} = 1.$$

Now $g(a) = 0$ and hence (from Theorem 13.3.6) $g(\mathbb{D}) = \mathbb{D}$. In the same way we can construct a Möbius h with $h(b) = 0$ and $h(\mathbb{D}) = \mathbb{D}$. Let $f = h^{-1}g$; then $f(a) = b$ and $f(\mathbb{D}) = \mathbb{D}$.

To prove the result for the points a and b in a general disc D, see Figure 13.3.1. Let t be any Möbius map of D onto $\mathbb{D}$, and let $u = t(a)$, $v = t(b)$. By the first part of this proof there is a Möbius map g with $g(u) = v$ and $g(\mathbb{D}) = \mathbb{D}$. Let $f = t^{-1}gt$; then $f(a) = t^{-1}gt(a) = t^{-1}g(u) = t^{-1}(v) = b$ and, similarly, $f(D) = D$. $\square$

13.4 Cross-ratios

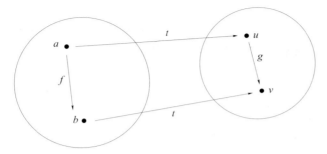

Figure 13.3.1

Exercise 13.3

1. In each of the following cases, find $f(C)$:
   (a) $f(z) = 1/z$ and $C = \{x + iy : x + y = 1\} \cup \{\infty\}$;
   (b) $f(z) = iz/(z-1)$ and $C = \mathbb{R} \cup \{\infty\}$;
   (c) $f(z) = iz/(z-1)$ and $C = \{z : |z| = 1\}$;
   (d) $f(z) = (z+1)/(z-1)$ and $C = \mathbb{R} \cup \{\infty\}$.
2. Show that the transformation $f(z) = (2z+3)/(z-4)$ maps the circle $|z - 2i| = 2$ onto the circle $|8z + (6 + 11i)| = 11$.
3. Find a Möbius map that maps the interior of the circle $|z - 1| = 1$ onto the exterior of the circle $|z| = 2$.
4. By considering the map $g(z) = (z-a)/(z-b)$, show that when $k > 0$ the set $\{z : |z-a|/|z-b| = k\}$ is a circle. What is this set when $k = 1$?
5. Let $f(z) = (z-i)/(iz-1)$. Show that f maps $\{x + iy : y > 0\}$ onto $\{z : |z| < 1\}$.
6. Find a Möbius transformation that maps $\{z : y > 0, \quad |z-1| < 1\}$ onto $\{z : x < 0, y > 0\}$.
7. Let C and C' be two Euclidean circles that are tangent at the point z_0, and let $f(z)$ be any Möbius map such that $f(z_0) = \infty$. Prove that $f(C)$ and $f(C')$ are parallel straight lines (with ∞ attached). Find a Möbius map f that maps the strip between the lines $y = 0$ and $y = 1$ onto the region between the circles $|z - 1| = 1$ and $|z - 2| = 2$.

13.4 Cross-ratios

By Theorem 13.2.1, there is a unique Möbius map that maps a given triple of distinct points onto another such triple. It follows that if we are given four distinct points z_1, z_2, z_3, z_4 and four distinct points w_1, w_2, w_3, w_4 there is at most one Möbius map f with $f(z_j) = w_j$ for $j = 1, 2, 3, 4$. We now give *a necessary and sufficient condition for the existence of such a map.*

Definition 13.4.1 The *cross-ratio* of four distinct points z_1, z_2, z_3, z_4 in $\mathbb{C}$ is defined to be

$$[z_1, z_2, z_3, z_4] = \frac{(z_1 - z_3)(z_2 - z_4)}{(z_1 - z_2)(z_3 - z_4)}, \qquad (13.4.1)$$

The value of the cross-ratio when, say, $z_j = \infty$ is the limiting value of $[z_1, z_2, z_3, z_4]$ as z_j tends to ∞. Explicitly,

$$[\infty, z_2, z_3, z_4] = \frac{z_2 - z_4}{z_3 - z_4};$$

$$[z_1, \infty, z_3, z_4] = -\left(\frac{z_1 - z_3}{z_3 - z_4}\right);$$

$$[z_1, z_2, \infty, z_4] = -\left(\frac{z_2 - z_4}{z_1 - z_2}\right);$$

$$[z_1, z_2, z_3, \infty] = \frac{z_1 - z_3}{z_1 - z_2}.$$

$\square$

The reader should note that some authors use different permutations of 1, 2, 3 and 4 on the right-hand side of (13.4.1); however, it does not matter which definition is used (provided, of course, that one is consistent). Our choice leads to the formula $[0, 1, w, \infty] = w$. We can now give the solution to the problem posed at the start of this section.

Theorem 13.4.2 *Given distinct points z_1, z_2, z_3, z_4, and distinct points w_1, w_2, w_3, w_4, a necessary and sufficient condition for the existence of a Möbius map f with $f(z_j) = w_j$, $j = 1, 2, 3, 4$, is*

$$[z_1, z_2, z_3, z_4] = [w_1, w_2, w_3, w_4]. \qquad (13.4.2)$$

In particular, for any Möbius transformation f,

$$[f(z_1), f(z_2), f(z_3), f(z_4)] = [z_1, z_2, z_3, z_4]. \qquad (13.4.3)$$

Proof Suppose first that there is a Möbius map f with $f(z_j) = w_j$, and suppose for the moment that none of the z_j, or the w_j, are ∞. If $f(z) = (az + b)/(cz + d)$ then $cz_j + d \neq 0$, and

$$w_j - w_k = f(z_j) - f(z_k) = \frac{(ad - bc)(z_j - z_k)}{(cz_j + d)(cz_k + d)}.$$

It is immediate from this and (13.4.1) that (13.4.2) holds. The general case, in which one of the z_j and one of the w_k may be ∞, follows in the same way

13.4 Cross-ratios

by using the appropriate formula for $f(\infty)$ and the cross-ratio (we omit the details).

Now suppose that (13.4.2) holds. Let g and h be Möbius maps such that $g(z_1) = 0$, $g(z_2) = 1$, $g(z_4) = \infty$, and $h(w_1) = 0$, $h(w_2) = 1$, $h(w_4) = \infty$. Then, from (13.4.2) and the invariance of cross-ratios that we have just established,

$$\begin{aligned} g(z_3) &= [0, 1, g(z_3), \infty] \\ &= [g(z_1), g(z_2), g(z_3), g(z_4)] \\ &= [z_1, z_2, z_3, z_4] \\ &= [w_1, w_2, w_3, w_4] \\ &= [h(w_1), h(w_2), h(w_3), h(w_4)] \\ &= [0, 1, h(w_3), \infty] \\ &= h(w_3). \end{aligned}$$

Now let $f = h^{-1}g$; then for each j, $f(z_j) = w_j$. $\square$

We give one (of many) applications of the cross-ratio.

Definition 13.4.3 The points $z_1, \ldots, z_n$ are said to be *concyclic* if they lie on some circle in $\mathbb{C}_\infty$.

Any three points are concyclic, but four points need not be. The cross-ratio of four points determines whether they are concyclic or not.

Theorem 13.4.4 *The four distinct points z_1, z_2, z_3, z_4 are concyclic if and only if $[z_1, z_2, z_3, z_4]$ is real.*

Proof Take any four distinct points z_j, let C be the circle through z_1, z_2 and z_4, and let g be the unique Möbius map with $g(z_1) = 0$, $g(z_2) = 1$ and $g(z_4) = \infty$. Then, by Theorem 13.3.2, $g(C) = \mathbb{R} \cup \{\infty\}$. As

$$\begin{aligned} [z_1, z_2, z_3, z_4] &= [g(z_1), g(z_2), g(z_3), g(z_4)] \\ &= [0, 1, g(z_3), \infty] \\ &= g(z_3), \end{aligned}$$

we see that $[z_1, z_2, z_3, z_4]$ is real if and only of $g(z_3) \in \mathbb{R}$, and this is so if and only if $z_3 \in C$. $\square$

Theorem 13.4.4 gives an alternative proof of Theorem 13.3.3. Given f and a circle C, choose z_1, z_2 and z_3 on C and let C' be the circle through $f(z_j)$, $j = 1, 2, 3$. If z is any other point of C, then $[z_1, z_2, z_3, z]$ is real and hence, by the invariance of the cross-ratio, the cross-ratio of the $f(z_k)$ is real. By

Theorem 13.4.4, the points $f(z)$, $f(z_1)$, $f(z_2)$ and $f(z_3)$ are concyclic; thus $f(z) \in C'$ and $f(C) \subset C'$.

Exercise 13.4

1. Show that if z_1, z_2, z_3 and z_4 are distinct points of $\mathbb{C}_\infty$, then the cross-ratio $[z_1, z_2, z_3, z_4]$ is not equal to 0, 1 or ∞.
2. Establish the invariance (14.4.3) of cross-ratios in the following way. Suppose that none of the z_j, and none of the $f(z_j)$ are ∞. Show first that (14.4.3) holds when $f(z) = az + b$. Now show that (14.4.3) holds when $f(z) = 1/z$. Finally, apply Theorem 13.1.6.
3. Let $z = x + iy$ and $w = u + iv$. By considering cross-ratios, show that the points $0, 1, z, w$ are concyclic if and only if $(|z|^2 - x)/y = (|w|^2 - u)/v$, and verify this directly by geometry.
4. Show that the circle through 0, 1 and w is given by $(t+1)w/(1+tw)$, where $t \in \mathbb{R} \cup \{\infty\}$.
5. Let C and C' be intersecting Euclidean circles, and suppose that $z_1, z_2, z_3 \in C$ and $z_1, z_3, z_4 \in C'$. Show that C and C' can be mapped by a Möbius map to an orthogonal pair of Euclidean lines if and only if the cross-ratio $[z_1, z_2, z_3, z_4]$ is purely imaginary (that is, has real part zero).

13.5 Möbius maps and permutations

Theorem 13.2.1 implies that there are exactly six Möbius maps that map the set $\{0, 1, \infty\}$ onto itself, and it is easy to see that these maps are

$$z, \quad \frac{1}{z}, \quad 1-z, \quad \frac{1}{1-z}, \quad \frac{z-1}{z}, \quad \frac{z}{z-1}. \qquad (13.5.1)$$

We leave the reader to check that these maps form a group, and that this group is isomorphic to the permutation group S_3. Indeed, if we regard S_3 as the group of permutations of $\{0, 1, \infty\}$ instead of $\{1, 2, 3\}$, we see that the six maps listed above correspond to the permutations $(0)(1)(\infty)$, $(0\,\infty)(1)$, $(0\,1)(\infty)$, $(0\,1\,\infty)$, $(0\,\infty\,1)$ and $(0)(1\,\infty)$, respectively.

More generally, given any permutation ρ of a finite subset X of $\mathbb{C}_\infty$, we might try to find a Möbius map f that agrees with ρ on X. However, Corollary 13.2.2 shows that this is not usually possible. Consider, for example, the permutation $(1\,2)(3\,4\,5)$. In this case ρ^3 has three and only three fixed points, and so if we were able to find a set X of five elements, and a Möbius f that acted on X in

13.5 Möbius maps and permutations

the same way that ρ acts on $\{1, \ldots, 5\}$, we would have $f^3(z) = z$ for three and only three z in X and this would violate Corollary 13.2.2.

We have seen that S_3, but not S_5, can be realised as a Möbius group, so it is natural to ask for a list of all permutations groups that can be realised by some Möbius group. In view of the comments just made, this list is likely to be short. In fact, the following more general statement is true: *any finite Möbius group is isomorphic to a cyclic or a dihedral group, or to one of the symmetry groups of the five Platonic solds.*

There is an important link between cross-ratios and permutations in S_4 which we shall now explore. Suppose that we start with a cross-ratio $[z_1, z_2, z_3, z_4]$ and a permutation ρ of $\{1, 2, 3, 4\}$. We can form a new cross-ratio $[w_1, w_2, w_3, w_4]$ by moving the entry z_k in the k-th place of the given cross-ratio to the $\rho(k)$-th place in the new cross-ratio. Thus $w_{\rho(k)} = z_k$ or, equivalently, $w_k = z_{\rho^{-1}(k)}$. For example, if $\rho = (1\ 2\ 3)$, then

$$[w_1, w_2, w_3, w_4] = [z_3, z_1, z_2, z_4] = [z_{\rho^{-1}(1)}, z_{\rho^{-1}(2)}, z_{\rho^{-1}(3)}, z_{\rho^{-1}(4)}].$$

We shall now investigate how, for a given ρ, the new cross-ratio is related to the original cross-ratio. Rather surprisingly, the value of the new cross-ratio is a function, say f_ρ, of *the value* of the original cross-ratio, but *not on the individual values of the z_j*. Moreover, the function f_ρ is always one of the Möbius maps listed in (13.5.1). We give a proof.

Take any permutation ρ (which will be fixed throughout this discussion). Take distinct z_j, let $\lambda = [z_1, z_2, z_3, z_4]$, and let g be the unique Möbius map with $g(z_1) = 0$, $g(z_2) = 1$ and $g(z_4) = \infty$. Then, by the invariance of the cross-ratio under g, we see that $g(z_3) = \lambda$. Now

$$[z_{\rho^{-1}(1)}, z_{\rho^{-1}(2)}, z_{\rho^{-1}(3)}, z_{\rho^{-1}(4)}]$$
$$= [g(z_{\rho^{-1}(1)}), g(z_{\rho^{-1}(2)}), g(z_{\rho^{-1}(3)}), g(z_{\rho^{-1}(4)})],$$

and this second cross-ratio is the cross-ratio of the points $0, 1, \lambda, \infty$ taken in some order. As the order of these points is determined by ρ, the value of the cross-ratio is of the form $f_\rho(\lambda)$, where (as is easily checked) f_ρ is one of the functions in (13.5.1). $\square$

An explicit example might help, so suppose that $\rho = (1\ 2)$. Then

$$[z_{\rho^{-1}(1)}, z_{\rho^{-1}(2)}, z_{\rho^{-1}(3)}, z_{\rho^{-1}(4)}] = [z_2, z_1, z_3, z_4]$$
$$= [g(z_2), g(z_1), g(z_3), g(z_4)]$$
$$= [1, 0, \lambda, \infty]$$
$$= 1 - \lambda$$
$$= 1 - [z_1, z_2, z_3, z_4].$$

Thus if $\rho = (1\,2)$, and $f_\rho(z) = 1 - z$, then

$$[z_2, z_1, z_3, z_4] = f_\rho\big([z_1, z_2, z_3, z_4]\big).$$

A similar argument can be used for any ρ in S_4, and we ask the reader to carry out the calculations in all cases in which ρ is a transposition; this leads to the following results:

(a) $f_{(12)}(z) = f_{(34)}(z) = 1 - z$;
(b) $f_{(13)}(z) = f_{(24)}(z) = z/(z-1)$;
(c) $f_{(14)}(z) = f_{(23)}(z) = 1/z$.

We can, of course, carry out the same calculation for any ρ, but the results follow more easily from the important formula

$$f_{\sigma\rho} = f_\sigma f_\rho \qquad (13.5.2)$$

which holds for any permutations σ and ρ in S_4. The proof of (13.5.2) is easy. Let $\mu = \sigma\rho$, $w_{\rho(k)} = z_k$ and $u_{\sigma(j)} = w_j$. Then $u_{\mu(k)} = u_{\sigma(\rho(k))} = w_{\rho(k)} = z_k$ and

$$\begin{aligned}
f_\sigma\big(f_\rho([z_1, z_2, z_3, z_4])\big) &= f_\sigma([w_1, w_2, w_3, w_4]) \\
&= [u_1, u_2, u_3, u_4] \\
&= [z_{\mu^{-1}(1)}, z_{\mu^{-1}(2)}, z_{\mu^{-1}(3)}, z_{\mu^{-1}(4)}] \\
&= f_\mu([z_1, z_2, z_3, z_4]),
\end{aligned}$$

and this completes the proof of (13.5.2). □

We can extract a lot of information from (13.5.2); for example,

$$f_{(1\,2)(3\,4)}(z) = f_{(1\,2)}\big(f_{(3\,4)}(z)\big) = 1 - (1 - z) = z \qquad (13.5.3)$$

as is evident from (13.4.1). More generally, we have seen that when ρ is a transposition, f_ρ is in the group, Γ say, of functions listed in (13.5.1). It follows from (13.5.2), and the fact that any ρ is a product of transpositions, that for every ρ in S_4, f_ρ is in Γ. Thus, if we think of Γ as S_3, the map $\rho \mapsto f_\rho$ is a homomorphism, say θ, of S_4 onto S_3.

The kernel K of θ is of interest, and (13.5.3) shows that $(1\,2)(3\,4)$ is in K. A similar argument shows that $(1\,3)(2\,4)$ and $(1\,4)(2\,3)$ are also in K, as (of course) is the identity permutation I. As $|K| = |S_4|/|S_3| = 4$, there can be no other elements in K. We summarize our results in the following theorem.

Theorem 13.5.1 *For each ρ in S_4 there is a Möbius map f_ρ in the group*

$$\Gamma = \left\{ z,\ \frac{1}{z},\ 1-z,\ \frac{1}{1-z},\ \frac{z-1}{z},\ \frac{z}{z-1} \right\}$$

of permutations of $\{0, 1, \infty\}$ such that, for any distinct z_1, z_2, z_3, z_4,
$$[z_{\rho^{-1}(1)}, z_{\rho^{-1}(2)}, z_{\rho^{-1}(3)}, z_{\rho^{-1}(4)}] = f_\rho([z_1, z_2, z_3, z_4]).$$
Moreover, the map $\rho \mapsto f_\rho$ is a homomorphism of the group S_4 onto the group Γ with kernel K, where $K = \{I, (1\,2)(3\,4), (1\,3)(2\,4), (1\,4)(2\,3)\}$.

Exercise 13.5

1. Verify (a), (b) and (c) in the text.
2. Find f_ρ when $\rho = (1\,2\,3)$. Express ρ as a product of transpositions and verify (13.5.2) in this case.

13.6 Complex lines

In this section we discuss an algebraic way, based on the complex vector space $\mathbb{C}^2$, to introduce the point ∞.

Definition 13.6.1 A *complex line* is a one-dimensional subspace of the vector space $\mathbb{C}^{2,t}$ of complex column vectors $(z_1, z_2)^t$. The set of all complex lines is denoted by $\mathcal{L}$.

A complex line L is the set of complex scalar multiples of some non-zero point in $\mathbb{C}^{2,t}$ and so is of the form
$$L = \{\lambda \begin{pmatrix} z_1 \\ z_2 \end{pmatrix} : \lambda \in \mathbb{C}\}. \tag{13.6.1}$$

Any 2×2 non-singular complex matrix
$$A = \begin{pmatrix} a & b \\ c & d \end{pmatrix} \tag{13.6.2}$$
acts as a linear transformation of $\mathbb{C}^{2,t}$ onto itself by the rule
$$\begin{pmatrix} z_1 \\ z_2 \end{pmatrix} \mapsto \begin{pmatrix} a & b \\ c & d \end{pmatrix} \begin{pmatrix} z_1 \\ z_2 \end{pmatrix} = \begin{pmatrix} az_1 + bz_2 \\ cz_1 + dz_2 \end{pmatrix}.$$
As A is non-singular, it maps a non-zero point to a non-zero point, and as it is linear it maps each complex line to a complex line. The same is true for A^{-1}, so we have the following result.

Lemma 13.6.2 *Any non-singular 2×2 complex matrix A is a bijection of $\mathcal{L}$ onto itself. The group $\mathrm{GL}(2, \mathbb{C})$ of non-singular 2×2 complex matrices acts as a group of permutations of $\mathcal{L}$.*

We want to see which lines map to which lines under a matrix A, and this is best done by considering the slope of a complex line. If $z_2 \neq 0$, we can form the quotients $(\lambda z_1)/(\lambda z_2)$ of the coordinates of the non-zero points on the line L in (13.6.1) and the common value of all of these quotients is the *slope* z_1/z_2 of L. The single complex line whose slope is not defined is

$$L(\infty) = \{\lambda \begin{pmatrix} 1 \\ 0 \end{pmatrix} : \lambda \in \mathbb{C}\},$$

and, by convention, we say that this line has slope ∞. Given a complex number w there is a unique complex line $L(w)$ with slope w, namely

$$L(w) = \{\lambda \begin{pmatrix} w \\ 1 \end{pmatrix} : \lambda \in \mathbb{C}\},$$

The following result is now clear.

Lemma 13.6.3 *The map $w \mapsto L(w)$ is a bijection from the set $\mathbb{C}_\infty$ onto the space $\mathcal{L}$ of all complex lines.*

This bijection is the key to understanding Möbius transformations from an algebraic point of view. Let A be the non-singular matrix in (13.6.2). Given a point w of $\mathbb{C}_\infty$, there is a unique line $L(w)$ with slope w, and this is mapped to, say $L(w')$, by A. It follows that we can regard A as the map $w \mapsto w'$ of $\mathbb{C}_\infty$ onto itself. Moreover, as A is a bijection of $\mathcal{L}$ onto itself, Lemma 13.6.3 says A acts as a bijection of $\mathbb{C}_\infty$ onto itself. In fact, as we shall now see, this action is identical to the action of the Möbius map $z \mapsto (az+b)/(cz+d)$.

We examine this action of A on $\mathbb{C}_\infty$ in greater detail. First, for any complex w,

$$\begin{pmatrix} a & b \\ c & d \end{pmatrix} \begin{pmatrix} w \\ 1 \end{pmatrix} = \begin{pmatrix} aw+b \\ cw+d \end{pmatrix}, \quad \begin{pmatrix} a & b \\ c & d \end{pmatrix} \begin{pmatrix} w \\ 0 \end{pmatrix} = \begin{pmatrix} aw \\ cw \end{pmatrix}.$$

If $c = 0$ then $ad = ad - bc \neq 0$, and we see that A maps the line of slope w to the line of slope $(aw+b)/d$. Also, A maps the line of slope ∞ to itself, so that the action of A on the set $\mathbb{C}_\infty$ (of slopes) is given by $z \mapsto (az+b)/d$. Now suppose that $c \neq 0$. If $w \neq -d/c$, then A maps the line of complex slope w to the line of complex slope $(aw+b)/(cw+d)$. If $w = -d/c$, then A maps the line of slope w to $L(\infty)$. Finally, A maps $L(\infty)$ to $L(a/c)$. Together these facts prove the following result.

Theorem 13.6.4 *Suppose that A is given by (13.6.2), and regard A as acting on the space $\mathbb{C}_\infty$ of slopes of complex lines. Then $A(z) = (az+b)/(cz+d)$.*

This result fully justifies the discussion of Möbius transformations given in Section 13.1.1 and, moreover, *it eliminates the need to consider as special* cases any arguments involving ∞ or 'division' by zero. Given a Möbius transformation f, we can choose a matrix of coefficients, say the matrix A in (13.1.6), and Theorem 13.1.2 implies that A is determined up to a non-zero scalar multiple. However, if we replace A by a scalar multiple, say λA, then this scalar multiple acts in exactly the same way as A does on the space $\mathcal{L}$ of complex lines, and hence it determines the same Möbius action on the space of slopes of lines.

Exercise 13.6

1. Prove that if two complex lines have the same slope, then they are the same line (thus the map $w \mapsto L(w)$ is properly defined).
2. Suppose that A is given by (13.6.2). Show that A maps $L(\infty)$ to itself if and only if $c = 0$.
3. Let $A = \begin{pmatrix} a & a \\ 0 & a \end{pmatrix}$, where $a \neq 0$. Show directly that the action of A on the space of slopes of complex lines is given by $z \mapsto z + 1$.
4. Discuss the action of the matrix A in (13.6.2) on the space $\mathcal{L}$ when $ad - bc = 0$.

13.7 Fixed points and eigenvectors

A point w in $\mathbb{C}_\infty$ is a *fixed point* of a Möbius map f if $f(w) = w$, and we recall that if a Möbius map is not the identity, then it has at most two fixed points. We shall now show that *the fixed points of a Möbius map correspond to the two lines of eigenvectors for the matrix corresponding to f*. Suppose that f is a Möbius map with a corresponding matrix A; thus A maps $L(w)$ to $L(w')$ if and only if $f(w) = w'$ (see Section 13.6). It follows that w is a fixed point of f if and only if A maps $L(w)$ to itself, and this is so if and only if each non-zero point on $L(w)$ is an eigenvector of A. We give a formal statement of this.

Theorem 13.7.1 *Let f be a Möbius map with corresponding matrix A. Then $f(w) = w$ if and only if $L(w)$ is a line of eigenvectors of A.*

Earlier we proved (by vector space methods) that every complex matrix has an eigenvector, and it follows from this that every Möbius transformation has a fixed point. If a Möbius transformation f has three distinct fixed points, then A has three distinct lines of eigenvectors (with different slopes). As A has at most two eigenvalues, it must have two eigenvectors, say v_1 and v_2, which lie on different complex lines but which correspond to the same eigenvalue μ, say. As $\{v_1, v_2\}$ is a basis of $\mathbb{C}^{2,t}$ this implies that $A = \mu I$, where I is the identity matrix, and this in turn implies that f is the identity Möbius map. This gives an algebraic proof of Corollary 13.2.2. We now give some examples to illustrate the link between fixed points and eigenvectors.

Example 13.7.2 The map $f(z) = z + 1$ has a single fixed point, namely ∞. The matrix

$$A = \begin{pmatrix} 1 & 1 \\ 0 & 1 \end{pmatrix}$$

represents f. It is clear that 1 is the only eigenvalue of A, so the eigenvectors of A are given by

$$\begin{pmatrix} 1 & 1 \\ 0 & 1 \end{pmatrix} \begin{pmatrix} z_1 \\ z_2 \end{pmatrix} = \begin{pmatrix} z_1 \\ z_2 \end{pmatrix}.$$

The general solution of this is $z_2 = 0$; thus, as predicted, $L(\infty)$ is the only line of eigenvectors of A. □

Example 13.7.3 It is easy to check that for any non-zero complex number k the Möbius map

$$f(z) = \frac{(k-2)z - 2(k-1)}{(k-1)z - (2k-1)}$$

fixes 1 and 2. The matrix

$$A = \begin{pmatrix} k-2 & 2-2k \\ k-1 & 1-2k \end{pmatrix}$$

represents f so that A must have eigenvectors $(1, 1)^t$ and $(2, 1)^t$. It is easy to check that this is indeed so, and that the corresponding eigenvalues are $-k$ and -1. □

Earlier we discussed the possibility of diagonalizing a matrix, so it is natural to consider what this means in the current context of Möbius maps. Suppose that the Möbius map f is represented by the matrix A, and that A has distinct eigenvalues, say λ and μ. Then there is a non-singular matrix B such that

$$BAB^{-1} = \begin{pmatrix} \lambda & 0 \\ 0 & \mu \end{pmatrix},$$

and BAB^{-1} has eigenvectors $(1, 0)^t$ and $(0, 1)^t$. Now B represents a Möbius map g, say, and as the transition from matrices to Möbius maps is a homomorphism, this means that gfg^{-1} fixes 0 and ∞. In fact, $gfg^{-1}(z) = (\lambda/\mu)z$, and

13.7 Fixed points and eigenvectors

we see that the diagonalization of A (which is achieved by taking the eigenvectors as a basis) is equivalent to conjugating f to gfg^{-1} thereby moving the fixed points to 0 and ∞. We can see this directly. If f has distinct fixed points z_1 and z_2, we let $g(z) = (z - z_1)/(z - z_2)$; them (by direct calculation) gfg^{-1} fixes 0 and ∞. If A has coincident eigenvalues, then there is a matrix B such that

$$BAB^{-1} = \begin{pmatrix} \lambda & \mu \\ 0 & \lambda \end{pmatrix},$$

where $\lambda \neq 0$ (as A is non-singular). In this case the corresponding gfg^{-1} is either a translation (when $\mu \neq 0$) or the identity map (when $\mu = 0$). Finally, if a Möbius map f has exactly one fixed point, say z_1, we let $g(z) = 1/(z - z_1)$ and then gfg^{-1} has a unique fixed point, namely ∞; thus gfg^{-1} is a translation. We summarize these results.

Theorem 13.7.4 *If a Möbius transformation f has exactly two fixed points then it is conjugate to some map $z \mapsto az$, where $a \neq 0$. If f has exactly one fixed point, it is conjugate to $z \mapsto z + 1$.*

Finally, we see how to compute the iterates of a Möbius map (or a 2×2 matrix). Suppose, for example, that f has two fixed points. Then there is some map g, and some constant k, such that $gfg^{-1}(z) = kz$. Then $gf^n g^{-1}(z) = (gfg^{-1})^n(z) = k^n z$, so that $f^n(z) = g^{-1}(k^n g(z))$. As g and k can be found explicitly, so can $f^n(z)$. This method is equivalent to that used for finding the powers of a matrix (that is diagonalize the matrix as in Section 10.3).

Exercise 13.7

1. Let $f(z) = (az + b)/(cz + d)$, where $ad - bc \neq 0$. Show that $f^2(z) = z$ for all z if and only if either f is the identity map I or $a + d = 0$. Is it true that for a 2×2 complex matrix A, $A^2 = I$, where I is now the identity matrix, if and only if $A = I$ or trace$(A) = 0$?
2. Let $f(z) = az + b$ and $g(z) = cz + d$, where neither is the identity map. Show that if f and g commute ($f(g(z)) = g(f(z))$ for all z) then either (i) f and g are translations, or (ii) f and g have a common finite fixed point. What does this say when interpreted in terms of 2×2 complex matrices?
3. Suppose that f is a Möbius map and that for some w in $\mathbb{C}$, $f(w) \neq w$ but $f(f(w)) = w$. Show that $f(f(z)) = z$ for all z in $\mathbb{C}_\infty$, and give a matrix proof of this result.

13.8 A geometric view of infinity

In order to study Möbius transformations properly we found it necessary to adjoin an 'abstract' point ∞ to $\mathbb{C}$. We have seen how to do this from an algebraic point of view, and we shall now discuss a construction in which the introduction of the additional point arises naturally in a geometric way. This method depends on some elementary, but very intuitive, ideas about continuity which we shall accept without question.

Let S denote the unit sphere $\{x \in \mathbb{R}^3 : ||x|| = 1\}$ in $\mathbb{R}^3$. We shall identify the complex number $x + iy$ with the point $(x, y, 0)$ in $\mathbb{R}^3$, and $\mathbb{C}$ cuts S in the circle $x^2 + y^2 = 1$ (the 'equatorial plane' of the sphere). Let $\zeta = (0, 0, 1)$ (the 'north pole' of S). Each point z of $\mathbb{C}$ can be projected linearly towards or away from ζ until it reaches the sphere S at some uniquely determined point w other than ζ. The map $\varphi : z \mapsto w$ is the *stereographic projection* of $\mathbb{C}$ into S, and it is illustrated in Figure 13.8.1.

It is easy to find an explicit formula for $\varphi(z)$. The general point on the line L through $(x, y, 0)$ and $(0, 0, 1)$ is

$$(0, 0, 1) + t[(x, y, 0) - (0, 0, 1)], \qquad (13.8.1)$$

where t is real, and this line meets the sphere S when

$$t^2 x^2 + t^2 y^2 + (1 - t)^2 = 1.$$

The two solutions of this are $t = 0$ (which corresponds to ζ), and $t = 2/(x^2 + y^2 + 1)$ which corresponds to $\varphi(z)$, where $z = x + iy$. If we now substitute this

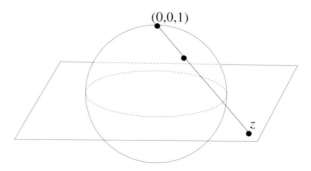

Figure 13.8.1

13.8 A geometric view of infinity

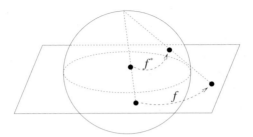

Figure 13.8.2

second solution in (13.8.1) we find that

$$\varphi(z) = \left(\frac{2x}{|z|^2+1}, \frac{2y}{|z|^2+1}, \frac{|z|^2-1}{|z|^2+1}\right). \tag{13.8.2}$$

It is evident from geometry that $\varphi(z) = z$ whenever $|z| = 1$, and this also follows from (13.8.2). Notice also that $\varphi(z)$ lies 'above' $\mathbb{C}$ when $|z| > 1$ and that it lies 'below' $\mathbb{C}$ when $|z| < 1$. In particular, $\varphi(0) = (0, 0, -1)$.

It is clear that φ is a bijection of $\mathbb{C}$ onto $S\setminus\{\zeta\}$. Moreover, as

$$\frac{2|x|}{|z|^2+1} \leq 2\frac{|z|}{|z|^2} = \frac{2}{|z|},$$

and similarly for y in place of x, (13.8.2) shows that $\varphi(z) \to \zeta$ as $|z| \to +\infty$. It is therefore natural to adjoin an 'abstract' point ∞ to $\mathbb{C}$ to form $\mathbb{C}_\infty$, and to define $\varphi(\infty) = \zeta$. Now φ is a bijection from $\mathbb{C}_\infty$ onto the sphere S.

The definition of a Möbius map f on $\mathbb{C}_\infty$ can now be explained in terms of a map from S to itself. Given a Möbius transformation f, we can construct a map $f^* : S \to S$ defined by $f^* = \varphi f \varphi^{-1}$; see Figure 13.8.2. The map f^* is defined at all points of S except at ζ and $\varphi(-d/c)$, and it is easy to see that the definitions $f(-d/c) = \infty$ and $f(\infty) = a/c$ are equivalent to making f^* a *continuous* map of S onto itself.

One of the benefits of stereographic projection is that it makes it clear why we should attach the point ∞ to a straight line L and regard the *extended line* $L \cup \{\infty\}$ as a circle. Indeed, it should be clear that lines in the plane correspond under stereographic projection to circles on S that pass through ζ, and conversely. Thus φ gives a bijection between the set of extended lines in $\mathbb{C}_\infty$ and the set of circles on S that pass through ζ.

It is also true (thought less obviously so) that a circle on the sphere that does not pass through ζ corresponds to a circle in $\mathbb{C}$. To see this note that a circle C on the sphere is the intersection of the sphere with a plane given, say, by

$\alpha_1 x_1 + \alpha_2 x_2 + \alpha_3 x_3 = \beta$, and that if $\zeta \notin C$ then $\alpha_3 \neq \beta$. If z in $\mathbb{C}$ is mapped to the circle C on S, then, from (13.8.2),

$$\alpha_1 \left(\frac{2x}{|z|^2 + 1} \right) + \alpha_2 \left(\frac{2y}{|z|^2 + 1} \right) + \alpha_3 \left(\frac{|z|^2 - 1}{|z|^2 + 1} \right) = \beta, \quad (13.8.3)$$

and as $\alpha_3 \neq \beta$, this is the equation of a circle in $\mathbb{C}$.

We have now seen that *the set of circles on S corresponds under φ^{-1} to the set of all circles and extended lines in $\mathbb{C}$.* This is further justification (if any is needed) for saying that $L \cup \{\infty\}$ a circle when L is a straight line.

Finally, suppose that z and w are in $\mathbb{C}$. Then their projections onto the sphere S are given by (13.8.2) and we can use this to compute the Euclidean distance beween the projections $\varphi(z)$ and $\varphi(w)$. A tedious (but elementary) calculation (which we leave as an exercise for the reader) shows that

$$||\varphi(z) - \varphi(w)|| = \frac{2|z - w|}{\sqrt{1 + |z|^2}\sqrt{1 + |w|^2}}, \quad (13.8.4)$$

and a similar (but simpler) argument shows that

$$||\varphi(z) - \varphi(\infty)|| = \frac{2}{\sqrt{1 + |z|^2}}. \quad (13.8.5)$$

The expression $||\varphi(z) - \varphi(w)||$ is known as the *chordal distance* between z and w (as it is the length of the chord of S that joins $\varphi(z)$ to $\varphi(w)$ in $\mathbb{R}^3$). It is geometrically evident that $||\varphi(z) - \varphi(w)|| \leq 2$; this also follows from the Cauchy–Schwarz inequality

$$|z - w|^2 = |1z + (-1)w|^2 \leq \sqrt{|z|^2 + (-1)^2}\sqrt{1^2 + |w|^2}.$$

Exercise 13.8

1. Find the image of the circle $\{z : |z| = r\}$ under stereographic projection onto S.
2. Show that the circle given in (13.8.3) is a straight line if and only if the plane $\alpha_1 x_1 + \alpha_2 x_2 + \alpha_3 x_3 = \beta$ passes through ζ.
3. Verify (13.8.4) and (13.8.5).
4. Show that if $(x_1, x_2, x_3) \in S$, then

$$\varphi^{-1}(x_1, x_2, x_3) = \frac{x_1 + ix_2}{1 - x_3}.$$

Show also that the projection of $\mathcal{S}$ onto $\mathbb{C}_\infty$ from the 'south pole' $(0, 0, -1)$ of $\mathcal{S}$ is given by

$$(x_1, x_2, x_3) \mapsto \frac{x_1 - ix_2}{1 + x_3}.$$

13.9 Rotations of the sphere

The stereographic projection φ maps $\mathbb{C}_\infty$ onto the sphere $\mathcal{S}$ so that every point is covered exactly once. Every map $f : \mathbb{C}_\infty \to \mathbb{C}_\infty$ corresponds to a map $f^* : \mathcal{S} \to \mathcal{S}$ defined by $f^* = \varphi f \varphi^{-1}$ and, in a similar way, every map $f^* : \mathcal{S} \to \mathcal{S}$ determines a map $f : \mathbb{C}_\infty \to \mathbb{C}_\infty$ by $f = \varphi^{-1} f \varphi$. The main result in this section is that any rotation of the sphere arises from a Möbius map in the sense that it is a map f^* for some choice of a Möbius map f. Note, however, that not all Möbius maps give rise to a rotation (for example, the Möbius maps with exactly one fixed point do not because f has one fixed point in $\mathbb{C}_\infty$ if and only if f^* has one fixed point in $\mathcal{S}$).

Theorem 13.9.1 *Every Möbius map of the form*

$$f(z) = \frac{az + b}{-\bar{b}z + \bar{a}}, \quad |a|^2 + |b|^2 = 1, \tag{13.9.1}$$

corresponds to a rotation f^* of the sphere, and every rotation of the sphere arises in this way.

The proof of this result involves several ideas, so we shall break the proof into several simpler steps.

Lemma 13.9.2 *Under the stereographic projection φ, two points z and w map to diametrically opposite points on the sphere $\mathcal{S}$ if and only if $w = -1/\bar{z}$.*

Proof The points $\varphi(z)$ and $\varphi(w)$ are diametrically opposite on $\mathcal{S}$ if and only if $\varphi(z) = -\varphi(w)$. We leave the reader to check from (13.8.2) that this is so if and only if $w = -1/\bar{z} = -z/|z|^2$. □

Lemma 13.9.3 *If a Möbius map f is such that f^* is a rotation of the sphere, then f is of the form* (13.9.1).

Proof Let $f(z) = (az + b)/(cz + d)$, where $ad - bc = 1$, and suppose that f^* is a rotation of the sphere. Then f^* preserves distances between points on the sphere, and so it must map diametrically opposite points to diametrically opposite points. This means that if $w = -1/\bar{z}$ then $f(w) = -1/\overline{f(z)}$; in other words, f must satisfy the relation

$$f(-1/\bar{z}) = -1/\overline{f(z)},$$

valid for all z. This relation is

$$\frac{-a+b\bar{z}}{-c+d\bar{z}} = \frac{-\bar{c}\bar{z}-\bar{d}}{\bar{a}\bar{z}+\bar{b}}$$

and (as z is arbitrary) it follows from Theorem 13.1.1 that there is some λ such that

$$\begin{pmatrix} b & -a \\ d & -c \end{pmatrix} = \lambda \begin{pmatrix} -\bar{c} & -\bar{d} \\ \bar{a} & \bar{b} \end{pmatrix}.$$

This shows that

$$1 = ad - bc = (\lambda\bar{d})(\lambda\bar{a}) - (-\lambda\bar{c})(-\lambda\bar{b}) = \lambda^2\overline{(ad-bc)} = \lambda^2,$$
$$1 = ad - bc = a(\lambda\bar{a}) - b(-\lambda\bar{b}) = \lambda(|a|^2 + |b|^2),$$

and together these show that $\lambda = 1$, and that (13.9.1) holds. $\square$

Lemma 13.9.4 *If f is of the form* (13.9.1), *then f^* is a rotation of the sphere.*

Proof Suppose that f is of the form (13.9.1). Then a calculation using (13.9.1) and (13.8.4) shows that for all z and w,

$$\|\varphi(f(z)) - \varphi(f(w))\| = \|\varphi(z) - \varphi(w)\|,$$

and this means that f^* is an isometry of the sphere S into itself. Now every isometry $g : S \to S$ extends to an isometry of $\mathbb{R}^3$ into itself by the rule $g(tx) = tg(x)$, where $x \in S$. Indeed, to see that this extension is an isometry of $\mathbb{R}^3$ we need only observe that the two larger triangles in Figure 13.9.1 are congruent. Moreover, it is clear that in this extension we have $g(0) = g(0x) = 0g(x) = 0$ for any x on S.

This means that f^* is an isometry of $\mathbb{R}^3$ onto itself and $f^*(0) = 0$; thus f^* is the action of an orthogonal matrix A on $\mathbb{R}^3$, and f^* is therefore either a reflection in some plane through the origin, or a rotation. However, f^* cannot be a reflection as this would imply that f^*, and hence also f, would have infinitely many fixed points. Thus f^* is a rotation. $\square$

Lemma 13.9.5 *Every rotation of S is of the form f^* for some Möbius map f of the form* (13.9.1).

Proof Let

$$f(z) = e^{i\theta}z = \frac{e^{i\theta/2}z + 0}{0z + e^{-i\theta/2}}, \quad g(z) = \frac{\bar{z}_0 z + 1}{-z + z_0}.$$

13.9 Rotations of the sphere

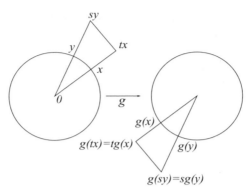

Figure 13.9.1

Then both f and g are of the form (13.9.1), and f^* is clearly a rotation of $\mathbb{R}^3$ by an angle θ about the 'vertical' axis in $\mathbb{R}^3$. Now let $F = g^{-1}fg$. As the reader can check (and we shall need this later) and composition of maps of the form (13.9.1) is again of this form, as is the inverse of any such map. Thus F is of the form (13.9.1), so that F^* is a rotation of $\mathcal{S}$. However,

$$\begin{aligned} F^* &= \varphi(g^{-1}fg)\varphi^{-1} \\ &= (\varphi g^{-1}\varphi^{-1})(\varphi f\varphi^{-1})(\varphi g\varphi^{-1}i) \\ &= (g^{-1})^* f^* g^* \\ &= (g^*)^{-1} f^* g^*. \end{aligned}$$

As g^* is a rotation, we see that F^* is a rotation of angle θ about some axis. Now F fixes z_0 and $-1/\bar{z}_0$; for example, $F(z_0) = g^{-1}fg(z_0) = g^{-1}f(\infty) = g^{-1}(\infty) = z_0$. Thus F^* is a rotation about an axis whose endpoints are $\varphi(z_0)$ and $-\varphi(z_0)$. As z_0 is arbitrary, we can take this axis to be any diameter of $\mathcal{S}$, and so we can choose z_0 and θ so that F^* is any pre-assigned rotation. □

These four lemmas compete our proof of Theorem 13.9.1. Clearly, though, there is more to these ideas. First, and as suggested in the proof of Lemma 13.9.5, we have the following result.

Theorem 13.9.6 *The Möbius maps of the form* (13.9.1) *for a group, which we denote by* $\mathcal{M}_0$.

Essentially the same argument proves the following result.

Theorem 13.9.7 *The set*

$$\mathcal{U} = \left\{ \begin{pmatrix} a & b \\ -\bar{b} & \bar{a} \end{pmatrix} : |a|^2 + |b|^2 = 1 \right\}$$

of 2×2 *complex matrices is a group, called the unitary group.*

It follows from our earlier work that the natural map $\Phi : \mathcal{U} \to \mathcal{M}_0$ is a homomorphism with kernel $\{\pm I\}$.

Finally, we can relate all these ideas to quaternions for, as we have seen earlier, quaternions and rotations of $\mathbb{R}^3$ are intimately linked. The group $\mathcal{U}$ defined in Theorem 13.9.7 is isomorphic to the multiplicative group of quaternions q with $||q|| = 1$ by the map

$$\begin{pmatrix} a & b \\ -\bar{b} & \bar{a} \end{pmatrix} \mapsto a_i + a_2 \mathbf{i} + b_1 \mathbf{j} + b_2 \mathbf{k},$$

where $a = a_1 + i a_2$ and $b = b_1 + i b_2$. If we identiy the complex number i with $\mathbf{i}$, this map can be written more concisely as $(a, b) \mapsto a + b\mathbf{j}$.

In conclusion, we can summarize these ideas in the following informal scheme, in which each 'link' has been discussed at some stage of this text:

$$\text{rotations} \longrightarrow \text{unit quaternions}$$
$$\text{unit quaternions} \longrightarrow \text{matrices in} \, \mathcal{U}$$
$$\text{matrices in} \, \mathcal{U} \longrightarrow \text{Möbius maps in} \, \mathcal{M}_0$$
$$\text{Möbius maps in} \, \mathcal{M}_0 \longrightarrow \text{rotations}$$

Exercise 13.9

1. Suppose that f is given by (13.9.1). Show that f fixes w if and only if it fixes $-1/\bar{w}$.
2. Show that the map $f(z) = e^{2i\theta} z$ can be written in the form (13.9.1), where θ is real. Let $z = x + iy$. Show that

$$\varphi(f(z)) = \left(\frac{2x \cos 2\theta - 2y \sin 2\theta}{|z|^2 + 1}, \frac{2x \sin 2\theta + 2y \cos 2\theta}{|z|^2 + 1}, \frac{|z|^2 - 1}{|z|^2 + 1} \right),$$

and deduce that f^* is a rotation of the sphere of angle 2θ about the vertical axis.
3. Show that if

$$f(z) = \frac{z \cos \theta + i \sin \theta}{iz \sin \theta + \cos \theta},$$

then f^* is a rotation of the sphere about the real axis. What is the angle of rotation of f^*? [Hint: consider $\varphi(i)$ and $\varphi(f(i))$.]
4. Let

$$\mathbf{1} = \begin{pmatrix} 1 & 0 \\ 0 & 1 \end{pmatrix}, \quad \mathbf{i} = \begin{pmatrix} i & 0 \\ 0 & -i \end{pmatrix}, \quad \mathbf{j} = \begin{pmatrix} 0 & 1 \\ -1 & 0 \end{pmatrix}, \quad \mathbf{k} = \begin{pmatrix} 0 & i \\ i & 0 \end{pmatrix},$$

13.9 Rotations of the sphere

so that if $z = x + iy$ and $w = u + iv$, then

$$\begin{pmatrix} z & w \\ -\bar{w} & \bar{z} \end{pmatrix} = x\mathbf{1} + y\mathbf{i} + u\mathbf{j} + v\mathbf{k}.$$

Show that in this identification, matrix multiplication corresponds to multiplication of quaternions.

14
Group actions

14.1 Groups of permutations

This chapter is devoted to the idea of a group acting on a set. Throughout this section, G will be a group of permutations of a set X, and we shall refer to this by saying that G *acts on* X. We shall discuss some of the geometric ideas that play an important role in the analysis of a group action, and we shall apply these to the study of the symmetry groups of regular solids.

Definition 14.1.1 Suppose that G acts on X. Then x is a *fixed point* of g in G if $g(x) = x$, and the set of fixed points of g is denoted by $\text{Fix}(g)$. Given x in X, the group $\{g \in G : g(x) = x\}$ of elements of G that fix x is called the *stabilizer* $\text{Stab}_G(x)$ of x. In plain language, $\text{Stab}_G(x)$ is the set of g that fix x, while $\text{Fix}(g)$ is the set of x that are fixed by g. □

Definition 14.1.2 Suppose that G acts on X and that $x \in X$. Then the subset $\{g(x) : g \in G\}$ of X is called the *orbit* of x under G, and it is denoted by $\text{Orb}_G(x)$. The group G is said to act *transitively* on X if $\text{Orb}_G(x) = X$ for one (or equivalently, for all) x in X. □

When the group G is understood from the context, we shall usually omit the suffix G and write $\text{Stab}(x)$, $\text{Fix}(g)$ and $\text{Orb}(x)$. It should be clear that G acts transitively on X if and only if for each x and y in X there is some g in G such that $g(x) = y$. Next, $\text{Orb}_G(x)$ is the set of points in X that x can be mapped to by some element of G. However, as

$$\{g^{-1}(x) : g \in G\} = \{g(x) : g \in G\} = \text{Orb}(x),$$

$\text{Orb}(x)$ is also the set of points that can be mapped to x by some g in G. It is obvious that X is the union of its orbits, and it is easy to see that *any two orbits are either equal or disjoint*. Indeed, suppose that there is some z in $\text{Orb}(x) \cap \text{Orb}(y)$, and take any w in $\text{Orb}(x)$. Then there are elements f, g

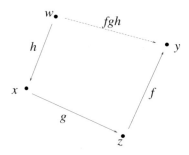

Figure 14.1.1

and h in G with $h(w) = x$, $g(x) = z$ and $f(z) = y$ (see Figure 14.1.1). Thus $fgh(w) = y$ so that $w \in \text{Orb}(y)$. We deduce that $\text{Orb}(x) \subset \text{Orb}(y)$, and the reverse inclusion holds by symmetry.

We illustrate these ideas with two examples (recall that $|E|$ is the number of elements in E).

Example 14.1.3 Let $X = \{1, 2, 3, 4, 5, 6, 7\}$, and let G be the cyclic group generated by (1 2)(3 4 5 6). Then G acts on X, and consists of the four elements I, (1 2)(3 4 5 6), (3 5)(4 6) and (1 2)(3 6 5 4). Here,

$$\text{Stab}(1) = \{I, (3\,5)(4\,6)\}, \qquad \text{Orb}(1) = \{1, 2\},$$
$$\text{Stab}(3) = \{I\}, \qquad \text{Orb}(3) = \{3, 4, 5, 6\},$$
$$\text{Stab}(7) = G, \qquad \text{Orb}(7) = \{7\}.$$

Notice that in each case, $|\text{Stab}(x)| \times |\text{Orb}(x)| = |G|$. □

Example 14.1.4 Let P be the regular n-gon whose vertices are at the n-th roots of unity. The symmetry group G of P acts on the complex plane $\mathbb{C}$, and we shall consider all points of $\mathbb{C}$. Each vertex v of P can mapped by a rotation in G to any other vertex. Thus $\text{Orb}(v) = V$, where V is the set of vertices of P, and $\text{Stab}(v)$ contains two elements, namely the identity I and the reflection in the line through v and 0. Clearly the same holds for any positive scalar multiple of a vertex. Next, consider the origin 0; here, $\text{Orb}(0) = \{0\}$ and $\text{Stab}(0) = G$. Finally, if z does not lie on any line of symmetry of P, then $\text{Stab}(z) = \{I\}$, and $\text{Orb}(z)$ contains $2n$ points. Again, in every case, $|\text{Stab}(z)| \times |\text{Orb}(z)| = |G|$. □

Suppose that G acts on X, and that x and y are in X. The next result says that the most general map that takes x to y can be written either as (i) the most general map f that fixes x followed by a single chosen map h of x to y, or as (ii) a chosen map h of x to y followed by the most general map g that fixes y

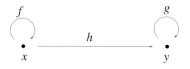

Figure 14.1.2

(see Figure 14.1.2). It also shows that if x and y are in the same orbit, then the set of h such that $h(x) = y$ is both a left coset, and a right coset, in G.

Theorem 14.1.5 *Suppose that G acts on X, and that $y = h(x)$, where $x, y \in X$ and $h \in G$. Then*

$$h\,\mathrm{Stab}(x) = \{g_0 \in G : g_0(x) = y\} = \mathrm{Stab}(y)\,h. \qquad (14.1.1)$$

Proof Let $G_0 = \{g_0 \in G : g_0(x) = y\}$. The general element of $h\,\mathrm{Stab}(x)$ is of the form hf, where $f(x) = x$, and as $hf(x) = h(x) = y$, this shows that $h\,\mathrm{Stab}(x) \subset G_0$. Now suppose that $g_0 \in G_0$. Then $g_0 = h(h^{-1}g_0)$ and this is in $h\mathrm{Stab}(x)$; thus $G_0 = h\,\mathrm{Stab}(x)$. The proof that $G_0 = \mathrm{Stab}(y)\,h$ is similar and is omitted. □

Note that in general there will be many maps, say $h_1, h_2, \ldots$, in G that map x to y and in this case we must have $h_1\,\mathrm{Stab}(x) = h_2\,\mathrm{Stab}(x) = \cdots$ and $\mathrm{Stab}(y)\,h_1 = \mathrm{Stab}(y)\,h_2 = \cdots$. The next result is an immediate consequence of (14.1.1); it says that the stabilizers of two points in the same orbit are *conjugate subgroups* of G. Thus they are isomorphic subgroups and so *have the same order*.

Corollary 14.1.6 *Suppose that G acts on X, and that $y = h(x)$, where $x, y \in X$ and $h \in G$. Then $\mathrm{Stab}(y) = h\,\mathrm{Stab}(x)\,h^{-1}$.*

We come now to a geometric form of Lagrange's theorem.

Theorem 14.1.7: the orbit-stabilizer theorem *Let G be a finite group acting on a finite set X. Then, for any x in X,*

$$|\mathrm{Orb}(x)| \times |\mathrm{Stab}(x)| = |G|. \qquad (14.1.2)$$

Proof Take any x in X, and let the orbit of x be $\{h_1(x), \ldots, h_r(x)\}$, where $h_i(x) \neq h_j(x)$ when $i \neq j$. According to Theorem 14.1.5, $h_j\,\mathrm{Stab}(x)$ is the set of elements in G that map x to $h_j(x)$, so we have

$$G = h_1\,\mathrm{Stab}(x) \cup \cdots \cup h_r\,\mathrm{Stab}(x),$$

where these cosets are pairwise disjoint (this is, in fact, the partitioning of G into left cosets with respect to the subgroup $\text{Stab}(x)$). Thus

$$|G| = \sum_{j=1}^{r} |\text{Stab}(x)| = r|\text{Stab}(x)|$$

as required. □

As an illustration of this result, we can see that the group of rotations of a cube has order 24. Indeed, the group acts as a permutation of the set of vertices, each vertex is fixed by a group of rotations of order three, and each vertex can be moved to any of the eight vertices by a rotation of the cube. Similarly, the group of rotations of a regular tetrahedron has 3×4 elements.

Results on group actions on finite sets can sometimes be proved by counting a finite set in different ways using the following simple principle. As $|E|$ is the number of elements in a set E, we have $|E| = \sum_{x \in E} 1$. Thus if E and F are finite sets, then

$$\sum_{x \in E} \left(\sum_{y \in F} 1 \right) = \sum_{y \in F} \left(\sum_{x \in E} 1 \right),$$

this being the number of pairs (x, y) with x in E and y in F.

The last result in this section provides a formula for the number of orbits in a group action and it is proved in this way. This result appeared in a text by Burnside in 1897, but it was known earlier by Cauchy, Frobenius and others.

Theorem 14.1.8: Burnside's lemma *Let G be a finite group acting on a finite set X. Then there are N orbits, where*

$$N = \frac{1}{|G|} \sum_{g \in G} |\text{Fix}(g)| = \frac{1}{|G|} \sum_{x \in X} |\text{Stab}(x)|. \quad (14.1.3)$$

In particular, N is the average number of fixed points that an element of G has.

Proof First,

$$\{(g, x) : g \in G, \; x \in \text{Fix}(g)\} = \{(g, x) : x \in X, \; g \in \text{Stab}(x)\},$$

as each set requires that $g(x) = x$. This implies that

$$\sum_{g \in G} \left(\sum_{x \in \text{Fix}(g)} 1 \right) = \sum_{x \in X} \left(\sum_{g \in \text{Stab}(x)} 1 \right),$$

or, equivalently, that

$$\sum_{g \in G} |\text{Fix}(g)| = \sum_{x \in X} |\text{Stab}(x)|,$$

and this shows that the two sums in (14.1.3) are equal to each other.

We now prove (14.1.3). The action of G on X partitions X into N pairwise disjoint orbits which we denote by $\mathcal{O}_1, \ldots, \mathcal{O}_N$. As summing over X is the same as summing over each orbit $\mathcal{O}_j$, and then summing over j, we obtain

$$\sum_{x \in X} |\text{Stab}(x)| = \sum_{j=1}^{N} \left(\sum_{x \in \mathcal{O}_j} |\text{Stab}(x)| \right). \qquad (14.1.4)$$

Now take any point y in $\mathcal{O}_j$. As the stabilizers of points in the same orbit have the same order, we see from (14.1.2) that

$$\sum_{x \in \mathcal{O}_j} |\text{Stab}(x)| = \sum_{x \in \mathcal{O}_j} |\text{Stab}(y)|$$
$$= |\mathcal{O}_j| \, |\text{Stab}(y)|$$
$$= |\text{Orb}(y)| \, |\text{Stab}(y)|$$
$$= |G|.$$

This combined with (14.1.4) gives (14.1.3). $\square$

The next example illustrates the use of Burnside's lemma and, as this is an important technique, we consider it in detail.

Example 14.1.9 Consider the problem of arranging two identical red beads, and four identical blue beads, on a circular ring of wire. We may assume that the beads are placed at the vertices of a regular hexagon, and we agree that two configurations are to be considered to be the same if one configuration can be mapped onto the other by a symmetry of the hexagon. We want to know how many different configurations there are. This number is three, for we can label the vertices 1, 2, 3, 4, 5, 6, and then each arrangement is equivalent to the two red beads being placed at one of the pairs (1, 2), (1, 3) or (1, 4) of vertices. Nevertheless, our aim is to illustrate the use of Burnside's Lemma rather than to solve the problem.

Each configuration of beads (without any identification) can be represented by a function $f : \{1, 2, 3, 4, 5, 6\} \to \{R, B\}$ with the property that $f(n) = R$ for two values of n and $f(n) = B$ for four values of n (this f represents the configuration in which the red beads appear at those vertices n for which $f(n) = R$). Let $\mathcal{F}$ be the set of such maps; clearly, $\mathcal{F}$ has $\binom{6}{2}$ ($= 15$) elements.

Now once we identify different configurations, different functions represent the same configuration of beads; in fact, the functions f_1 and f_2 represent the same configuration (after identification) if and only if there is some symmetry, say ρ, of the hexagon such that $f_2 = f_1 \rho$. The symmetries ρ belong to the dihedral group D_{12}, which we shall now regard as acting on $\mathcal{F}$ by the rule $\rho : f \mapsto f\rho$, or, equivalently, $\rho(f) = f\rho$. Now we see that f_1 and f_2 represent the same configuration of beads if and only if there is some ρ such that $f_2 = \rho(f_1)$; thus *the number N of different configurations is the same as the number of orbits in the action of D_{12} on $\mathcal{F}$.* As $|D_{12}| = 12$, Burnside's Lemma gives

$$N = \frac{1}{12} \sum_{\rho \in D_{12}} |\text{Fix}(\rho)|$$

$$= \frac{1}{12} \sum_{\rho \in D_{12}} |\{f \in \mathcal{F} : f = f\rho\}|.$$

To find N we need to find, for each ρ in D_{12}, how many functions f in $\mathcal{F}$ satisfy $f = f\rho$, and we consider each ρ in turn. First, when $\rho = I$, all fifteen functions f satisfy $f = f\rho$. If ρ is a rotation for which there is some f with $f = f\rho$ then ρ must have a two-cycle (corresponding to the two positions of the red beads). The only rotation in D_{12} with a two cycle is $(1\,4)(2\,5)(3\,6)$, and there are exactly three functions f for which $f = f\rho$ (for example, the function f with $f(2) = f(5) = R$).

Now suppose that ρ is a reflection across a line joining two vertices. There are three such ρ, and for each of these there are three functions f with $f = f\rho$. For example, if $\rho = (2\,6)(3\,5)$, then the three functions take the value R on the sets $\{2, 6\}$, $\{3, 5\}$ and $\{1, 4\}$, respectively. Finally, suppose that ρ is a reflection across a line joining the mid-points of two sides. Again there are three such ρ and three functions f for each ρ. For example, if $\rho = (1\,4)(2\,3)(5\,6)$, then the three sets on which a function can take the value R are $\{1, 4\}$, $\{2, 3\}$ and $\{5, 6\}$. We conclude that N is $[15 + 3 + (3 \times 3) + (3 \times 3)]/12$, which is 3. □

Exercise 14.1

1. Let G be the group $\{1, i, -1, -i\}$. Show that for each g in G the map $x \mapsto gx$ is a permutation of G.
2. Let G be a finite group of permutations of a finite set X. Show that if G is abelian and transitive, then $|G| = |X|$.
3. Let Q be a plane quadrilateral and let $G(Q)$ be its symmetry group (that is the group of Euclidean isometries mapping Q onto itself). Show that $G(Q)$ has at most 8 elements (so that Q has the largest symmetry group when it is a square). For each n in $\{1, 2, \ldots, 8\}$ determine whether or not there is a quadilateral Q with $G(Q)$ of order n. Is it true that if $G(Q)$ has order eight then Q is a square?
4. Work through the problem considered in Example 14.1.9 in the case when there are three identical red beads and three identical blue beads.

14.2 Symmetries of a regular polyhedron

The following is a list of all regular polyhedra (see Section 5.5):

(1) the *tetrahedron* (four triangular faces, six edges, four vertices, and three faces meeting at each vertex);
(2) the *cube* (six square faces, twelve edges, eight vertices, and three faces meeting at each vertex);
(3) the *octahedron* (eight triangular faces, twelve edges, six vertices, and four faces meeting at each vertex);
(4) the *dodecahedron* (twelve pentagonal faces, thirty edges, twenty vertices, and three faces meeting at each vertex);
(5) the *icosahedron* (twenty triangular faces, thirty edges, twelve vertices, and five faces meeting at each vertex).

These polyhedra were first listed by Theatus in about 400 BC, they are often referred to as the *Platonic solids*, and they are illustrated in Figure 5.5.1. We shall take the existence of these polyhedra for granted, together with the fact that the vertices of each of these polyhedra lie on a sphere, which we may assume is centred at the origin in $\mathbb{R}^3$. For each polyhedron, we consider the *group* G^+ *of rotations* of $\mathbb{R}^3$ that leaves the polyhedron invariant (and permutes its vertices). Observe that as any such rotation permutes the set of vertices, it must leave the (unique) sphere through the vertices invariant, and hence it must fix the origin. Next, we recall that, say, q regular p-gons meet at each vertex v, and we accept (again without proof) that there is a rotation of order q (about an axis through v and the centre of the polyhedron) which fixes v and leaves the polyhedron invariant. This rotation will cyclically permute the q edges emanating from v, and the stabilizer of each vertex is therefore a cyclic group of order q. Finally, we accept without proof that, by repeated rotations of the polyhedron, we can move any vertex to any other; thus the orbit of any given vertex v is the set of all V vertices (and G^+ acts transitively on V). If we now apply the Orbit-stabilizer theorem (Theorem 14.1.7) we conclude that G^+ has order qV; thus, from (5.5.2),

$$|G^+| = \frac{4pq}{2q + 2p - pq} = 2E.$$

This gives $|G^+|$ to be twelve for the tetrahedron, twenty four for the cube and the octahedron, and sixty for the dodecahedron and the icosahedron. Note that we could have found that $|G^+| = 2E$ from the plausible assumption that a given *directed* edge of the polyhedron can be moved by a rotation of the polyhedron

14.2 Symmetries of a regular polyhedron

onto any other edge with either direction assigned to it. In any event, we have established the following fact.

Theorem 14.2.1 *The group of rotations of a regular polyhedron with E edges has order $2E$.*

Each regular polyhedron is left invariant by a reflection σ across some plane passing through the origin. As the composition of two indirect isometries is a direct isometry, the group G of isometries that leave the polyhedron invariant has the coset decomposition $G^+ \cup \sigma G^+$, so that $|G| = 2|G^+|$. Note that whereas each rotation (in G^+) is physically realizable as a rotation of $\mathbb{R}^3$, the reflections that leave the polyhedron invariant cannot be realized physically. Mathematically, this distinction is whether the symmetry lies in the orthogonal group (of matrices A with $\det(A) = \pm 1$) or the special orthogonal group (of matrices A with $\det(A) = 1$).

The rest of this section contains a discussion of each Platonic solid and its symmetry group. Throughout, G^+ will denote the group of rotations, and G the group of isometries, that leave the solid invariant. We shall take each solid in turn, but we draw attention to the fact that the cube and the octahedron are *dual* solids, as are the dodecahedron and the icosahedron, while the tetrahedron is *self-dual*. Duality implies that if a polyhedron P has V vertices, E edges and F faces, then the dual polyhedron has F vertices, E edges and V faces, but it means more than this.

Example 14.2.2: the cube The existence of the cube is not in doubt. The symmetry group G^+ has twenty four elements, and we shall now identify the twenty three rotations which, together with I, constitute G^+. First, let L_1 be a line through the centres of opposite faces. There are three choices for L_1 and, for each choice, G^+ contains a cyclic subgroup of order four of rotations about L_1. This provides us with nine non-trivial rotations in G^+. Next, let L_2 be a line through the midpoints of opposite edges. There are six choices of L_2 and, for each choice, G^+ contains a subgroup of order two of rotations about L_2. This provides us with six more non-trivial rotations. Finally, let L_3 be a line through a pair of opposite vertices. There are four choices of L_3 and, for each choice, G^+ contains a cyclic subgroup of order three of rotations about L_3. This provides us with eight more non-trivial rotations in G, making a total of twenty three non-trivial (distinct) rotations in G.

We shall now show that *the group G^+ of rotations of a cube C is isomorphic to S_4*. Let d_1, d_2, d_3, d_4 be the four diagonals of C. As each rotation is an isometry, a rotation of C must map a pair of diametrically opposite vertices to another such pair, and this means that we can regard G^+ as acting on the

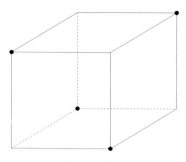

Figure 14.2.1

set $\{d_1, d_2, d_3, d_4\}$ of diagonals. Now consider a rotation of order two whose axis is the line through the midpoints of a pair of opposite edges, say E and E'. It is easy to see that this induces a transposition of one pair of diagonals while leaving the other two diagonals invariant (the two invariant diagonals are those which do not have a common endpoint with E or E'). As S_4 is generated by its transpositions, we see that G^+ (regarded as the permutation group S_4 of the four diagonals) contains S_4. As $|G^+| = 24 = |S_4|$, we see that $G^+ = S_4$.

Example 14.2.3: the tetrahedron The existence of the regular tetrahedron is not in doubt, but perhaps the simplest construction of a regular tetrahedron is as follows. Mark a vertex v of a cube, and then mark the three other vertices of the cube that share a diagonal of a face with v. The four marked vertices are then the vertices of a regular tetrahedron (see Figure 14.2.1). Notice that each pair of opposite edges of the tetrahedron arises as a pair of skew diagonals of opposite faces of the cube.

It is easy to describe the eleven non-trivial rotations in the symmetry group G^+ of a tetrahedron T. First, let L_1 be the line through a vertex and the centroid of the opposite face. There are four choices of L_1 and, for each choice, there is a cyclic group of order three of rotations about L_1 that leaves T invariant. This provides eight non-trivial rotations of T. The remaining three rotations are the rotations of order two whose axis is on a lines though the mid-points of opposite edges of T (for example, in Figure 14.2.1 one such axis is the vertical line through the centre of the cube). If we label the vertices, say a, b, c and d, and simply write down the twelve elements of G^+ as permutations of $\{a, b, c, d\}$, we see immediately that *the group G^+ of rotational symmetries of a regular tetrahedron is the alternating group A_4*. In fact, it is clear (geometrically) that G^+ cannot contain a rotation that acts as a transposition on $\{a, b, c, d\}$.

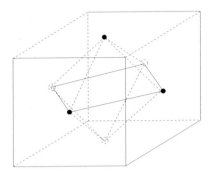

Figure 14.2.2

Now consider opposite edges e and e' of T (for example, the horizontal edges of the tetrahedron in Figure 14.2.1) We know that e lies in a plane Π that is orthogonal to e' and passing through the midpoint of e'. Let α be the reflection in Π. Then α is a symmetry of T that interchanges the two endpoints of e' while fixing each endpoint of e. If we view G as acting on the set of four vertices of T, then G contains every transposition of two vertices, and hence every permutation of these vertices. As $|G| = 24 = |S_4|$, we see that $G = S_4$.

Example 14.2.4: the octahedron Take a cube C; then the six midpoints of the faces of the cube form the vertices of a regular octahedron (see Figure 14.2.2 in which the 'solid' points are the mid-points of the visible faces of the cube). Similarly, the eight centroids of the equilateral triangular faces of a regular octahedron are the vertices of a cube. This shows that the cube and octahedron are dual solids (roughly speaking, the vertices of one are at the centres of the faces of the other). If we rotate the cube we rotate the inscribed octahedron; likewise, a rotation of an octahedron is also a rotation of the inscribed cube. It is clear, then, that the two symmetry groups of the octahedron coincide with those of the cube.

Example 14.2.5: the icosahedron The icosahedron and the dodecahedron are more difficult to construct, and the geometry of these figures reveals that the golden ratio

$$\tau = \tfrac{1}{2}\left(\sqrt{5}+1\right),$$

appears as the ratio of certain lengths in the construction. We note for future use that $\tau^2 = \tau + 1$, and that $\tau^3 = \tau^2 + \tau = 2\tau + 1$. We now claim that the twelve points whose coordinates are

$$(\pm 1, 0, \pm\tau), \quad (0, \pm\tau, \pm 1), \quad (\pm\tau, \pm 1, 0) \qquad (14.2.1)$$

form the twelve vertices of a regular icosahedron of edge length 2 that lies on the sphere S_0 centred at the origin with radius $\tau + 2$. To see this, let us focus on the point $(\tau, 1, 0)$, and let S_1 be the sphere with centre $(\tau, 1, 0)$ and radius 2. The points (x, y, z) on $S_0 \cap S_1$ are the common solutions of

$$x^2 + y^2 + z^2 = \tau + 2,$$
$$(x - \tau)^2 + (y - 1)^2 + z^2 = 4,$$

and this intersection is easily seen to be the plane Π given by $x\tau + y = \tau$. Let us now see which of the points in (14.2.1) lie on Π. If $x = 0$ then $y = \tau$ so the points $(0, \tau, \pm 1)$ lie on Π. If $x = 1$ then $y = 0$ and the points $(1, 0, \pm\tau)$ lie on Π. If $x = -1$ then $y = 2\tau$ so there are no points in (14.2.1) with $x = 1$ that lie on Π. Similarly, we cannot have $x = -\tau$. Finally, if $x = \tau$, then $y = 1$ so the single point $(\tau, 1, 0)$ lies on Π. We have now shown that the five points $(0, \tau, \pm 1)$, $(1, 0, \pm\tau)$ and $(\tau, 1, 0)$ in (14.2.1) lie at a distance two from $(\tau, 1, 0)$, and it is easily checked that any two of these points are at a distance two apart. The reader may care to draw a diagram (with three axes) in which these five points, the point $(\tau, 1, 0)$ and the plane Π are illustrated.

The determination of the symmetry group of the icosahedron requires more work, and here is one way to approach the problem. Briefly, the icosahedron has twenty triangular faces which can be subdivided into five sets of four triangles in such a way that the centroids of the four triangles in a set form the vertices of a regular tetrahedron. For example, the following sets of vertices of the icosahedron determine a triangle T_j, and the four centroids of the faces T_j form a regular tetrahedron:

$$T_1 : (\tau, -1, 0), (\tau, 1, 0), (1, 0, -\tau);$$
$$T_2 : (-1, 0, \tau), (0, \tau, 1), (1, 0, \tau);$$
$$T_3 : (0, -\tau, 1), (0, -\tau, -1), (-\tau, -1, 0);$$
$$T_4 : (-\tau, 1, 0), (-1, 0, -\tau), (0, \tau, -1).$$

A similar argument holds for the other sets of triangular faces, and in this way we can construct five regular tetrahedra whose vertices lie at the centroids of the faces of the icosahedron. Any symmetry of the icosahedron will permute these five tetrahedra, and it can be shown that the symmetry group G^+ acts on these five tetrahedra as an even permutation. Thus G^+ may be regarded as a subgroup of A_5 (the even permutations of five objects) and as $|G^+| = 60 = |A_5|$, we see that $G^+ = A_5$. We state (without proof) that the full symmetry group of the icosahedron is $A_5 \times C_2$ (and not S_5).

Example 14.2.6: the dodecahedron The dodecahedron and the icosahedron are dual solids; that is the centroids of the faces of one of these solids are at the vertices of a solid of the other type. Accordingly, their symmetry groups are the same and, with a little work) one can write down the coordinates of the vertices of a regular dodecahedron.

Exercise 14.2

1. Let G be the group of permutations of $\{a, b, c, d\}$, and let $\alpha = (b\,c\,d)$ and $\beta = (c\,d\,a)$. Show that A_4 is generated by α and β, and interpret this in terms of the symmetries of a regular tetrahedron.
2. Verify the following construction of a regular octahedron. Consider a vertex v_j and its opposite face f_j of a regular tetrahedron T. Let Π_j be the plane, parallel to f_j, and through the mid-points of the three edges that contain v_j. Then Π_j cuts the tetrahedron into two pieces, one of which is a smaller tetrahedron T_j, say, with vertex v_j. If we remove the four tetrahedra T_j from T the remaining solid is a regular octahedron.
3. Let C be the cuboid given by

$$\{(x, y, z) \in \mathbb{R}^3 : |x| \leq a, \ |y| \leq b, \ |z| \leq c\},$$

where a, b and c are distinct, and let G^+ be the group of rotations about the origin that leave C invariant. Show that $|G^+| = 4$, and that $|G| = 8$. Identify (geometrically) all eight elements of G, and show that G consists of the group of eight diagonal matrices, all of whose diagonal elements are ± 1.
4. Show that if a rotation of a cube leaves each diagonal invariant then it is the trivial rotation.

14.3 Finite rotation groups in space

We have identified the finite groups of rotations that arise as the symmetry group of one of the regular polygons, or one of the polyhedra, and we now raise the question of whether there are any other finite groups of rotations. The answer is 'no', and the sole aim of this section is to prove the first of the following two results, and to state the second. A very brief sketch of a proof of the second result is given in Section 15.6.

Theorem 14.3.1 *Any finite group of rotations of $\mathbb{R}^3$ is cyclic, dihedral, or the symmetry group of a regular solid.*

Theorem 14.3.2 *Any finite group of Möbius maps is isomorphic to a finite group of rotations of $\mathbb{R}^3$.*

The proof of Theorem 14.3.1 Let G be a finite, non-trivial, group of rotations of $\mathbb{R}^3$. Then each element g of G leaves the unit sphere $||x|| = 1$ invariant, and g has exactly two (diametrically opposite) fixed points on the sphere (unless $g = I$). Let X be the set of points on the sphere that are fixed by some non-trivial rotation in G. If $x \in X$, then there is some h in G with $h(x) = x$, and then $g(x)$ is fixed by ghg^{-1}. It follows that each g in G maps X onto itself, so that G acts as a group of permutations of the finite set X.

Now X is the union of, say N orbits, which we denote by $\mathcal{O}_1, \ldots, \mathcal{O}_N$. Associated to each orbit $\mathcal{O}_j$ is an integer n_j which is the order of the stabilizer of any x in $\mathcal{O}_j$. Theorem 14.1.7 implies that $n_j |\mathcal{O}_j| = |G|$, and as we obviously have $|X| = \sum_j |\mathcal{O}_j|$, we find that

$$|X| = |G| \sum_{j=1}^{N} 1/n_j. \tag{14.3.1}$$

Next, each g in G, $g \neq I$, has exactly two fixed points in X, and I has $|X|$ fixed points in X. Thus, from Theorem 14.1.8,

$$N|G| = \sum_{g \in G} |\text{Fix}(g)| = 2(|G| - 1) + |X|.$$

If we use (14.3.1) to eliminate $|X|$, we find that

$$2 - \frac{2}{|G|} = \sum_{j=1}^{N} \left(1 - \frac{1}{n_j}\right). \tag{14.3.2}$$

As $1/2 \leq 1 - 1/n_j < 1$ and $|G| \geq 2$, this shows that N is 2 or 3.

Case 1: $N = 2$. In this case (14.3.2) reduces to

$$\frac{2}{|G|} = \frac{1}{n_1} + \frac{1}{n_2}$$

which, as $n_j \leq |G|$, gives $n_1 = n_2 = |G|$. As $N = 2$ there are exactly two orbits $\mathcal{O}_1$ and $\mathcal{O}_2$ of fixed points, and as $|\mathcal{O})j| = |G|/n_j$, each orbit contains only one point. These points must be diametrically opposite points on $\mathcal{S}$, and we deduce that G is a cyclic group of rotations of the sphere.

Case 2: $N = 3$. In this case we have

$$1 + \frac{2}{|G|} = \frac{1}{n_1} + \frac{1}{n_2} + \frac{1}{n_3}. \tag{14.3.3}$$

We may assume that $n_1 \leq n_2 \leq n_3$; then $1 < 1 + 2/|G| \leq 3/n_1$ so that $n_1 = 2$. Using this in (14.3.3) we find that

$$\frac{1}{2} + \frac{2}{|G|} = \frac{1}{n_2} + \frac{1}{n_3} \leq \frac{2}{n_2},$$

so that n_2 is 2 or 3. We can now use these values of n_2 in (14.3.3), with $n_1 = 2$, to find all solutions of (14.3.3). Thus with $N = 3$ we obtain the following possibilities:

(1) $(n_1, n_2, n_3) = (2, 2, n)$ and $|G| = 2n$;
(2) $(n_1, n_2, n_3) = (2, 3, 3)$ and $|G| = 12$;
(3) $(n_1, n_2, n_3) = (2, 3, 4)$ and $|G| = 24$;
(4) $(n_1, n_2, n_3) = (2, 3, 5)$ and $|G| = 60$.

We shall not go into details, but we now have enough information about the possibilities for G to show that in (1) G is the dihedral group D_{2n}, while in (2), (3) and (4) G is one of the symmetry groups of the regular polyhedra. □

14.4 Groups of isometries of the plane

Each isometry of $\mathbb{C}$ is either a direct isometry $z \mapsto az + b$, or an indirect isometry $z \mapsto a\bar{z} + b$, where, in each case, $|a| = 1$. The group of all isometries is denoted by $I(\mathbb{C})$, and the group of direct isometries by $I^+(\mathbb{C})$. Both of these groups act on $\mathbb{C}$, and our first task is to find all finite subgroups of $I(\mathbb{C})$. As always, I denotes the identity map.

Theorem 14.4.1 *The only finite subgroups of $I(\mathbb{C})$ are the cyclic and dihedral groups.*

Proof Let G be a finite subgroup of $I(\mathbb{C})$, let $\{z_1, \ldots, z_n\}$ be an orbit, and let ζ be the centre of gravity of this orbit. If g is in G then g permutes the points in the orbit, so that if $g(z) = az + b$, then

$$g(\zeta) = a\left(\frac{1}{n}\sum_j z_j\right) + b = \frac{1}{n}\sum_j (az_j + b) = \frac{1}{n}\sum_j g(z_j) = \zeta.$$

The same is true if $g(z) = a\bar{z} + b$, so that the elements of G have a common fixed point ζ. By choosing coordinates appropriately we may assume that $\zeta = 0$;

then every element of G is of the form $z \mapsto az$ or $z \mapsto a\bar{z}$, where $|a| = 1$. Now consider the subgroup G^+ of direct isometries in G. This is a finite group of rotations about the origin, and it is easy to see that this is a cyclic group. If G contains a map $z \mapsto a\bar{z}$ then G is a dihedral group. □

The next task is to describe all abelian subgroups of $I(\mathbb{C})$.

Theorem 14.4.2 *Each of the following is an abelian subgroup of* $I(\mathbb{C})$:

(a) *the group* $\mathcal{T}$ *of all translations;*
(b) *a group of rotations* $\mathcal{R}$ *about a given point;*
(c) *a group* $\mathcal{K}$ *(of order four) generated by the reflections in two orthogonal lines;*
(d) *a group* $\mathcal{G}$ *generated by a reflection across a line and all translations along that line.*

Further, any abelian subgroup of $I(\mathbb{C})$ *is a subgroup of one of these groups (for a suitable choice of point or lines).*

Proof We suppose first that G contains only direct isometries. We may suppose that G is not the trivial group $\{I\}$; then G contains a translation, or a rotation, or both.

(i) *Suppose that G contains a translation*, say $f(z) = z + t$, where $t \neq 0$. Let g be any element of G, say $g(z) = az + b$. As $fg(0) = gf(0)$ we see that $a = 1$, so that g is a translation. Thus G is a subgroup of $\mathcal{T}$.

(ii) *Suppose that G contains a rotation*. Then, from (i), G contains only rotations and I. Let f and g be any rotations in G with (finite) fixed points ζ_f and ζ_g, respectively. As G is abelian, $f(g(\zeta_f)) = g(f(\zeta_f)) = g(\zeta_f)$, so that f fixes $g(\zeta_f)$. As f has a unique fixed point, $g(\zeta_f) = \zeta_f$. However, g also has a unique fixed point; thus $\zeta_f = \zeta_g$. We conclude that G is a group of rotations about a single point, and this is case (b).

We now consider an abelian group G that contains an indirect isometry. There are three cases to consider, namely when the subgroup G^+ of direct isometries is $\{I\}$, or contains translations, or contains rotations.

(iii) *Suppose that $G^+ = \{I\}$*. If g is an indirect isometry in G, then g^2 is a direct isometry so that $g^2 = I$ and $g = g^{-1}$. Thus every indirect isometry in G is a reflection. If g and h are indirect isometries in G, then $gh \in G^+$ so that $gh = I$. Thus $h = g^{-1} = g$, and then $G = \{I, g\}$, where g is a reflection.

(iv) *Suppose that G^+ contains a rotation*. We may assume that the rotation is $f(z) = \mu z$, where $|\mu| = 1$ and $\mu \neq 1$. Let $g(z) = a\bar{z} + b$ be any indirect isometry in G. As $fg = gf$ we see that for all z, $\mu a \bar{z} + \mu b = a \bar{\mu} \bar{z} + b$. This shows first that $\mu b = b$, so that $b = 0$, and second, that $\mu = \bar{\mu}$. As $|\mu| = 1$ and $\mu \neq 1$ we see that $\mu = -1$. It follows that G contains only one rotation, namely $f(z) = -z$, and that every indirect isometry in G is of the form $z \mapsto c\bar{z}$, where $|c| = 1$, and hence is a reflection. Now suppose that g and h are reflections in G. Then gh is either I or f; hence h is g or gf, and $G = \{I, f, g, gf\}$. The reflections g and gf must be reflections across *orthogonal lines* because their product is the rotation f by an angle of π.

(v) *Suppose that G^+ contains a translation*, say $f(z) = z + t$; we may assume that t is real and non-zero. In addition, G contains an indirect isometry, say $g(z) = a\bar{z} + b$. As $fg = gf$ we see that $a = 1$, so that G contains $f(z) = z + t$ and $g(z) = \bar{z} + b$. Next, suppose that h is any direct isometry in G; then (by what we have shown above) h is a translation, say $h(z) = z + s$. As $gh = hg$ we see that s is real; thus the only direct isometries in G are real translations. Now consider any indirect isometry k in G; then, as above, $k(z) = \bar{z} + c$, say. As $gk = kg$ we see that b and c have the same imaginary part, say ρ. It follows that G is a subgroup of the group generated by all real translations and the map $z \mapsto \bar{z} + i\rho$. Finally, this larger group is a group of the form $\mathcal{G}$ in (d) with the invariant line given by $y = \rho/2$. □

The most well known groups of isometries of $\mathbb{C}$ are the seventeen 'wallpaper groups' and the seven 'frieze groups'. We shall discuss the frieze groups as *these provide a good illustration of the use of quotient groups, and of conjugacy classes*. A 'frieze' is a decorative strip with a repeating pattern, and a frieze group is the symmetry group of some frieze. Frieze groups are often described by drawing repeated motifs along a line, but here we prefer a more analytical approach (which is free of motifs) in order to illustrate the use of group theory. Reader wishing to 'see' the seven possibilities can find pictures of them in many texts (or create pictures from our list of seven groups).

Given any group G of isometries of $\mathbb{C}$, the set T of translations in G is a subgroup of G. This is the *translation subgroup* of G, and it is always *a normal subgroup of G*. To see this, take any translation in G, say $f(z) = z + t$. Now any direct isometry in G is of the form $g(z) = az + b$, and any indirect isometry is of the form $h(z) = c\bar{z} + d$. A simple calculation shows that both gfg^{-1} and hfh^{-1} are translations, and this shows that T is a normal subgroup of G. The most important consequence of this fact is that we can now consider the quotient group G/T.

Definition 14.4.3 A *frieze group* is a group $\mathcal{F}$ of isometries of $\mathbb{C}$ that leaves the real line $\mathbb{R}$ invariant, and whose translation subgroup $\mathcal{T}$ is an infinite cyclic group.

Our aim is to classify the frieze groups, and we shall do this by showing that the quotient group $\mathcal{F}/\mathcal{T}$ has at most four elements, and then considering all possibilities. However, before we can list the possibilities, we need to decide when two frieze groups are to be considered as the 'same' group. Two frieze groups may be isomorphic but have quite different geometric actions; for example, the symmetry group of a normal pattern of footprints is an infinite cyclic group generated by a glide reflection, and this has a quite different geometric action to the (isomorphic) infinite cyclic group generated by a translation. This means that a classification by isomorphism classes is inadequate (this is because we are interested in the geometry as well as the algebra), and we shall classify frieze groups *geometrically* by considering conjugacy in $I(\mathbb{C})$. Thus two frieze groups $\mathcal{F}_1$ and $\mathcal{F}_2$ are to be 'identified' if and only if there is an isometry g such that $\mathcal{F}_2 = g\mathcal{F}_1 g^{-1}$. Informally, conjugate frieze groups have similar geometric actions.

If $\mathcal{T}_1$ and $\mathcal{T}_2$ are cyclic groups of translations, generated by $z \mapsto z + t_1$ and $z \mapsto z + t_2$, respectively, then $\mathcal{T}_2 = g\mathcal{T}_1 g^{-1}$, where $g(z) = (t_2/t_1)z$. Thus any frieze group is conjugate to another frieze group whose translation subgroup $\mathcal{T}$ is generated by $z \mapsto z + 1$. It follows that from now on we may (and shall) restrict our attention to frieze groups whose translation subgroup $\mathcal{T}$ is the group of integer translations $z \mapsto z + n$, where $n \in \mathbb{Z}$. It is convenient to call such a frieze group a *standard frieze group*. We shall prove the following result.

Theorem 14.4.4 *Any frieze group is conjugate to exactly one of the seven groups*

$$\langle z + 1 \rangle, \tag{14.4.1}$$

$$\langle z + 1, -z \rangle, \ \langle z + 1, -\bar{z} \rangle, \ \langle z + 1, \bar{z} \rangle, \ \langle z + 1, \bar{z} + \tfrac{1}{2} \rangle, \tag{14.4.2}$$

$$\langle z + 1, -z, \bar{z} \rangle, \ \langle z + 1, -z, \bar{z} + \tfrac{1}{2} \rangle. \tag{14.4.3}$$

where $\langle a_1, \ldots, a_k \rangle$ denotes the group generated by $a_1, \ldots, a_k$.

A summary of the proof The first step is to show that, apart from translations, there are only four types of elements in a frieze group. We then show that two elements of the same type yield the same coset with respect to $\mathcal{T}$; thus the quotient group $\mathcal{F}/\mathcal{T}$ has order at most five. The next step is to show that every non-trivial element in the quotient group has order two, and this leads to the following result. □

14.4 Groups of isometries of the plane

Lemma 14.4.5 *Let $\mathcal{F}$ be a standard frieze group. Then $\mathcal{F}/\mathcal{T}$ is either the trivial group, a cyclic group of order two, or isomorphic to the Klein 4-group.*

Clearly, (14.4.1) corresponds to the case when $\mathcal{F}/\mathcal{T}$ is the trivial group (that is, $\mathcal{F} = \mathcal{T}$). A little more analysis shows that (14.4.2) and (14.4.3) correspond to the cases when $\mathcal{F}/\mathcal{T}$ has order two and four, respectively.

We shall now give the details, and we begin by finding a general form of an element g in a standard frieze group $\mathcal{F}$. First, $g(z)$ is either $az+b$ or $a\bar{z}+b$, where $b = g(0)$ and $a = g(1) - g(0)$. As $g(\mathbb{R}) = \mathbb{R}$ we see that a and b are real. As $|a| = 1$ this gives $a = \pm 1$. Finally, if $g(z) = \bar{z} + b$ then g^2 is a translation so that $2b \in \mathbb{Z}$. This shows that every element of a standard frieze group is of one of the following forms:

(1) $z \mapsto z + m$, $m \in \mathbb{Z}$;
(2) $z \mapsto -z + b$, $b \in \mathbb{R}$;
(3) $z \mapsto -\bar{z} + b$, $b \in \mathbb{R}$;
(4) $z \mapsto \bar{z} + m$, $m \in \mathbb{Z}$.
(5) $z \mapsto \bar{z} + \frac{1}{2} + m$, $m \in \mathbb{Z}$.

We call these the five different *types* of elements. Note that $\mathcal{F}$ cannot contain elements of type (4) and elements of type (5) as otherwise, $\mathcal{F}$ would contain $z \mapsto z + \frac{1}{2}$ (which is not in $\mathcal{T}$).

The crucial observation is that if g and h are of the same type, then $g^{-1}h$ is a translation; thus we have the equality $g\mathcal{T} = h\mathcal{T}$ of cosets. This implies that each of the five types provides *at most one coset* to $\mathcal{F}/\mathcal{T}$; thus $\mathcal{F}/\mathcal{T}$ has order at most five. Next, if g is any element of $\mathcal{F}$, then $g^2 \in \mathcal{T}$ so that in the quotient group, $(g\mathcal{T})(g\mathcal{T}) = g^2\mathcal{T} = \mathcal{T}$. Thus *every element of $\mathcal{F}/\mathcal{T}$ has order two*, and, as $\mathcal{F}/\mathcal{T}$ has order at most five, this implies that $\mathcal{F}/\mathcal{T}$ must have order one, two or four. Moreover, if it has order four, it cannot be cyclic and so it must be isomorphic to the Klein four-group. We have now proved Lemma 14.4.5, and we consider each of the three cases.

Case 1: $\mathcal{F}/\mathcal{T}$ is the trivial group.
In this case $\mathcal{F} = \mathcal{T}$, and $\mathcal{F}$ is the group given in (14.4.1).
Case 2: $\mathcal{F}/\mathcal{T}$ has order two.
In this case $\mathcal{F} = \mathcal{T} \cup g\mathcal{T}$, where g is one of the four types (2)–(5), and $\mathcal{F}$ is generated by g and t, where $t(z) = z + 1$.
If g is of type (2), say $g(z) = -z + b$, let $h(z) = z - b/2$. Then $hgh^{-1} = -z$ and $hth^{-1} = t$, so that $h\mathcal{F}h^{-1} = \langle z + 1, -z \rangle$.
If g is of type (3), say $g(z) = -\bar{z} + b$, we take h as above, and then $h\mathcal{F}h^{-1} = \langle z + 1, -\bar{z} \rangle$.
If g is of type (4), say $g(z) = \bar{z} + m$, where $m \in \mathbb{Z}$, then $\mathcal{F} = \langle z + 1, \bar{z} \rangle$.

Finally, if g is of type (5), say $g(z) = \bar{z} + \frac{1}{2} + m$, where $m \in \mathbb{Z}$, then $\mathcal{F} = \langle z+1, \bar{z} + \frac{1}{2} \rangle$. All these cases are listed in (14.4.2).

Case 3: $\mathcal{F}/\mathcal{T}$ has order four.

In this case $\mathcal{F}/\mathcal{T}$ is the quotient group consisting of exactly four cosets, with each coset containing an element of one of the types listed above. As $\mathcal{T}$ is one of these cosets, and as $\mathcal{F}$ cannot contain both elements of type (4) and elements of type (5), we see that there are only two possibilities for $\mathcal{F}/\mathcal{T}$, namely

$$\mathcal{T} \cup g_2 \mathcal{T} \cup g_3 \mathcal{T} \cup g_4 \mathcal{T}, \quad \mathcal{T} \cup g_2 \mathcal{T} \cup g_3 \mathcal{T} \cup g_5 \mathcal{T}, \tag{14.4.4}$$

where g_j is of type j. In both cases, $\mathcal{F}$ contains $g_2(z) = -z + b$ and by replacing $\mathcal{F}$ by $h\mathcal{F}h^{-1}$, where $h(z) = z - b/2$, we may assume that $g_2(z) = -z$. Note that as h is a translation, it commutes with $t(z) = z + 1$. Also, $hg_j h^{-1}$ has the same type as g_j so that the description (14.4.4) of the two cases remains valid.

In the first case in (14.4.4), $\mathcal{F}$ contains $g_3(z) = -\bar{z} + b$, say, and $g_2(z) = -z$ so it also contains $\bar{z} - b$. As this element is of type (4) (the first case in (14.4.4) has no elements of type (5)), we see that $b \in \mathbb{Z}$, so that $\mathcal{F} = \langle z+1, -z, \bar{z} \rangle$.

Finally, consider the second case in (14.4.4). As before, $\mathcal{F}$ contains $g_3(z) = -\bar{z} + b$ and hence also $\bar{z} - b$. This time, this element must be of type (5) so we see that $b - \frac{1}{2} \in \mathbb{Z}$. It is now clear that $\mathcal{F}$ contains $\bar{z} + \frac{1}{2}$, and that. $\mathcal{F} = \langle z+1, -z, \bar{z} + \frac{1}{2} \rangle$. This completes the proof of Theorem 14.4.4. □

A full discussion of the seventeen *crystallographic groups*, or 'wallpaper groups' as they are popularly called, is long, and we shall content ourselves with a brief description of the starting point of such an investigation. Of course, it is not difficult to list the seventeen groups (and again, there are many texts that contains pictures of such groups); the hard work goes into showing that these are the only such groups. Consider a group G of isometries acting on $\mathbb{C}$. Each point z of $\mathbb{C}$ gives rise to an orbit $\{g(z) : g \in G\}$, and it is evident (because the elements of G are isometries) that if one orbit accumulates at some point in $\mathbb{C}$, then so does every orbit. Thus every orbit accumulates somewhere in $\mathbb{C}$ or none do.

Definition 14.4.6 A group G of isometries of $\mathbb{C}$ is *discrete* if no orbit accumulates in $\mathbb{C}$.

A frieze group is a discrete group of isometries of $\mathbb{R}$ whose translation subgroup is cyclic. A *crystallographic group* is a discrete group of isometries of $\mathbb{C}$

whose translation subgroup is generated by two translations in different directions. Thus, in some sense, the crystallographic groups are the two-dimensional versions of the one-dimesional frieze groups, and it is clear that there are crystallographic groups in all dimensions. We end by noting that, up to a suitable identification of groups, there are exactly seventeen plane crystallographic groups and 230 crystallographic groups in $\mathbb{R}^3$.

Exercise 14.4

1. Find (explicitly) the group of isometries g of $\mathbb{C}$ which satisfy $g(\mathbf{Z}) = \mathbf{Z}$. Is this group abelian?
2. If $f(z) = az + b$, or if $f(z) = a\bar{z} + b$, then $a = f(1) - f(0)$. Now write $a_f = a = f(1) - f(0)$. Show that the map $f \mapsto a_f$ is a homomorphism of $I(\mathbb{C})$ onto the multiplicative group $\{z : |z| = 1\}$, and that the kernel of this homomorphism is the group of all translations. Deduce that if G is any group of isometries of $\mathbb{C}$, then the translation subgroup T of G is a normal subgroup of G.
3. Show that no two of the seven groups listed in Theorem 14.4.4 are conjugate to each other.
4. Let G be a discrete group of isometries of $\mathbb{C}$. Show that the translation subgroup T of G is of the form $\{z \mapsto ma + nb : m, n \in \mathbb{Z}\}$, where a and b are non-zero complex numbers such that a/b is not real.

14.5 Group actions

In this section we take a closer look at some of the ideas discussed earlier in this chapter. We begin with two simple examples which show the need for a deeper investigation.

Example 14.5.1 Consider the group G consisting of isometries $I(z) = z$, $f(z) = \bar{z}$, $g(z) = -z$ and $h(z) = -\bar{z}$. Clearly, G acts on $\mathbb{C}$. However, we can also view G as a group acting on $\mathbb{R}$. Now $I = f$ and $g = h$ throughout $\mathbb{R}$ so that although G has order four, it appears to have order two when we restrict its action to $\mathbb{R}$. □

Example 14.5.2 We have already used the idea that a group G can be thought of as acting on a particular set if we are prepared to change our view of G; for example, we considered G as a group of rotations of $\mathbb{R}^3$, but then decided to view G as a permutation of the vertices of a polyhedron. In this example, we point out that a given group G can (in this sense) act on a set X in many ways. For example, the cyclic group $\{1, -1\}$ can act on the complex plane either as the group generated by a rotation of order two, or as the group generated by a reflection in a line. □

These (and other) ideas are best dealt with in the following way. At present, we have decided that a group G acts on a set X if G is a group of permutations of X. However, we obtain a greater flexibility (and rigour) if we now extend this definition to say that a group G acts on X if G is isomorphic to some group Γ of permutations of X. For example, if $\theta : G \to \Gamma$ is an isomorphism, we can think of G as acting on X by thinking of $g(x)$ when we really mean $(\theta(g))(x)$. In Example 14.5.2, for example, the group $\{1, -1\}$ acts on $\mathbb{C}$ either through the isomorphism $\theta_1(1) = I$ and $\theta_1(-1) = g$, where $g(z) = -z$, or through the isomorphism θ_2 defined by $\theta_2(1) = I$ and $\theta_2(-1) = h$, where $h(z) = \bar{z}$.

The apparent 'collapse' of order illustrated in Example 14.4.1 can be explained by noting that there the map θ is a *homomorphism* rather than an isomorphism. Let us be quite explicit. Using the same notation as in Example 14.4.1, let $G = \{I, f, g, h\}$ and $\Gamma = \{I, g\}$. Then G acts on $\mathbb{C}$ in the usual way. Now let θ be the homomorphism from G onto Γ defined by $\theta(I) = \theta(f) = I$ and $\theta(g) = \theta(h) = g$. Note that θ is a homomorphism; for example, $\theta(fg) = \theta(h) = g = Ig = \theta(f)\theta(g)$. The important point to note here is that two functions in G have the same θ-image if and only if the two functions agree on $\mathbb{R}$; for example, $g(x) = h(x)$ for all real x, and $\theta(g) = \theta(h)$, even though $g \neq h$. Thus the apparent 'collapse' of the order of the group in Example 14.4.1 is associated with, and explained by, the 'collapse' of the group under a homomorphism. These comments lead us to the following more general notion of the action of a group on a set.

Definition 14.5.3 Let G be a group and X be a set. An *action of G on X* is a homomorphism θ of G onto some group Γ of permutations of X. In this case we say that *G acts on X*. □

Of course, if G happens to be a group of permutations of X, then we can take θ to be the identity map from G to itself, and this recaptures the earlier, and most obvious, way that G can act on X.

In Example 14.5.1 we have an action of a group G on a set X (namely $\mathbb{R}$) in which it is impossible to tell the difference between two elements in G by their action on X alone. Now it is usually important to know whether or not the elements of G can be distinguished by their action on X alone, so we introduce the following terminology.

Definition 14.5.4 Let the action of a group G on a set X be given by the homomorphism $\theta : G \to \Gamma$. Then the action of G is said to be *faithful*, or G is said to *act faithfully on X*, if θ is an isomorphism. □

If this action is *not* faithful, then there are two elements f and g in G, with $f \neq g$ but $\theta(f) = \theta(g)$. In particular, $\theta(f)$ and $\theta(g)$ are the same permutation of X.

Our immediate concern is to examine the impact that this discussion has on the results in Section 14.1, and we shall see that the results there remain valid for any group G that acts on X in this wider sense. Suppose that θ is an action of a group G on X, and let $\Gamma = \theta(G)$ (so Γ is a group of permutations of X). For brevity, we write g_θ for $\theta(g)$, where $g \in G$. Naturally, for x in X and $g \in G$ we define

$$\text{Stab}_G(x) = \{g \in G : g_\theta(x) = x\},$$
$$\text{Orb}_G(x) = \{g_\theta(x) : g \in G\} = \{\gamma(x) : \gamma \in \Gamma\} = \text{Orb}_\Gamma(x),$$
$$\text{Fix}_G(g) = \{x \in X : g_\theta(x) = x\}.$$

Now $\text{Orb}_G(x) = \text{Orb}_\Gamma(x)$, and we need to find analogous relations for the stabilizer and fixed point set. Let K be the kernel of the homomorphism θ. Then every element of Γ is the image of exactly $|K|$ elements in G, so that $|G| = |K|\,|\Gamma|$. Moreover, as $g \in \text{Stab}_G(x)$ if and only if $g_\theta \in \text{Stab}_\Gamma(x)$ we see that

$$|\text{Stab}_G(x)| = |K|\,|\text{Stab}_\Gamma(x)|.$$

Now Theorem 14.1.7 shows that

$$|\Gamma| = |\text{Orb}_\Gamma(x)| \times |\text{Stab}_\Gamma(x)|,$$

and from the observations just made we now have the following result.

Theorem 14.5.5 *Suppose that the finite group G acts on X as in Definition 14.5.3. Then for any x in X, $|G| = |\text{Orb}_G(x)| \times |\text{Stab}_G(x)|$.*

We began our study of groups with groups of permutations, and then moved on to study 'abstract groups'. The last result in this section is due to Cayley, and it shows that 'abstract' groups are, in fact, no more abstract than the apparently more concrete permutation groups.

Theorem 14.5.6 *Every group is isomorphic to a group of permutations of some set.*

Let us illustrate the main idea in the proof with an explicit example.

Example 14.5.7 Let G be the additive group $\mathbb{R}$. For each real a, let t_a be the translation by a (so that $t_a(x) = x + a$), and let $T = \{t_a : a \in \mathbb{R}\}$. Then each t_a is a permutation of $\mathbb{R}$, and T is a group under composition. Moreover, as $t_{a+b} = t_a t_b$, the map $a \mapsto t_a$ is a homomorphism of G onto T. In fact, this homomorphism is an isomorphism because if $t_a = t_b$ then $a = b$. It is worthwhile to pause and extract the main idea from this example: we have used each element a of G to create a permutation t_a of G (as a set) in such a way that G itself is isomorphic to the group of these permutations. The proof of Theorem 14.5.6 is similar. □

The proof of Theorem 14.5.6 Let G be any group. For each g in G let $\theta_g : G \to G$ be the map defined by $\theta_g(h) = gh$ (informally, θ_g is the instruction 'multiply on the left by g'). We begin by showing that each θ_g is a permutation of G. Certainly, θ_g is a map of G into itself. Next, if $\theta_g(f) = \theta_g(h)$ then $gf = gh$ so that $f = h$; thus θ_g is injective. Finally, take any h in G and notice that $\theta_g(g^{-1}h) = gg^{-1}h = h$ so that θ_g is surjective. Thus each θ_g is a permutation of G.

It follows that we have just constructed a map θ that takes g to θ_g, and this is a map from G to the group, say $\mathcal{P}$, of permutations of G. Now it is easy to see that θ is a homomorphism from G to $\mathcal{P}$. Indeed, given g and h in G,

$$\theta_{gh}(f) = (gh)f = g(hf) = \theta_g(hf) = \theta_g\theta_h(f),$$

and as this holds for all f, we see that $\theta_{gh} = \theta_g\theta_h$. Thus $\theta : G \to \mathcal{P}$ is a homomorphism. Finally, let Γ be the image of G under θ. As G is a group and θ is a homomorphism, we see that Γ is a subgroup of $\mathcal{P}$. By definition, θ maps G onto Γ; thus θ is an isomorphism from G onto the subgroup Γ of $\mathcal{P}$. □

Exercise 14.5

1. Let G be the group of transformations $\{I, f, g, h\}$, where $f(z) = -z$, $g(z) = \bar{z}$ and $h(z) = -\bar{z}$. Show that the map $v \mapsto fv$ is a permutation of G.
2. Let G be any group, and let G act on itself as described in the proof of Cayley's theorem. Show that G acts faithfully on itself.
3. Let G be a group and H a subgroup of G. Show that G acts on the set of left cosets by the rule that g (in G) takes hH to ghH; equivalently, $g(hH) = ghH$. Now H itself is a left coset ($= eH$), so we can ask for the subgroup of elements of G that fix H. Show that this subgroup is H; thus any subgroup of any group arises as the stabilizer of some group action.

15
Hyperbolic geometry

15.1 The hyperbolic plane

In the earlier chapters we have discussed both Euclidean geometry and spherical (non-Euclidean) geometry, and in this last chapter we discuss a second type of non-Euclidean geometry, namely hyperbolic geometry. Gauss introduced the term *non-Euclidean geometry* to describe a geometry which does *not* satisfy Euclid's axiom of parallels, namely that if a point P is not on a line L, then there is exactly one line through P that does not meet L. In spherical geometry, the 'lines' are the great circles, and in this case any two lines meet. Hyperbolic geometry is a geometry in which there are infinitely many lines through the point P that do not meet the line L, and it was developed independently by Gauss (in Germany), Bolyai (in Hungary) and Lobatschewsky (in Russia) around 1820.

We begin by describing the points and lines of hyperbolic geometry without any reference to distance. We shall take the hyperbolic plane to be the upper half-plane $\mathcal{H} = \{x + iy : y > 0\}$ in $\mathbb{C}$. Notice that the real axis $\mathbb{R}$ is not part of $\mathcal{H}$. A *hyperbolic line* (that is, a line in the hyperbolic geometry) is a semicircle in $\mathcal{H}$ whose centre lies on $\mathbb{R}$; such semi-circles are orthogonal to $\mathbb{R}$. However, as our concept of circles includes 'straight lines' (see Chapter 14), we must also regard those straight lines that are orthogonal to $\mathbb{R}$ as hyperbolic lines. Figure 15.1.1 illustrates the hyperbolic lines in $\mathcal{H}$, and we remark that the two 'different' types of line are only different because we are viewing them from a Euclidean perspective.

We notice immediately that Euclid's Parallel Axiom fails; indeed, the two semi-circles have a common point P that does not lie on the line L; moreover, it is easy to see that there are infinitely many hyperbolic lines through P and not meeting L. It is clear, however, that *any two hyperbolic lines meet in at most one point*, and that *there is is a unique hyperbolic line through any two distinct points in $\mathcal{H}$*.

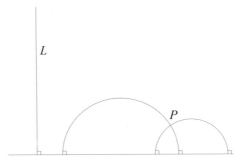

Figure 15.1.1

Now let

$$\Gamma = \left\{ z \mapsto \frac{az+b}{cz+d} : a, b, c, d \in \mathbb{R}, \ ad - bc > 0 \right\}.$$

First, we leave the reader to check that Γ is a group. Next, we note that if g is in Γ then it maps $\mathcal{H}$ into itself. Indeed, if we write $g(z) = (az+b)/(cz+d)$, and $z = x + iy$, then

$$\operatorname{Im}[g(x+iy)] = \frac{(az+b)\overline{(cz+d)}}{(cz+d)\overline{(cz+d)}} \qquad (15.1.1)$$
$$= \frac{(ad-bc)\,y}{|cz+d|^2} > 0.$$

Exactly the same reason shows that g also maps the lower half-plane (given by $y < 0$) into itself; thus g must also map the circle $\mathbb{R} \cup \{\infty\}$ into itself. As Γ is a group, the same holds for g^{-1}; thus

$$g(\mathbb{R} \cup \{\infty\}) = \mathbb{R} \cup \{\infty\} = g^{-1}(\mathbb{R} \cup \{\infty\}).$$

This implies that the coefficients a, b, c and d in g may be chosen to be real (note that we cannot assert that they *are* real, for they are only determined to within a complex scalar multiple). The case $c = 0$ is easy, so we may assume that $c \neq 0$. Then, by scaling the coefficients by the factor $1/c$, we can choose these coefficients so that, in effect, $c = 1$. Then, as $g(\infty) = a$, and $g^{-1}(\infty) = -d$, we see that a and d are real. Finally, if $a = 0$ then $-b = ad - bc > 0$ so that b is real. If, however, $a \neq 0$, then $-b/a = g^{-1}(0)$ so that b is again real. To summarize: if $g(z) = (az+b)/(cz+d)$, and $g \in \Gamma$, then we may assume that a, b, c and d are real. This implies that, for each z,

$$\overline{g(z)} = \overline{\left(\frac{az+b}{cz+d}\right)} = \frac{\bar{a}\bar{z}+\bar{b}}{\bar{c}\bar{z}+\bar{d}} = \frac{a\bar{z}+b}{c\bar{z}+d} = g(\bar{z}). \qquad (15.1.2)$$

15.1 The hyperbolic plane

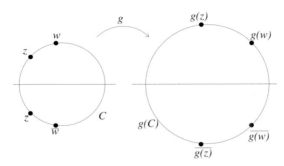

Figure 15.1.2

Let us now see how g in Γ acts on hyperbolic lines. Let z and w be distinct points of $\mathcal{H}$, and let us consider the hyperbolic line L through z and w. Now L is part of the (unique) Euclidean circle C that passes through z, w, $\bar{z}$ and $\bar{w}$, so that $g(C)$ passes through $g(z)$, $g(w)$, $g(\bar{z})$ and $g(\bar{w})$. However, (15.2.1) now implies that $g(C)$ passes through $g(z)$, $\overline{g(z)}$, $g(w)$ and $\overline{g(w)}$, so that $g(C)$ is orthogonal to $\mathbb{R}$ (see Figure 15.1.2). It follows that $g(L) = \mathcal{H} \cap g(C)$, and this proves the following result.

Theorem 15.1.1 *If $g \in \Gamma$ and L is a hyperbolic line, then $g(L)$ is a hyperbolic line.*

Theorem 15.1.1 suggests that the elements of Γ might be regarded as the rigid motions of hyperbolic geometry. This suggestion is strengthened by the fact (which will not be proved here) that any bijective map of $\mathbb{C}_\infty$ onto itself that maps circles to circles is a Möbius map of z or of $\bar{z}$; this is a type of converse of Theorem 13.3.2. Further, any Möbius map that preserves $\mathcal{H}$ must be in Γ (Exercise 15.1.1). In the next section we shall introduce a distance in $\mathcal{H}$, and we shall then see that the elements of Γ are indeed the isometries of $\mathcal{H}$.

There is a second model of the hyperbolic plane which is useful, and often preferable to the model $\mathcal{H}$. In this model the hyperbolic plane is the unit disc $\mathbb{D}$, namely $\{z : |z| < 1\}$ (see Figure 15.1.3). The Möbius map $g(z) = (z - i)/(z + i)$ maps $\mathcal{H}$ onto $\mathbb{D}$ (because $\mathcal{H}$ is given by $|z - i| < |z + i|$), and so we may take the hyperbolic lines in the model $\mathbb{D}$ to be the images under g of the hyperbolic lines in $\mathcal{H}$. Thus the hyperbolic lines in $\mathbb{D}$ are the arcs of circles in $\mathbb{D}$ whose endpoints lie on the cirle $|z| = 1$ and which are orthogonal to this circle at their endpoints. The two models $\mathcal{H}$ and $\mathbb{D}$ may be used interchangeably, and any result about one may be transferred to the other by any Möbius map that maps $\mathcal{H}$ to $\mathbb{D}$, or $\mathbb{D}$ to $\mathcal{H}$.

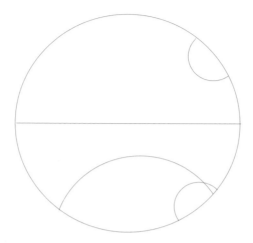

Figure 15.1.3

Exercise 15.1

1. Show that if f is a Möbius map which maps $\mathbb{R} \cup \{\infty\}$ onto itself, then f can be written in the form $f(z) = (az+b)/(cz+d)$, where a, b, c, d are real and $ad - bc \neq 0$. Show further that $f(\mathcal{H}) = \mathcal{H}$ if and only if $ad - bc > 0$.
2. Let z_1 and z_2 be distinct points in the hyperbolic plane. Show that there is a unique hyperbolic line that passes through z_1 and z_2.
3. Suppose that w_1 and w_2 are in $\mathcal{H}$. Show that there is some g in Γ such that $g(w_1) = w_2$. This shows that *the stabilizer of any point in $\mathcal{H}$ is conjugate to the stabilizer of any other point in $\mathcal{H}$.*
4. Verify the steps in the following argument. Let $g(z) = (z-i)/(z+i)$; then g maps $\mathcal{H}$ onto $\mathbb{D}$, where $\mathbb{D} = \{z : |z| < 1\}$. Note that $g(i) = 0$ and $g(-i) = \infty$. Now suppose that f maps $\mathcal{H}$ onto itself and fixes i; then f also fixes $-i$. It follows that gfg^{-1} maps $\mathbb{D}$ onto itself, and fixes 0 and ∞. Thus gfg^{-1} is a Euclidean rotation about the origin, and hence *the group of hyperbolic isometries that fix a given point w is isomorphic to the group of Euclidean rotations that fix the origin.*

15.2 The hyperbolic distance

We shall now introduce a distance in $\mathcal{H}$, and then show that the elements of Γ are isometries for this distance (in fact, they are the only orientation-preserving isometries). There are two ways to define this *hyperbolic distance*, and we shall

15.2 The hyperbolic distance

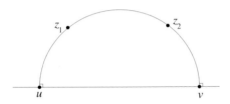

Figure 15.2.1

start with the more elementary way. Consider distinct points z_1 and z_2 in $\mathcal{H}$, and let L be the hyperbolic line through these points. Then L has endpoints u and v, say, chosen so that u, z_1, z_2 and v occur in this order along L (see Figure 15.2.1). We can find a Möbius map g in Γ such that $g(u) = 0$ and $g(v) = \infty$ (see Exercise 15.2.1); then $g(z_1) = ia$ and $g(z_2) = ib$, say, where $0 < a < b$. If we now recall that cross-ratios are invariant under Möbius maps, we have

$$[u, z_1, z_2, v] = [0, ia, ib, \infty] = b/a > 1.$$

This allows us to make the following definition.

Definition 15.2.1 The *hyperbolic distance* between z_1 and z_2 in $\mathcal{H}$ is $\log [u, z_1, z_2, v]$ when $z_1 \neq z_2$, and zero otherwise. We denote this distance by $\rho(z_1, z_2)$. □

Theorem 15.2.2 *The elements of Γ preserve the hyperbolic distance between two points in $\mathcal{H}$.*

Proof This is immediate because the hyperbolic distance is defined as a cross-ratio, cross-ratios are invariant under Möbius maps, and each g in Γ is a Möbius map. □

Notice that if, in Definition 15.2.1, we have $z_1 = ia$ and $z_2 = ib$, where $0 < a < b$, then

$$\rho(ia, ib) = \log[0, ia, ib, \infty] = \log b/a$$

(see Definition 13.4.1). This leads to the following more general result.

Theorem 15.2.3 *The hyperbolic distance is additive along hyperbolic lines.*

Proof Suppose that z_1, z_2, z_3 lie on a hyperbolic line L with end-points u and v such that u, z_1, z_2, z_3, v occur in this order along L. We can find some g in Γ such that $g(u) = 0$ and $g(v) = \infty$, and then $g(z_j) = ia_j$, where $0 < a_1 < a_2 < a_3$.

As $\rho(z_i, z_j) = \log a_j/a_i$ when $i < j$ we see that
$$\rho(z_1, z_3) = \log a_3/a_1$$
$$= \log a_3/a_2 + \log a_2/a_1$$
$$= \rho(z_1, z_2) + \rho(z_2, z_3).$$
$\square$

We are now in a position to give explicit formuale for the hyperbolic distance (see Exercise 15.2.2).

Theorem 15.2.4 *For z and w in $\mathcal{H}$,*
$$\sinh^2 \frac{1}{2}\rho(z, w) = \frac{|z - w|^2}{4\operatorname{Im}[z]\operatorname{Im}[w]}, \quad (15.2.1)$$
$$\cosh^2 \frac{1}{2}\rho(z, w) = \frac{|z - \bar{w}|^2}{4\operatorname{Im}[z]\operatorname{Im}[w]}. \quad (15.2.2)$$

Proof First, choose g in Γ (as above) so that $g(z) = ia$ and $g(w) = ib$, where $0 < a < b$. By applying the map $z \mapsto z/a$ (which is in Γ), we may assume that $a = 1$. Then $\rho(z, w) = \rho(i, ib) = \log b$, so that
$$\sinh^2 \frac{1}{2}\rho(z, w) = \sinh^2\left(\log\sqrt{b}\right) = \frac{(b-1)^2}{4b}. \quad (15.2.3)$$

Next, let
$$F(z, w) = \frac{|z - w|^2}{4\operatorname{Im}[z]\operatorname{Im}[w]}.$$

Then, from (13.1.2) and (15.1.1), we see that F is invariant under any g in Γ; that is,
$$F\bigl(g(z), g(w)\bigr) = F(z, w).$$

Thus
$$F(z, w) = F(i, ib) = \frac{(b-1)^2}{4b}, \quad (15.2.4)$$

and this together with (15.2.3) gives (15.2.1). The second formula (15.2.2) follows from the fact that, for all z, $\cosh^2 z = 1 + \sinh^2 z$. $\square$

We remark that, as Theorem 15.2.4 suggests, in calculations involving the hyperbolic distance it is almost always advantageous to use the functions sinh or cosh of $\rho(z, w)$ or $\frac{1}{2}\rho(z, w)$; *only rarely is $\rho(z, w)$ used by itself.*

We end with a brief discussion of an alternative (but equivalent) way to define distance. First, we define the *hyperbolic length* of a curve γ in $\mathcal{H}$ to be

the line integral
$$\int_\gamma \frac{|dz|}{y},$$
where, as usual $z = x + iy$. Now let L be the hyperbolic line through two points z and w in $\mathcal{H}$, and let σ be the arc of L that lies between z and w. It can be shown that σ has hyperbolic length $\rho(z, w)$ and, moreover, that any other curve joining z to w has a greater hyperbolic length than σ. Thus the hyperbolic line through two points does indeed give the shortest path between these points.

Finally, if g is in Γ, and $g(z) = (az + b)/(cz + d)$, then, from (13.1.2), we see that
$$|g'(z)| = \frac{|ad - bc|}{|cz + d|^2},$$
where $g'(z)$ is the usual derivative of g. In conjunction with (15.1.1) this gives
$$\frac{|g'(z)|}{\text{Im}[g(z)]} = \frac{1}{\text{Im}[z]}.$$
This (together with the formula for a change of variable in a line integral) shows that for each g in Γ, and each curve γ, $g(\gamma)$ has the same hyperbolic length as γ.

Exercise 15.2

1. Show that for any u and v in $\mathbb{R} \cup \{\infty\}$ with $u \neq v$, there is a g in Γ with $g(u) = 0$ and $g(v) = \infty$. [Hint: apply $z \mapsto -1/(z - v)$ and then a translation.]
2. The functions sinh and cosh are defined by $\sinh z = (e^z - e^{-z})/2$ and $\cosh z = (e^z + e^{-z})/2$. Show that (a) $\cosh^2 z - \sinh^2 z = 1$, and (b) $\cosh 2z = 2\cosh^2 z - 1 = 1 + 2\sinh^2 z$.
3. Find the hyperbolic distance between the points $1 + iy$ and $-1 + iy$ as a function of y. Show that for a given positive t there is a value of y such that this distance is t.
4. Let L be the Euclidean line given by $\text{Im}[z] = 2$. Show that $2i$ is the point on L that is closest (as measured by the hyperbolic distance) to the point i.

15.3 Hyperbolic circles

Suppose that $w \in \mathcal{H}$, and $r > 0$. The *hyperbolic circle* with *hyperbolic centre* w and *hyperbolic radius* r is the set $\{z \in \mathcal{H} : \rho(z, w) = r\}$.

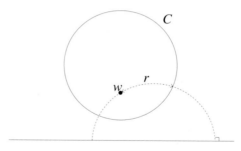

Figure 15.3.1

Theorem 15.3.1 *Each hyperbolic circle is a Euclidean circle in $\mathcal{H}$.*

Proof Let C be the hyperbolic circle with centre w and radius r. There is a map g in Γ with $g(w) = i$ (see Exercise 15.1.2), so that $g(C)$ is the hyperbolic circle with centre i and radius r. Now by Theorem 15.2.4, $z \in g(C)$ if and only if $|z - i|^2/4y = \sinh^2 \frac{1}{2}r$, where $z = x + iy$. This equation simplifies to give $x^2 + (y - \cosh r)^2 = \sinh^2 r$, so that $g(C)$ is a Euclidean circle in $\mathcal{H}$. As g^{-1} maps circles to circles, and $\mathcal{H}$ to itself, we see that $g^{-1}(g(C))$, namely C, is a Euclidean circle in $\mathcal{H}$. □

Notice that the hyperbolic centre of a hyperbolic circle is *not* the same as its Euclidean centre (and similarly for the radii); indeed, the hyperbolic circle $g(C)$ in the proof of Theorem 15.3.1 has hyperbolic centre i, and Euclidean centre $i \cosh r$ (and $\cosh r > 1$). A hyperbolic circle with centre w and hyperbolic radius r is illustrated in Figure 15.3.1. Finally, it can be shown that the length of a hyperbolic circle of hyperbolic radius r is $2\pi \sinh r$, and that its hyperbolic area (which we have not defined) is $4\pi \sinh^2(\frac{1}{2}r)$. Notice that the hyperbolic radius of a hyperbolic circle of radius r grows roughly like πe^r; in the Euclidean case, it is $2\pi r$. Finally, we mention (but do not prove) the hyperbolic counterpart of the fact that the area of a spherical triangle is π less than its angle sum.

Theorem 15.3.2 *The area of a hyperbolic triangle with angles α, β and γ is $\pi - (\alpha + \beta + \gamma)$. In particular, this area cannot exceed π.*

Exercise 15.3

1. Find the equation of the hyperbolic circle with centre $2i$ and radius e^2. Suppose that this circle meets the imaginary axis at ia and ib, where $0 < a < 2 < b$. Find a and b, and verify directly that $\rho(ia, 2i) = e^2 = \rho(2i, ib)$.

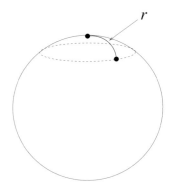

Figure 15.3.2

2. Consider the unit sphere in $\mathbb{R}^3$ as a model of spherical geometry in which distances are measured on the surface of the sphere. What is the circumference of the circle whose centre is at the 'north pole' (see Figure 15.3.2) and whose radius is r? Now compare the circumference of a circle of radius r in Euclidean geometry, spherical geometry and hyperbolic geometry.

15.4 Hyperbolic trigonometry

A hyperbolic triangle is a triangle whose sides are arcs of hyperbolic lines. We begin with the hyperbolic version of Pythagoras' theorem (see Figure 15.4.1).

Theorem 15.4.1 *Suppose that a hyperbolic triangle has sides of hyperbolic lengths a, b and c, and that the two sides of lengths a and b are orthogonal. Then* $\cosh c = \cosh a \, \cosh b$.

In our proof of this result we shall need to use the fact that a Möbius map is *conformal*; that is, it preserves the angles between circles. In particular, this implies that if two circles C and C' are orthogonal, and if g is any Möbius map, then $g(C)$ and $g(C')$ are orthogonal. We shall not give a proof of this (although the proof is not difficult).

Proof Let the vertices of the hyperbolic triangle be v_a, v_b and v_c, where v_a is opposite the side of length a, and so on. There is some g in Γ such that $v_c = i$ and $v_b = ik$, where $k > 1$. As g preserves the orthogonality of circles we see that g maps v_a to some point $s + it$, where $s^2 + t^2 = 1$ (see Figure 15.4.2). As g preserves hyperbolic distances, this means that we may assume that $v_a = s + it$,

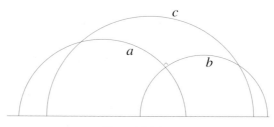

Figure 15.4.1

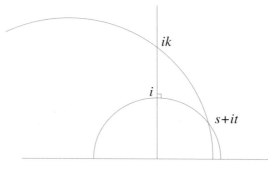

Figure 15.4.2

$v_b = ik$ and $v_c = i$. Now $\rho(i, ik) = a$, $\rho(i, s + it) = b$ and $\rho(ik, s + it) = c$. As

$$\cosh a = \frac{k^2 + 1}{2k}, \quad \cosh b = \frac{1}{t}, \quad \cosh c = \frac{k^2 + 1}{2tk},$$

the given formula follows. □

It is interesting to examine Pythagoras' theorem for small triangles, and for large triangles. As $\cosh z = 1 + z^2/2! + z^4/4! + \cdots$, we see that when a, b and c are very small, the formula is, up to the second-order terms, $c^2 = a^2 + b^2$. Thus, infinitesimally, the hyperbolic version of Pythagoras' theorem agrees with the Euclidean version. This is because the hyperbolic distance is obtained form the Euclidean distance by applying a 'local scaling factor' of $1/y$ at z. As this scaling factor is essentially constant on an infinitesimal neighbourhood of a point, the 'infinitesimal hyperbolic geometry is just a scaled version of the Euclidean geometry. However, as the scaling factor varies considerably over large distances, the global hyperbolic geometry is very different from the Euclidean geometry. For example, if in Pythagoras' theorem, a, b and c are all very large, then, as $\cosh x$ is approximately $e^x/2$ when x is large, we have (approximately) $4e^c = e^a e^b$ so that $c = a + b - \log 2$. In other words, in a 'large' hyperbolic right-angled triangle, the length of the hypotenuse is almost the sum of the lengths of the other two sides! If this were the case in Euclidean geometry, then the triangle would be very 'flat', but this is not so in hyperbolic geometry.

Finally, we remark that hyperbolic trigonometry is as rich and well understood as Euclidean trigonometry (and spherical trigonometry) is. For example, there is a sine rule, and cosine rules in hyperbolic geometry. In most applications it is the hyperbolic trigonometry that is important, and hyperbolic geometry by itself has relatively few applications.

Exercise 15.4

1. Suppose that a, b and c are the sides of a right-angled hyperbolic triangle with the right-angle opposite the side of length c. Prove that $c \leq a + b$; this is a special case of the triangle inequality.
2. Consider a right-angled hyperbolic triangle with both sides ending at the right angle having length a. Let the height of this triangle be h (the distance from the right angle to the third side). Find h as a function of a. What is the limiting behaviour of h as $a \to +\infty$?

15.5 Hyperbolic three-dimensional space

We end this text with a very brief description of three-dimensional hyperbolic geometry, and a sketch of the proof of Theorem 14.3.2. These are given in this and the next section, and they combine many of the ideas that have been introduced in this text. We take *hyperbolic space* to be the upper-half of $\mathbb{R}^3$, namely

$$\mathcal{H}^3 = \{(x, y, t) \in \mathbb{R}^3 : t > 0\}.$$

It is convenient to identify the point (x, y, t) with the quaternion $x + y\mathbf{i} + t\mathbf{j}$, and also to identify the quaternion $\mathbf{i}$ with the complex number i. Thus we can write (x, y, t) as $z + t\mathbf{j}$, where z is the complex number $x + iy$. Note that in this notation we have the convenient formula

$$z\mathbf{j} = (x + yi)\mathbf{j} = x\mathbf{j} + y\mathbf{k} = x\mathbf{j} - y\mathbf{ji} = \mathbf{j}\bar{z}. \tag{15.5.1}$$

Suppose now that $g(z) = (az + b)/(cz + d)$, where $ad - bc \neq 0$. We can now let g act on hyperbolic space $\mathcal{H}^3$ by the rule

$$g : z + t\mathbf{j} \mapsto \bigl[a(z + t\mathbf{j}) + b\bigr]\bigl[c(z + t\mathbf{j}) + d\bigr]^{-1}, \tag{15.5.2}$$

where this computation is to be carried out in the algebra of quaternions. This lengthy (but elementary) exercise shows that

$$g(z+t\mathbf{j}) = \frac{(az+b)(\bar{c}\bar{z}+\bar{d}) + a\bar{c}t^2 + |ad-bc|t\mathbf{j}}{|cz+d|^2 + |c|^2 t^2}. \quad (15.5.3)$$

Notice that as quaternions are not commutative, we have to choose (and then be consistent about) which side we shall write the inverse in (15.5.2). However, in (15.5.3), the denominator is real (and positive), and as every real number commutes with every quaternion, we can write it in the usual form for a fraction without any ambiguity. Notice also that if we put $t=0$ in (15.5.3), we recapture the correct formula for the action of g on $\mathbb{C}$.

The consequences of (15.5.3) are far-reaching. First, if we consider g to be a translation, say $g(z) = z+b$, then we find that $g(z+t\mathbf{j}) = (z+b) + t\mathbf{j}$; thus g is just the 'horizontal' translation by b. If $g(z) = az$, then we find that

$$g(z+t\mathbf{j}) = az + |a|t\mathbf{j}.$$

If $|a|=1$, so that g is a rotation of the complex plane, then g acts on $\mathcal{H}^3$ as a rotation about the vertical axis through the origin. If $a>0$, so that g acts as a 'stretching' from the origin by a factor a in $\mathbb{C}$, then g also acts as a stretching (from the origin, and by the same factor) in $\mathcal{H}^3$. Of course, the more interesting case is when $g(z) = 1/z$; here

$$g(z+t\mathbf{j}) = \frac{\bar{z}+t\mathbf{j}}{|z|^2 + t^2}.$$

We define the lines in $\mathcal{H}$ to be the 'vertical' semi-circles, and the 'vertical' rays (exactly as in the two-dimensional case; see Figure 15.5.1), and we can define the hyperbolic distance between two points again by a cross-ratio (thinking of the vertical plane through the two points as the complex plane), or by integrating $|dx|/x_3$, where $x = (x_1, x_2, x_3)$, over curves. When all this has been done, we arrive at the following beautiful result.

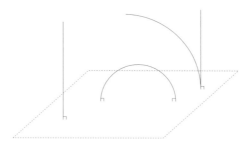

Figure 15.5.1

Theorem 15.5.1 *Every Möbius map acts on hyperbolic space $\mathcal{H}^3$ as a hyperbolic isometry, and every isometry that preserves orientation is a Möbius map.*

15.6 Finite Möbius groups

Finally, we give only the briefest sketch of the ideas behind a proof of Theorem 14.3.2. The aim of this sketch is to give the reader a glimpse of some beautiful interaction between algebra and geometry, and it is far from being complete. First, the Möbius maps (that act on $\mathbb{C}_\infty$) can be extended (either as a composition of reflections and inversions, or in terms of the quaternion algebra) to act on all of $\mathbb{R}^3$. The upper-half $\mathcal{H}^3$ of $\mathbb{R}^3$ with the hyperbolic metric $ds = |dx|/x_3$ is a model of three-dimensional hyperbolic geometry, and the Möbius group is the group of orientation-preserving isometries of this space.

Now let G be a finite Möbius group; then G may be regarded as a finite group of isometries of $\mathcal{H}^3$, so that each point in $\mathcal{H}^3$ has a finite orbit. Take any orbit and let B be the smallest hyperbolic ball that contains the orbit. Analytic arguments show that B is unique, and as the chosen orbit is invariant under G, so is B, and hence (finally) so too is the hyperbolic centre of B. This argument proves that the elements of the finite group G have a common fixed point ζ in $\mathcal{H}^3$. There is now a Möbius map (which acts on all of $\mathbb{R}^3 \cup \{\infty\}$) that converts the upper-half space model of three-dimesional hyperbolic space into the unit ball model (much as there is a Möbius map that takes the upper half-plane to the unit disc). This can be chosen so that ζ is carried to the origin; thus the finite Möbius G is conjugate to a Möbius group G' of hyperbolic isometries that act on the unit ball in $\mathbb{R}^3$ with the extra property that every element of G' fixes 0. It is not difficult to show that every such isometry is a Euclidean rotation of $\mathbb{R}^3$ and the sketch of the proof is complete. □

Index

A^t, 146
A_n, 13
D_{2n}, 237
G^+, 290
I_X, 17
$L(\infty)$, 272
$L \cup \{\infty\}$, 277
$M^{m \times n}(\mathbb{F})$, 149
$O(k)$, 9
q-cycle, 9
$R(f, g)$, 171
$SO(n)$, 202
S_n, 8
$x \star y$, 212
$\arg z$, 36
$\bar{z}$, 33
cosh, 312, 313
$\Delta(f)$, 171
$\gcd(m, n)$, 233
∞, 255
$\ker(\theta)$, 246
$\mu(E)$, 77
$\oplus$, 116
$\oplus_n$, 28
$\otimes_n$, 28
$\rho(z_1, z_2)$, 311
$\rho_{\text{col}}(A)$, 140
$\rho_{\text{row}}(A)$, 140
sinh, 312, 313
$\varepsilon(\rho)$, 13
$\mathbb{C}$, 31
$\mathbb{C} \cup \{\infty\}$, 256
$\mathbb{C}^*$, 247
$\mathbb{C}^n$, 103
$\mathbb{C}^{n,t}$, 104
$\mathbb{H}$, 96

$\mathbb{H}_0$, 97
$\mathbb{N}$, 22
$\mathbb{Q}$, 2, 5
$\mathbb{R}$, 2, 5, 22, 26
$\mathbb{R}^+$, 247
$\mathbb{R}^n$, 103
$\mathbb{R}^\#$, 27
$\mathbb{R}^{n,t}$, 104
$\mathbb{Z}$, 2, 4, 22
$\mathbb{Z}_n$, 28
$\mathcal{F}$, 300
$\mathcal{H}^3$, 317
$\mathcal{L}$, 271, 272
$\mathcal{L}(V, W)$, 132
$\mathcal{M}$, 256
$\mathcal{M}_0$, 281
$\mathcal{P}(X)$, 7, 18
$\mathcal{P}_A$, 182
$\mathcal{S}$, 74, 276
$\mathcal{U}$, 281
$\det(A)$, 144, 145
$\dim(V) = 0$, 106
Fix(g), 284
GL$(2, \mathbb{C})$, 257, 271
I$(\mathbb{C})$, 297
I$^+(\mathbb{C})$, 297
$\ker(\alpha)$, 127
Orb$_G(x)$, 284
P$(\mathbb{C}, d)$, 120
SL$(2, \mathbb{C})$, 258
SL$(2, \mathbb{Z})$, 258
Stab$_G(x)$, 284
tr(X), 152
Im, 34
lcm, 233
Re, 34

Index

Abel, 216
abelian group, 216, 221, 236
abelian group; see also group, commutative, 4
act, 304
act transitively, 284
action
    faithful, 304
action of a group, 304
action of a matrix, 154
addition modulo n, 28
additive function, 125
algebraic structure, 22
alternating group, 13, 292
altitude, 70
angle, 198
anti-trace, 152
Argand, 31
argument, 36
associative law, 2
associativity, 16
augmented matrix, 141
auxiliary equation, 119
axiom of parallels, 307
axis of rotation, 90

basis, 106, 158
    change of, 168
    orthonormal, 198
    standard, 108
bell-ringing, 15
bijection, 17, 18
bijective, 17
binary operation, 2
binomial theorem, 36
Bolyai, 307
Burnside, 2, 287
Burnside's lemma, 287, 288

Cancellation Law, 3
Cardan, 31, 47
cardinality, 216
Cauchy, 2, 287
Cauchy–Schwarz inequality, 197, 198
Cayley, 305
Cayley's theorem, 305
Cayley–Hamilton theorem, 190
centre of a group, 222
characteristic equation, 181, 182, 208
characteristic polynomial, 182
Chebychev polynomial, 40
chordal distance, 278

circle, 262
    complement, 263
    Euclidean, 261
    unit, 264
circle group, 253
closure axiom, 2
coefficients, 149
column, 149
column rank, 140
column vector, 149
    j-th, 139
common divisor, 232
commutative group; see group, commutative, 4, 216
commute, 4, 216
complement of a circle, 263
complementary components of a circle, 264
complex conjugate, 33
complex line, 271
complex number, 31
complex plane
    extended, 256
complex polynomials, 120, 123
    even, 105
    odd, 105
composition, 6, 16
concyclic points, 267, 268
conformal, 315
conjugacy class, 242, 243, 299
conjugate, 242
conjugate subgroup, 242, 286
convex polygon, 79
convex polyhedron, 81
convex spherical polygon, 80
coordinates, 53
coplanar, 53, 54, 61
coprime, 25, 233
corkscrew, 66
corkscrew rule, 63
    right-handed, 59
coset, 220, 237, 243
    left, 222, 243, 253, 286
    right, 222, 243, 286
coset decomposition, 222
cosine rule
    for a spherical triangle, 77
Cotes, 39, 50
Cramer's rule, 143
cross-ratio, 266, 311, 318
crystallographic group; see also wallpaper group, 302

cube, 291
cubic equation, 46
cubic polynomial, 171
cycle, 9, 13
   length, 9
cycle type, 244
cyclic group, 230–232, 269, 296, 297

d'Alembert, 48
de Moivre, 39
de Moivre's theorem, 39
deficiency
   total, 87, 88
deficiency of a vertex, 87
degree of a polynomial, 48
del Ferro, 47
deltahedron, 88
DeMorgan, 24
Descartes, 31
Descartes' theorem, 88
determinant, 64, 65, 144, 165, 169
diagonal elements, 149
diagonal matrix, 153, 184
diagonalizable matrix, 185, 188, 189, 191
diagonalizing, 274
difference equation, 118, 186
dihedral group, 237, 238, 241, 244, 259, 269, 296, 297
dimension, 106
   finite, 106
direct isometry, 206
direct sum of two subspaces, 115
directed line segment, 54
discrete group, 302
discriminant, 171
disjoint permutations, 9
distributive laws, 27
divisor, 24, 232
dodecahedron, 290, 295
dual, 291
dynamics, 193

edges
   of a polyhedron, 83
   of a spherical triangle, 79
eigenvalue, 175, 178, 179, 201, 273
eigenvector, 175, 178, 273
   generalized, 179
entries; see coefficients, 149
equivalence class, 228

equivalence relation, 227, 243
Erlangen Programme, 1
Euclidean circle, 261
Euclidean line, 261
Euler, 2, 38, 39, 48, 80
Euler's formula for triangulations, 80, 81, 85, 87
Euler's function $\varphi(n)$, 234
Euler's theorem, 234
even complex polynomials, 105
even function, 151
even permutation, 13
extended complex plane, 256
extended line, 277
extends, 126

faces
   of a polyhedron, 83
   of a triangle, 80
faithful action, 304
Fermat's theorem, 234
Ferrari, 47
Fibonacci sequence, 120
field, 22, 27, 211
   skew, 98
fixed point, 8, 260, 273, 284
frieze group, 299, 300, 302
   standard, 300
Frobenius, 287
function, 6, 16
   linear, 56
Fundamental Theorem of Algebra, 48, 51, 120, 176, 180, 182, 208
Fundamental Theorem of Arithmetic, 24

Gauss, 2, 31, 48, 307
Gaussian integers, 253
General Linear group, 257
generated, 105, 114, 220, 230
Girard, 77
glide reflection, 42, 236
golden ratio, 40, 293
great circle, 74
greatest common divisor, 172, 233
group, 1, 2
   abelian, 216, 236
   alternating, 13
   circle, 253
   commutative or abelian, 4
   crystallographic, 302
   cyclic, 269, 296, 297

Index

dihedral, 237, 238, 241, 244, 259, 269, 296, 297
discrete, 302
frieze, 300, 302
General, 257
Modular, 258
Möbius, 256, 269, 319
non-abelian, 241
non-commutative, 4
permutation, 268
quotient, 252
Special Linear, 258
symmetric, 8, 14
unitary, 281
group action, 304

half-space, 81
Hamilton, 95
homogeneous linear equation, 136, 261
homogeneous quadratic form, 206
homomorphism, 246
homomorphism, kernel of a, 246
Hurwitz, 214
hyperbolic centre, 313, 314
hyperbolic circle, 313, 314
hyperbolic distance, 311
hyperbolic geometry, 307
hyperbolic length, 312
hyperbolic line, 307, 309
hyperbolic radius, 313
hyperbolic space, 317
hyperbolic triangle, 314, 315
hyperbolic trigonometry, 315
hyperplane, 135

icosahedron, 290, 293
identity element, 3, 17
identity function, 6
identity map, 91
identity matrix, 163
imaginary part of a complex number, 31
index
  cyclic, 230–232
indirect isometry, 206
Induction, first Principle of, 22
inhomogeneous linear equations, 141
injective, 17
integers, set of, 22
invariant planes, 193
invariant subspace, α-, 175, 177
invariant under, 175

inverse element, 3
inverse function, 6, 17
inverse matrix, 164
inverse transformation, 164
invertible, 6, 17, 18
invertible matrix, 164, 200, 204
irrational number, 26
isometry, 41, 42, 89, 91, 92, 204, 242
  direct, 94, 206
  indirect, 94, 206
isomorphic groups, 226
isomorphism, 131, 226
iterate, 17, 133, 260

join of two subspaces, 112

kernel, 127
kernel of a homomorphism, 246, 248, 252
Klein 4-group, 301
Klein 4-group, 240
Klein, F., 1

Lagrange's interpolation formula, 121
Lagrange's theorem, 223, 232, 286
Laplace, 48
least common multiple, 233
least upper bound, 22
left coset, 222, 243, 253, 286
Legendre, 80
Legendre polynomials, 122
Leibniz, 48
length, 122
length of a cycle, 9
line, 135, 175
  complex, 271
  Euclidean, 261
linear function, 56
linear independence, 107
linear map, 124
linear operator, 125
linear transformation; see linear map, 124
Lobatschewsky, 307
lower-triangular matrix, 153
lune, 77

magic square, 152, 153
map, 16, 124
  linear, 124
mapping, 16
matrix, 136, 149
  action, 154

matrix (*cont.*)
  complex, 149
  diagonal, 153, 184
  diagonalizable, 185, 188, 189, 191
  identity, 163
  inverse, 164
  invertible, 164, 200, 204
  orthogonal, 200
  product, 154
  real, 149
  skew-symmetric, 150, 204
  square, 149
  symmetric, 150
  transpose, 146, 157
  upper-triangular, 153
  zero, 149
matrix coefficients, 149
matrix polynomials, 189
matrix product, 156
matrix representation, 158, 160, 207
Maurolico, 23
median, 70
modular arithmetic, 28, 221
Modular group, 258
modulus, 34
monoid, 234
multiplication modulo n, 28
multiplicity, 182
Möbius group, 256, 269, 319
Möbius inverse, 257
Möbius map, 254, 256, 257, 259, 260, 266, 272, 273, 279, 282, 296, 309, 315, 319

natural numbers, set of, 22
non-abelian group, 241
non-Euclidean geometry, 307
non-singualr matrix; see invertible, 164
normal subgroup, 243, 248, 251–253
nullity, 127

octahedron, 290, 293
odd complex polynomials, 105
odd function, 151
odd permutation, 13
one-to-one; see injective, 17
onto; see surjective, 17
operator
  linear, 125
orbit, 9, 284
orbit-decomposition, 10
orbit-stabilizer theorem, 286

order, 22, 216, 227, 237
  finite, 216
  infinite, 216
order of a permutation, 15
orientation, 310, 319
  negative, 63, 64, 66
  positive, 63, 64, 66
orthogonal, 55, 59, 122, 135
orthogonal group, 202
  special, 202
orthogonal map, 199
orthogonal matrix, 200
orthogonal projection onto a plane, 155
orthonormal basis, 198, 199

parallel, 54
Parallelogram Law of Addition, 53
parity, 12
Peano, 24
Pell sequence, 120
permutation, 11, 13, 18
  disjoint, 9
  even, 13
  odd, 13
permutation group, 268
perpendicular; see orthogonal, 55
Plato, 83
Platonic solids, 83, 84, 269, 290
point at infinity, 255
polar coordinates, 37
polar form of a complex number, 38
polygon
  convex, 79
  convex spherical, 80
  regular, 237
  spherical, 79, 80
polyhedron, 81
  convex, 81
  edges of, 83
  faces of, 83
  general, 87
  regular, 83, 84
  vertices of, 83
polynomial
  real, 50
polynomial of degree n, 48
polynomials
  complex, 110
  trigonometric, 104
prime number, 24
primitive root of unity, 46

product; see also composition, 6
proper subgroup, 219
proper subspace, 111
Ptolemy's theorem, 40
Pythagoras' theorem
    for a spherical triangle, 76
    for a tetrahedron, 72
Pythagoras' theorem, hyperbolic version, 315

quartic equation, 47
quaternion, 95, 96, 282, 318
    addition, 96
    conjugate of a, 98
    multiplication, 96
    norm of a, 98
    pure, 97, 99
    unit, 100
quintic equation, 47
quotient group, 252, 299

range, 127
rank, 127
rational number, 26
real line, 22
real numbers, set of, 22
real part of a complex number, 31
reciprocal vectors, 62
recurrence relation; see difference equation, 118
reflection, 42, 89, 93, 99, 205
regular n-gon, 45, 285
regular polygon, 45, 237
regular polyhedra, 290
regular polyhedron, 83, 84
representation
    standard, 10
resultant, 171
right coset, 222, 243, 286
root of a of polynomial, 48
root of unity, 44, 218, 237
    primitive, 46
rotation, 42, 93, 100
    of $\mathbb{R}^3$, 90
row, 149
row rank, 140
row vector, 137, 149
    j-th, 139

scalar product, 55, 93, 122, 135, 210
scalar triple product, 60

screw-motion, 94
self-conjugate subgroup, 243
self-dual, 291
set
    closed with respect to an operation, 2
shear, 176
sign, 13
sine rule, 71
    for a spherical triangle, 77
skew, 69
skew field, 98
skew-symmetric matrix, 150, 204
span, 107
Special Linear group, 258
special orthogonal group, 202
spherical area, 77
spherical distance, 74
spherical geometry, 74
spherical polygon, 79, 80
    area of, 79
spherical triangle, 79
    edges of a, 79
    vertices of a, 79
stabilizer, 284, 310
standard basis, 108
standard frieze group, 300
standard representation, 10
stereographic projection, 276, 279
subgroup, 218
    conjugate, 242, 286
    non-trivial, 219
    normal, 243, 248, 251–253
    proper, 219
    self-conjugate, 243
    translation, 299
    trivial, 219
subspace, 111
    α-invariant, 175, 177
    proper, 111
sum of two subspaces; see join, 112
surjective, 17
symmetric difference, 5
symmetric group, 8, 14
symmetric linear map, 207
symmetric matrix, 150
symmetry, 237

Tartaglia, 47
tetrahedron, 71, 290, 292
Theatus, 290
topological invariant, 81

torus, 236
total deficiency, 87, 88
trace, 152
translation, 42, 89, 236
translation subgroup, 299
transpose, 146, 157
transposition, 11
triangle inequality, 35, 56, 197
triangulation, 79
trigonometric polynomials, 104
trivial subgroup, 219

unit, 234
unit circle, 264
unit disc, 264, 309
unit quaternion, 100
unit sphere, 276
unit vector, 54
unitary group, 281
upper-triangular matrix, 153

valency, 85, 88
vector, 52, 53
   column, 149
   components of, 53
   length, 54
   norm, 54
   row, 137
   unit, 54
vector product, 57, 58, 93, 212
vector space, 102, 103
   complex, 103
   finite-dimensional, 106
   infinite-dimensional, 106
   real, 103
vector triple product, 62
vectors
   reciprocal, 62
vertices
   of a polyhedron, 83
   of a spherical triangle, 79
von Dyck, 2

Wallis, 31
wallpaper group; see also crystallographic group, 299, 302
Well-Ordering Principle, 22
Wessel, 31